VOLUME ONE HUNDRED AND TWENTY SEVEN

INTERNATIONAL REVIEW OF NEUROBIOLOGY

Controversies In Diabetic Neuropathy

INTERNATIONAL REVIEW OF NEUROBIOLOGY

VOLUME 127

SERIES EDITORS

R. ADRON HARRIS

Waggoner Center for Alcohol and Drug Addiction Research
The University of Texas at Austin
Austin, Texas, USA

PETER JENNER

Division of Pharmacology and Therapeutics
GKT School of Biomedical Sciences
King's College, London, UK

VOLUME ONE HUNDRED AND TWENTY SEVEN

INTERNATIONAL REVIEW OF NEUROBIOLOGY

Controversies In Diabetic Neuropathy

Edited by

NIGEL A. CALCUTT
University of California, San Diego, La Jolla, CA, United States

PAUL FERNYHOUGH
University of Manitoba and St. Boniface Hospital Research Centre, Winnipeg, MB, Canada

ELSEVIER

AMSTERDAM • BOSTON • HEIDELBERG • LONDON
NEW YORK • OXFORD • PARIS • SAN DIEGO
SAN FRANCISCO • SINGAPORE • SYDNEY • TOKYO
Academic Press is an imprint of Elsevier

Academic Press is an imprint of Elsevier
50 Hampshire Street, 5th Floor, Cambridge, MA 02139, USA
525 B Street, Suite 1800, San Diego, CA 92101-4495, USA
The Boulevard, Langford Lane, Kidlington, Oxford OX5 1GB, UK
125 London Wall, London, EC2Y 5AS, UK

First edition 2016

ISBN: 978-0-12-803915-1
ISSN: 0074-7742

Publisher: Zoe Kruze
Acquisition Editor: Kirsten Shankland
Editorial Project Manager: Hannah Colford
Production Project Manager: Surya Narayanan Jayachandran
Designer: Maria Ines Cruz

Typeset by SPi Global, India

CONTENTS

Section II
New Models of Diabetic Neuropathy

Section III
Mechanisms and Therapies

CONTRIBUTORS

V. Bril
Ellen and Martin Prosserman Centre for Neuromuscular Diseases, University Health Network, University of Toronto, Toronto, ON, Canada

N.A. Calcutt
University of California, San Diego, La Jolla, CA, United States

C. Casellini
Eastern Virginia Medical School, Strelitz Diabetes and Neuroendocrine Center, Norfolk, VA, United States

R.T. Dobrowsky
The University of Kansas, Lawrence, KS, United States

S.M. Emery
The University of Kansas, Lawrence, KS, United States

P. Fernyhough
University of Manitoba; St. Boniface Hospital Research Centre, Winnipeg, MB, Canada

O.J. Freeman
University of Manchester, Manchester, United Kingdom

N.J. Gardiner
University of Manchester, Manchester, United Kingdom

R.A. Malik
Weill Cornell Medicine-Qatar, Qatar Foundation, Education City, Doha, Qatar

C. Martin
University of Michigan Medical School, Ann Arbor, MI, United States

M.-L. Névoret
Impeto Medical Inc., San Diego, CA, United States

R. Pop-Busui
University of Michigan Medical School, Ann Arbor, MI, United States

S.M. Todorovic
School of Medicine, University of Colorado Anschutz Medical Campus, Aurora, CO, United States

A.I. Vinik
Eastern Virginia Medical School, Strelitz Diabetes and Neuroendocrine Center, Norfolk, VA, United States

S. Yagihashi*
Hirosaki University Graduate School of Medicine, Hirosaki, Japan

---

*Present address: The Nukada Institute of Medical and Biological Research, 4-16 Inage-machi, Inage-ku, Chiba, Japan. E-mail: soroku@nukada.ac.jp.

M.A. Yorek
Iowa City Health Care System; Fraternal Order of Eagles Diabetes Research Center, University of Iowa, Iowa City, IA, United States

D.W. Zochodne
Neuroscience and Mental Health Institute and Alberta Diabetes Institute, University of Alberta, Edmonton, AB, Canada

PREFACE

I edited the previous version of this book—*The Neurobiology of Diabetic Neuropathy* (*Int. Rev. Neurobiol.*, volume 50), which was published in 2002. At that time the estimated number of people worldwide with diabetes mellitus was about 170 million and the projection was that it would rise to 366 million by the year 2030. That estimate was just a little conservative. The global number of sufferers had reached 387 million by 2014 and the projection for 2035 is now close to 600 million (figures from the WHO and IDF).

As Radica Pop-Busui and Catherine Martin point out, the prevalence of neuropathy in a diabetic population is a function of disease duration and efficacy of diabetes management, but a reasonable estimate is that approximately 50% of patients with diabetes will eventually develop neuropathy. So, the topic of this book will eventually impact on approximately 300 million people and, if the rates of morbid obesity continue to rise, that is an underestimate. It represents a lot of morbidity, a lot of misery, and a huge cost to health-care providers. Quite apart from the inherent interest and challenge to the scientists involved, this epidemiology is reason enough for this book to be published and, hopefully, widely read.

To put the present volume in context, we need to examine where the 2002 version left us, and to see whether the topics covered in 2015 volume have sprung from its predecessor or whether they reflect new initiatives. First, I have to admit that the text of the new book is not available to me at the time of writing; I am working only from the authors' abstracts.

In 2002 the primary metabolic progenitors for neuropathy were considered to be oxidative stress, nonenzymic glycation of proteins, and flux through the polyol pathway, all interconnected direct consequences of raised plasma glucose. Is glucose still viewed as the major metabolic insult and are these three secondary dysmetabolic consequences still viewed as important? There still seems to be a consensus that glucose remains the chief villain of the piece, with the added possibility from Doug Zochodne, that insulin per se may have a relevant trophic role, whose benefit is lacking in diabetes. The three secondary villains, oxidative stress, glycation, and polyol pathway flux, are still seen as important members of the dramatis personae, at least as far as Paul Fernyhough is concerned.

The 2002 volume went on to consider secondary targets of these aberrations of glucose metabolism en route to neuropathy. Five such were appraised: protein kinase C, MAP kinases, neurofilaments, apoptosis, and nerve blood flow. These 2002 chapters and the subsequent passage of time lead this author to the following conclusions. Alterations in activation of protein kinase C are by-products in animal models; there is no involvement in the development of human neuropathy. The same applies to apoptosis; indeed evidence for its existence in the nervous system in diabetes is equivocal. On the other hand, activation of MAP kinases occurs in peripheral nerves in both animal models and human diabetic patients. Whether this is early enough to be a prime driver is less certain, but MAP kinases remain an established mechanism whereby extracellular consequences of glucose dysmetabolism can alter the phenotype of nerve and Schwann cells to promote the detriments seen in neuropathies. The arguments for an involvement of neurofilaments in neuropathy were and remain strong for both animal models and humans. Furthermore, these changes may be promoted by MAP kinase activation. In contrast, the exquisitely careful and even-handed analysis of nerve blood flow involvement, written by Doug Zochodne in the 2002 book, shows that the changes depend on the animal model and the method of measurement, and their candidacy for an early causative factor in human sufferers is dubious.

Against this background, where does the new book take us? Soroku Yagihashi has reviewed primary metabolic insults from raised glucose and emphasizes the point that all these changes are well established in (at least some) animal models, but their involvement in development of the human condition is not proven. None of the contributors to the present volume appear to shoot down MAP kinases or aberrant neurofilament phosphorylation as potential stages in the development of neuropathy. Nor do any contributors appear to resurrect protein kinase C or reduced nerve blood flow as instrumental events. We have a very interesting concept from Rick Dobrowsky and Sean Emery, who argue that it is likely that the body mounts defenses against the consequences of glucose toxicity and that a potential therapeutic avenue would be to assist those defenses. This has the significant advantage that a novel therapeutic agent, based on this premise, might have fewer side effects than agents designed to arrest events downstream of glucose toxicity. They cite assisting heat-shock protein 70 as a feasible example to achieve neuroprotection.

Neuropathic pain is usually the earliest symptom of peripheral neuropathy in diabetes. It has proved to be a difficult end point for inclusion in clinical trials, but we need to understand the mechanism behind it. If the factors underlying the pain were common to those precipitating the later neurodegeneration, then it would be a valuable therapeutic focus. An intervention that had a powerful effect on neuropathic pain via its underlying cause, rather than nonspecific analgesia, might well be useful in trials and it would have the added advantage of ameliorating symptoms, rather than impacting on signs, such as electrophysiology, which might be unrelated to the condition or develop too late, when the pathology has become irreversible. Thus, the chapter by Slobodan Todorovic, dissecting the pain mechanisms, is of great value to our understanding.

This new book returns to the consideration of the value of modeling the disease and/or its components. Nigel Calcutt suggests that clinicians, as superintendents of failed clinical trials, are not entitled to blame inaccuracies of animal models for their failures. We are long past the notion that animal models can holistically represent human diabetic neuropathy. Instead the focus must be to use animals to model specific components of the dysmetabolisms inherent in diabetes and to gauge their impact on the nervous system. In this context, Mark Yorek has introduced a rat model with the defects associated with Type II diabetes. This will give a metabolic profile that differs from the conventional streptozotocin-diabetic rat and may include or rule out some of the candidate metabolic precursors for neuropathy.

The same questions and constraints should be applied to experiments based on neurons in culture. These are entirely valid as a means of tracking subcellular changes related to glucose and its sequelae. Indeed, as Natalie Gardiner and Ollie Freeman illustrate in their chapter, studies on nerve cultures can give a uniquely clean identification of neuronal targets for glucose toxicity. That done, the requirement is to show that these targets are features of the human condition.

Interpretation of findings from such models demands two questions. (1) Does the model register a deviant or exaggerated metabolic anomaly that is relevant to human diabetes? (2) Are the direct consequences of this dysmetabolism present in or relevant to human diabetes? The polyol pathway serves as an excellent example. Widely studied in the rat model, the answer to (1) is yes, but the answer to (2) is maybe, which explains the uncertainty that still surrounds the efficacy of aldose reductase inhibitors. Thus, in vitro and in vivo models are acceptable approaches to mechanisms, but that is as far as their predictive values go. I suggest, therefore, that debates of the "rat vs mouse" type are best left to back-stage discussions and not dragged out in front of the audience.

It follows from the above that, unless these two questions have positive answers for an end point in a model, efficacy for a potential treatment in animals cannot be a certain springboard for clinical trials. Again, aldose reductase inhibitors authenticate this assertion. Thus, for many clinical trials that have been run, their failure might be blamed on interpretation of data from the model, at least in part.

So, we come to the issue of clinical trials. Both the old and the new volumes have addressed this. A systematic and scientific approach to understanding diabetic neuropathy was kick-started in 1979 by a review published by Rex Clements entitled "Diabetic neuropathy: new concepts of its etiology" (*Diabetes* 28: 604–611), so we are not far from 50 years of research with no therapeutic outcomes. It would not be unreasonable for the finger of blame to be looking for targets. It might hover over clinical trials. I do not know how many compounds have been through clinical trials for diabetic neuropathy, but it is well into double figures, and the sad fact is that not only have no successes emerged, but we cannot be certain that the agents tested were failures. To generalize, the trials have been too short and we cannot be sure that the right end points were selected. As Vera Bril points out in this volume, trials lasting 5–8 years will be required to demonstrate clinical benefits. Rayaz Malik reviews potential end points for identification of patients at risk or for inclusion in trials. None of these is entirely satisfactory for either purpose. Clear signs of true neuropathy probably indicate the condition is irreversible.

There are additional problems in finding suitable treatment. A successful agent may have to be given prophylactically for neuropathy, perhaps starting with the first signs of background retinopathy as an indicator of susceptibility to diabetes complications. This would mean that significant side effects might make the treatment unacceptable for asymptomatic patients. Thus, I can think of no more difficult challenges to pharmacologists and clinical trial designers than diabetic neuropathy, especially when duration of trials and numbers of patients are constrained by finance and the acceptable pace of the return on investment. Let us hope that this book spurs the scientists to reveal the appropriate drug targets and that the industry keeps faith enough to rise to the challenge.

D. Tomlinson
University of Manchester, Manchester, United Kingdom

SECTION I

Clinical Context

CHAPTER ONE

A Brief Introduction to the History and Controversies of Clinical Trials in Diabetic Neuropathy

N.A. Calcutt*[,1], P. Fernyhough[†,‡]
*University of California, San Diego, La Jolla, CA, United States
[†]University of Manitoba, Winnipeg, MB, Canada
[‡]St. Boniface Hospital Research Centre, Winnipeg, MB, Canada
[1]Corresponding author: e-mail address: ncalcutt@ucsd.edu

The earliest clinical trials in subjects with diabetic neuropathy targeted pain as the primary end point (Kastrup, Angelo, Petersen, et al., 1986; Kvinesdal, Molin, Froland, et al., 1984; Rull, Quibrera, Gonzalez-Millan, et al., 1969; Saudek, Werns, & Reidenberg, 1977). This approach recognized that pain was a particularly disruptive feature of diabetic neuropathy and took the pragmatic view that, in the absence of known pathogenic mechanisms, drugs that produced pain relief in other conditions might show benefit in diabetic patients. Sadly, the limited advances made in understanding the pathogenesis of pain in diabetes and difficulties inherent in translating preclinical research to therapeutic use mean that this speculative approach of almost half a century ago is not dissimilar to many current drug development programs for painful diabetic neuropathy. Recent progress in identifying the pathogenesis of painful diabetic neuropathy is discussed in detail elsewhere in this volume. It is also noteworthy that other aspects of nerve function, such as large fiber conduction velocity and action potential amplitudes, were initially included in clinical trials primarily to establish that any pain relief achieved by therapeutic intervention was not due to general neurotoxicity of the drugs under investigation. However, reports that glucose metabolism through the polyol pathway contributed to large fiber nerve conduction velocity (NCV) slowing in diabetic rodents (Tomlinson, Holmes, & Mayer, 1982) supported the evolution of large fiber electrophysiological dysfunction into the de facto biomarker for diabetic peripheral neuropathy. A statistically significant improvement of conduction velocity slowing persists as the primary

International Review of Neurobiology, Volume 127
ISSN 0074-7742
http://dx.doi.org/10.1016/bs.irn.2016.03.014

objective of therapeutic intervention for many regulatory authorities to this day. However, the ascendancy of such a narrow focus on large fiber function in diabetic neuropathy has been questioned (Greene, Brown, Braunstein, et al., 1981; Malik, 2014) and is coming under increasing scrutiny in the face of therapies designed to primarily protect small sensory fibers. Ongoing validation of new and sensitive methods for measuring small fiber function and structure (Chen, Graham, Dabbah, et al., 2015) may further strengthen the case for a more flexible, therapy-specific, approach.

Reports that glucose metabolism by aldose reductase and subsequent nerve myo-inositol depletion caused NCV slowing in diabetic rats (Greene, De Jesus, & Winegrad, 1975; Tomlinson et al., 1982) prompted a number of small-scale clinical trials in subjects with diabetic neuropathy that used NCV as an arbiter of drug efficacy. Myo-inositol supplementation was not successful (Gregersen, Bertelsen, Harbo, et al., 1983; Salway, Whitehead, Finnegan, et al., 1978), was not obviously pertinent in species other than rats (Calcutt, Willars, & Tomlinson, 1988; Dyck, Sherman, Hallcher, et al., 1980; Miwa, Kanbara, & Okuda, 1989), and has largely faded from investigation as a contributor to the pathogenesis of diabetic neuropathy. In contrast, aldose reductase inhibitors (ARIs) showed some efficacy against NCV slowing (Fagius & Jameson, 1981; Judzewitsch, Jaspan, Polonsky, et al., 1983) and sensory disorders such as pain (Jaspan, Maselli, Herold, et al., 1983; Young, Ewing, & Clarke, 1983). These small successes identified statistically significant improvements in NCV of around 1 m/s, representing about 10% of the deficit but outside what was considered meaningful when extrapolated to clinical features (Dyck & O'Brien, 1989). Even so, these results prompted a number of extensive development and clinical trial programs that began with the ARIs sorbinil, ponalrestat (statil), and tolrestat, as reviewed elsewhere (Oates, 2008). Improvements of NCV in the 1 m/s range following ARI treatment continue to be reported (Polydefkis, Arezzo, Nash, et al., 2015), despite refinement of drug efficacy and clinical trial design.

Thioctic acid, an antioxidant also known as alpha-lipoic acid, was used in early clinical trials (Sachse & Willms, 1980) and showed efficacy against pain and other symptoms but not against NCV slowing. This treatment has reemerged in the past 15 years following extensive clinical trials that were not entirely weighted toward NCV slowing for demonstration of efficacy (Papanas & Ziegler, 2014; Ziegler, Low, Freeman, et al., 2016). The success of clinical trials of alpha-lipoic acid appears to depend on how efficacy is defined by investigators and regulatory agencies such that the opinions of authors contributing to this volume are divided. Alpha-lipoic acid is available by prescription to treat diabetic neuropathy in Germany and oral formulations are considered an over-the-counter nutritional supplement in the United States but do not have FDA approval for treating diabetic neuropathy.

Other agents used in early clinical trials included vitamin B6 as pyridoxine (Levin, Hanscom, Fisher, et al., 1981), which can perhaps be viewed as a precursor to a later and recurrent therapeutic theme of testing dietary supplements with antioxidant properties. Gangliosides, glycosphingolipid constituents of neuronal plasma membranes, were also used (Abraham, Abraham, & Wynn, 1984; Crepaldi, Fedele, Tiengo, et al., 1983), despite an ill-defined mechanistic rationale. Subsequent preclinical data have shown that gangliosides modulate content and signaling of endogenous neurotrophic factors and their receptors (Fukuda, Fukui, Hikichi, et al., 2015; Nishio, Fukumoto, Furukawa, et al., 2004; Vieira, de Almeida e Silva Lima Zollner, Malaguti, et al., 2008). This places use of gangliosides as the harbinger of a sustained interest in the development of neurotrophic and other growth factors as a therapeutic approach for diabetic neuropathy (Calcutt, Jolivalt, & Fernyhough, 2008). Attempts to replace diminished endogenous neurotrophic support in diabetes or to promote growth factor-driven neuronal survival and regrowth continue to the present. Indeed, it is striking how many of the early therapeutic approaches have persisted in one guise or another, with increased polyol pathway flux, oxidative stress, and loss of neurotrophic support still being targets in recently reported clinical trials (Kessler, Smith, Cha, et al., 2015; Polydefkis et al., 2015; Ziegler et al., 2016). Whether this reflects an early recognition of the dominant pathogenic mechanisms of diabetic neuropathy that has simply required refinement of drugs, trial design, and end point measurements or else represents an introspective intellectual stagnation is a controversy that is frequently raised by those outside the field.

The other intervention studied during this early period of clinical trials in diabetic neuropathy was improved glycemic control, usually by instituting more intensive insulin regimes in type 1 diabetic subjects (Dahl-Jorgensen, Brinchmann-Hansen, Hanssen, et al., 1986; Service, Rizza, Daube, et al., 1985). A notable feature of such trials was that they followed subjects for periods of months–years rather than the weeks–months that were the norm in early drug intervention trials. Indeed, the short duration of clinical trials, along with low numbers of subjects, recruitment of subjects with severe and probably irreversible neuropathy, and an inability to assess efficacy against the proposed pathogenic mechanism (nerve polyol pathway flux, oxidative damage, etc.), have been widely acknowledged as weaknesses that could have limited the success of early clinical trials. Improving glycemic control allows for a clear and well-validated readout of subject compliance and treatment efficacy via $HbA1_C$ levels. The initial successes against NCV slowing led to instigation of large multicenter clinical trials of intensive glycemic control such as the Oslo (Amthor, Dahl-Jorgensen, Berg, et al., 1994), DCCT (1988), and EDIC studies. These studies were careful to address trial design concerns and emphasized standardized objective measures of nerve function. Their influence on clinical trial design has persisted to the present.

The DCCT and EDIC trials codified many good practices for clinical trials. These include long durations of treatment, use of multiple trial sites, recruitment of large numbers of subjects with mild to moderate neuropathy, implementation of standardized protocols with objective measurements, and the tracking of subject compliance. However, their very success may also have introduced an unhelpful rigidity to subsequent decades of preclinical and clinical research in diabetic neuropathy, specifically in their focus on hyperglycemia as the primary pathogenic mechanism and large fiber NCV slowing as the primary readout of efficacy. Many of the current controversies in diabetic neuropathy arise from an evolving appreciation that the prolonged intensive insulin therapy of the DCCT was only partially effective and that risk factors such as hypertension and dyslipidemia may contribute independently to neuropathy. The capacity of insulin to act as a direct trophic factor for peripheral nerve and the absence of C-peptide replacement during insulin therapy may also contribute to both the apparent efficacy of intensive glycemic control and its incomplete efficacy in studies such as the DCCT. Further, the focus on large fiber NCV slowing as the primary readout of treatment efficacy may constrain use of technological advances in the objective quantification of small fiber neuropathy to preferentially determine efficacy of agents that specifically target this aspect of diabetic neuropathy.

The persistent failure of clinical trials of agents to treat diabetic neuropathy is frequently attributed to poor drug design and inappropriate animal models that identify pathogenic mechanisms that are not pertinent to humans. However, at the other end of the pipeline, recognition that one size (of trial design) may not fit all (therapeutic approaches) should allow a flexibility that embraces advances in how diabetic neuropathy is assessed so that both new and old therapies can be evaluated in a scientifically rational and clinically meaningful manner.

REFERENCES

Factors in development of diabetic neuropathy. Baseline analysis of neuropathy in feasibility phase of Diabetes Control and Complications Trial (DCCT). The DCCT Research Group. (1988). *Diabetes, 37*, 476–481.

Abraham, R. R., Abraham, R. M., & Wynn, V. (1984). A double blind placebo controlled trial of mixed gangliosides in diabetic peripheral and autonomic neuropathy. *Advances in Experimental Medicine & Biology, 174*, 607–624.

Amthor, K. F., Dahl-Jorgensen, K., Berg, T. J., et al. (1994). The effect of 8 years of strict glycaemic control on peripheral nerve function in IDDM patients: The Oslo Study. *Diabetologia, 37*, 579–584.

Calcutt, N. A., Jolivalt, C. G., & Fernyhough, P. (2008). Growth factors as therapeutics for diabetic neuropathy. *Current Drug Targets, 9*, 47–59.

Calcutt, N. A., Willars, G. B., & Tomlinson, D. R. (1988). Statil-sensitive polyol formation in nerve of galactose-fed mice. *Metabolism, 37*, 450–453.

Chen, X., Graham, J., Dabbah, M. A., et al. (2015). Small nerve fiber quantification in the diagnosis of diabetic sensorimotor polyneuropathy: Comparing corneal confocal microscopy with intraepidermal nerve fiber density. *Diabetes Care, 38*, 1138–1144.

Crepaldi, G., Fedele, D., Tiengo, A., et al. (1983). Ganglioside treatment in diabetic peripheral neuropathy: A multicenter trial. *Acta Diabetologica Latina, 20*, 265–276.

Dahl-Jorgensen, K., Brinchmann-Hansen, O., Hanssen, K. F., et al. (1986). Effect of near normoglycaemia for two years on progression of early diabetic retinopathy, nephropathy, and neuropathy: The Oslo study. *British Medical Journal (Clinical Research Ed), 293*, 1195–1199.

Dyck, P. J., & O'Brien, P. C. (1989). Meaningful degrees of prevention or improvement of nerve conduction in controlled clinical trials of diabetic neuropathy. *Diabetes Care, 12*, 649–652.

Dyck, P. J., Sherman, W. R., Hallcher, L. M., et al. (1980). Human diabetic endoneurial sorbitol, fructose, and myo-inositol related to sural nerve morphometry. *Annals of Neurology, 8*, 590–596.

Fagius, J., & Jameson, S. (1981). Effects of aldose reductase inhibitor treatment in diabetic polyneuropathy—A clinical and neurophysiological study. *Journal of Neurology, Neurosurgery & Psychiatry, 44*, 991–1001.

Fukuda, Y., Fukui, T., Hikichi, C., et al. (2015). Neurotropin promotes NGF signaling through interaction of GM1 ganglioside with Trk neurotrophin receptor in PC12 cells. *Brain Research, 1596*, 13–21.

Greene, D. A., Brown, M. J., Braunstein, S. N., et al. (1981). Comparison of clinical course and sequential electrophysiological tests in diabetics with symptomatic polyneuropathy and its implications for clinical trials. *Diabetes, 30*, 139–147.

Greene, D. A., De Jesus, P. V., Jr., & Winegrad, A. I. (1975). Effects of insulin and dietary myoinositol on impaired peripheral motor nerve conduction velocity in acute streptozotocin diabetes. *Journal of Clinical Investigation, 55*, 1326–1336.

Gregersen, G., Bertelsen, B., Harbo, H., et al. (1983). Oral supplementation of myoinositol: Effects on peripheral nerve function in human diabetics and on the concentration in plasma, erythrocytes, urine and muscle tissue in human diabetics and normals. *Acta Neurologica Scandinavica, 67*, 164–172.

Jaspan, J., Maselli, R., Herold, K., et al. (1983). Treatment of severely painful diabetic neuropathy with an aldose reductase inhibitor: Relief of pain and improved somatic and autonomic nerve function. *Lancet, 2*, 758–762.

Judzewitsch, R. G., Jaspan, J. B., Polonsky, K. S., et al. (1983). Aldose reductase inhibition improves nerve conduction velocity in diabetic patients. *New England Journal of Medicine, 308*, 119–125.

Kastrup, J., Angelo, H., Petersen, P., et al. (1986). Treatment of chronic painful diabetic neuropathy with intravenous lidocaine infusion. *British Medical Journal (Clinical Research Ed.), 292*, 173.

Kessler, J. A., Smith, A. G., Cha, B. S., et al. (2015). Double-blind, placebo-controlled study of HGF gene therapy in diabetic neuropathy. *Annals of Clinical and Translational Neurology, 2*, 465–478.

Kvinesdal, B., Molin, J., Froland, A., et al. (1984). Imipramine treatment of painful diabetic neuropathy. *JAMA, 251*, 1727–1730.

Levin, E. R., Hanscom, T. A., Fisher, M., et al. (1981). The influence of pyridoxine in diabetic peripheral neuropathy. *Diabetes Care, 4*, 606–609.

Malik, R. A. (2014). Which test for diagnosing early human diabetic neuropathy? *Diabetes, 63*, 2206–2208.

Miwa, I., Kanbara, M., & Okuda, J. (1989). Improvement of nerve conduction velocity in mutant diabetic mice by aldose reductase inhibitor without affecting nerve myo-inositol content. *Chemical & Pharmaceutical Bulletin (Tokyo), 37*, 1581–1582.

Nishio, M., Fukumoto, S., Furukawa, K., et al. (2004). Overexpressed GM1 suppresses nerve growth factor (NGF) signals by modulating the intracellular localization of NGF receptors and membrane fluidity in PC12 cells. *Journal of Biological Chemistry, 279*, 33368–33378.

Oates, P. J. (2008). Aldose reductase, still a compelling target for diabetic neuropathy. *Current Drug Targets, 9*, 14–36.

Papanas, N., & Ziegler, D. (2014). Efficacy of alpha-lipoic acid in diabetic neuropathy. *Expert Opinion on Pharmacotherapy, 15*, 2721–2731.

Polydefkis, M., Arezzo, J., Nash, M., et al. (2015). Safety and efficacy of ranirestat in patients with mild-to-moderate diabetic sensorimotor polyneuropathy. *Journal of the Peripheral Nervous System, 20*, 363–371.

Rull, J. A., Quibrera, R., Gonzalez-Millan, H., et al. (1969). Symptomatic treatment of peripheral diabetic neuropathy with carbamazepine (Tegretol): Double blind crossover trial. *Diabetologia, 5*, 215–218.

Sachse, G., & Willms, B. (1980). Efficacy of thioctic acid in the therapy of peripheral diabetic neuropathy. *Hormone and Metabolic Research Supplement Series, 9*, 105–107.

Salway, J. G., Whitehead, L., Finnegan, J. A., et al. (1978). Effect of myo-inositol on peripheral-nerve function in diabetes. *Lancet, 2*, 1282–1284.

Saudek, C. D., Werns, S., & Reidenberg, M. M. (1977). Phenytoin in the treatment of diabetic symmetrical polyneuropathy. *Clinical Pharmacology and Therapeutics, 22*, 196–199.

Service, F. J., Rizza, R. A., Daube, J. R., et al. (1985). Near normoglycaemia improved nerve conduction and vibration sensation in diabetic neuropathy. *Diabetologia, 28*, 722–727.

Tomlinson, D. R., Holmes, P. R., & Mayer, J. H. (1982). Reversal, by treatment with an aldose reductase inhibitor, of impaired axonal transport and motor nerve conduction velocity in experimental diabetes mellitus. *Neuroscience Letters, 31*, 189–193.

Vieira, K. P., de Almeida e Silva Lima Zollner, A. R., Malaguti, C., et al. (2008). Ganglioside GM1 effects on the expression of nerve growth factor (NGF), Trk-A receptor, proinflammatory cytokines and on autoimmune diabetes onset in non-obese diabetic (NOD) mice. *Cytokine, 42*, 92–104.

Young, R. J., Ewing, D. J., & Clarke, B. F. (1983). A controlled trial of sorbinil, an aldose reductase inhibitor, in chronic painful diabetic neuropathy. *Diabetes, 32*, 938–942.

Ziegler, D., Low, P. A., Freeman, R., et al. (2016). Predictors of improvement and progression of diabetic polyneuropathy following treatment with alpha-lipoic acid for 4 years in the NATHAN 1 trial. *Journal of Diabetes and Its Complications, 30*, 350–356.

CHAPTER TWO

Neuropathy in the DCCT/EDIC—What Was Done Then and What We Would Do Better Now

R. Pop-Busui[1], C. Martin
University of Michigan Medical School, Ann Arbor, MI, United States
[1]Corresponding author: e-mail address: rpbusui@umich.edu

Contents

Abstract

The Diabetes Control and Complications Trial (DCCT) and its epidemiological follow-up, the Epidemiology of Diabetes Interventions and Complications (EDIC) provide important insight on the natural history of distal symmetrical polyneuropathy and cardiovascular autonomic neuropathy in patients with type 1 diabetes and on the impact of intensive treatment of hyperglycemia on disease progression. This chapter summarizes the design and methods used for neuropathy evaluations both in the DCCT and in EDIC, the characteristics of the DCCT/EDIC patient population, and summarizes the findings of the DCCT/EDIC relative to neuropathic complications of type 1 diabetes. Lessons learned from the DCCT and EDIC experiences of longitudinal assessments of neuropathic complications are also reviewed.

International Review of Neurobiology, Volume 127
ISSN 0074-7742
http://dx.doi.org/10.1016/bs.irn.2016.02.020

1. INTRODUCTION

Diabetic neuropathies are among the most prevalent of the chronic complications of diabetes (Boulton et al., 2005; Tesfaye et al., 2010), with a broad spectrum of forms and clinical manifestations. The most studied in clinical trials are distal (length-dependent) symmetrical sensorimotor polyneuropathy (DSPN) and cardiovascular autonomic neuropathy (CAN). Outcome measures specific to DSPN and CAN forms were included in the Diabetes Control and Complications Trial (DCCT) design and will be discussed in this chapter.

The DCCT enrolled 1441 patients with type 1 diabetes between 1983 and 1989 at 28 clinical sites across United States and Canada (DCCT, 1993). The DCCT included a primary prevention cohort and a secondary intervention cohort. The primary prevention cohort had diabetes for 1–5 years (mean 2.6 years) and no retinopathy at baseline. The secondary intervention cohort had diabetes for 1–15 years (mean 8.7 years) and mild-to-moderate retinopathy at baseline (DCCT, 1993). Patients were randomly assigned to intensive (INT) or conventional (CONV) insulin therapy. Briefly, the intensive therapy group used either insulin pumps, or three or more daily injections of insulin, together with frequent daily self-blood glucose monitoring in order to achieve blood glucose levels as close as possible to normal. The conventional therapy group used one or two injections of insulin daily, without routine blood glucose monitoring with a goal of minimal symptoms related to hyper- or hypoglycemia. In 1993, after an average of 6.5 years of follow-up, HbA1c was 7.4% in the intensive group and 9.1% in the conventional group (DCCT, 1993). The HbA1c improvement in the intensive group was associated with significantly reduced incidence of diabetic retinopathy, nephropathy, and neuropathy and the randomized, clinical trial was therefore stopped (DCCT, 1993, 1995, 1998). Participants from both the INT and CONV groups were encouraged to continue or to adopt an intensive diabetes treatment regimen and DCCT CONV group subjects were instructed in intensive therapy by the DCCT treatment teams. By the end of 1993, after an average of 6.5 years of follow-up, all DCCT participants were returned to their prior health care providers for ongoing care.

The DCCT finding that intensive insulin therapy of type 1 diabetes to improve glycemia profoundly reduced complications of type 1 diabetes triggered fundamental changes in the management of diabetes in the United

States and worldwide. Intensive management of hyperglycemia was adopted as the standard of care for most patients with type 1 diabetes (American Diabetes Association, 2015).

The observational Epidemiology of Diabetes Interventions and Complications (EDIC) study began in 1994, with a majority (94%) of former DCCT subjects agreeing to participate (Epidemiology of Diabetes Interventions and Complications (EDIC), 1999; DCCT, 1993). The EDIC was established to monitor the long-term effects of prior INT compared to prior CONV treatment on the development and progression of more advanced complications and of cardiovascular disease among the DCCT cohort.

Subsequent EDIC evaluations demonstrated long-term benefits of prior intensive glycemic control on complications (DCCT, 2002; The Diabetes Control and Complications Trial/Epidemiology of Diabetes Interventions and Complications Research Group, 2000; Writing Team for DCCT/EDIC, 2003) and cardiovascular disease (Nathan et al., 2005). The persistent beneficial effect of prior intensive glucose control has been termed "metabolic memory" (Writing Team for DCCT/EDIC, 2003).

2. NEUROPATHY OUTCOMES ASSESSMENTS

Diabetic neuropathies are one of the most prevalent chronic complications of diabetes, with multiple manifestations, consistent risk factors, and complex pathogenetic mechanisms. The specific presentation of diabetic neuropathy reflects the distribution and size of nerve fibers involved. By far, the most prevalent neuropathies observed in diabetes are DSPN and diabetic autonomic neuropathy, especially CAN (American Diabetes Association, 2015; Ang, Jaiswal, Martin, & Pop-Busui, 2014).

2.1 DCCT and EDIC Design

DSPN and CAN were evaluated both in DCCT and in EDIC. Table 1 lists the measures obtained in DCCT and EDIC. The institutional review boards of all participating centers approved the DCCT and EDIC study protocols, and all participants provided written informed consent. Participation in DSPN, CAN, and other neuropathy-related outcome assessments during DCCT and EDIC are shown in Fig. 1.

2.1.1 Assessment of DSPN in DCCT

DSPN was assessed at DCCT baseline and after 5 years of DCCT participation and/or at the end of the DCCT study. The assessment included

Table 1 Neuropathy Evaluations During DCCT and During EDIC

	DCCT	EDIC
DSPN evaluations	Symptoms of DSPN Signs (neurological examination by board-certified neurologist) Electrophysiology (median, peroneal, sural nerves)	Symptoms of DSPN Signs (neurological examination by board-certified neurologist) Electrophysiology (median, peroneal, sural nerves) Michigan Neuropathy Screening Instrument (MNSI) Quantitative sensory testing (VPT) NeuroQuol
CAN evaluations	Deep breathing test Valsalva maneuver Postural changes in blood pressure	Deep breathing test Valsalva maneuver Postural changes in blood pressure Autonomic symptom profile
Other		Urologic complications Gastroparesis (pilot only)

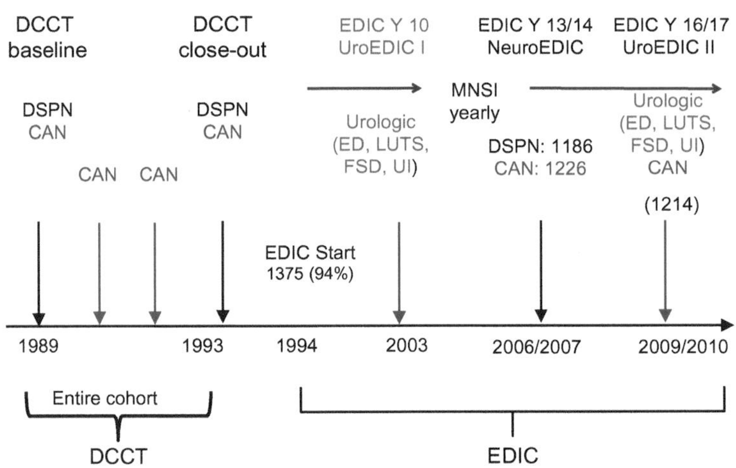

Fig. 1 Timeline of main and other neuropathy evaluations in the DCCT/EDIC. *CAN*, cardiovascular autonomic neuropathy; *DSPN*, distal symmetrical polyneuropathy; *ED*, erectile dysfunction; *FSD*, female sexual dysfunction; *LUTS*, lower urinary tract symptoms; *MNSI*, Michigan Neuropathy Screening Instrument; *UI*, urinary incontinence.

standardized medical history and physical examination by board-certified neurologists, and electrophysiology. Nerve conduction studies were performed on the dominant side and included the median (motor and sensory), peroneal (motor), and sural (sensory) nerves using standard techniques and specified anatomical landmarks or stimulation-to-recording electrode distances for each study of motor and sensory nerves as has been amply described (DCCT, 1993, 1995). Absolute threshold levels for the individual attributes were defined as the median of the upper or lower limits provided by participating laboratories (DCCT, 1995). All participating neurologists were required to complete protocol training and certification to assure uniform implementation of the study protocol across all participating sites, with certification again required for the EDIC year 13/14 measures, including ensuring uniform temperature match (DCCT, 1993, 1995; Martin, Albers, & Pop-Busui, 2014). Neurologists were masked to participant treatment assignment during the DCCT.

Three DSPN outcomes were defined: clinical neuropathy, nerve conduction study (NCS) abnormalities, and confirmed clinical neuropathy. The diagnosis of *clinical neuropathy* was based on a structured history and physical examination performed by the DCCT/EDIC-certified neurologist and required at least two positive responses among symptoms, sensory signs, or ankle reflexes (diminished or absent) consistent with a distal symmetrical polyneuropathy and without causal explanation aside from diabetes (Albers et al., 2010; Martin et al., 2014). The outcome of *NCS abnormality* required an absolute abnormality of amplitude, conduction velocity, or F-wave latency in at least two anatomically distinct nerves (DCCT, 1995). Both clinical neuropathy and nerve conduction studies were secondary outcome measures in DCCT (DCCT, 1993, 1995). The primary DSPN outcome of *confirmed clinical neuropathy* required both the presence of clinical neuropathy and NCS abnormalities as previously defined (Albers et al., 2010; DCCT, 1993, 1995; Martin et al., 2014).

2.1.2 Assessment of DSPN in EDIC

All the comprehensive DSPN measures performed in DCCT described earlier were repeated once during the 13th and 14th years of EDIC follow-up (the Neurology Protocol or NeuroEDIC).

A new measure, the Michigan Neuropathy Screening Instrument (MNSI), was introduced at the start of the EDIC study and was performed annually as a measure of DSPN before the full DCCT evaluations could be repeated (Martin et al., 2006). The MNSI is a two-part tool that includes a 15 question self-administered symptom assessment plus a focused examination of the feet to assess skin and structural abnormalities, distal vibration perception, and ankle reflexes (Feldman et al., 1994; Martin et al., 2006). The MNSI is validated to have a good sensitivity and specificity to detect the presence of DSPN (Feldman et al., 1994; Herman et al., 2012). Using the MNSI, DSPN was defined as either an MNSI questionnaire score of ≥ 7 or an MNSI exam score of ≥ 2.5; these scores being based on the initial validation of the MNSI instrument (Feldman et al., 1994). The MNSI evaluations through EDIC year 8 suggested that the metabolic memory phenomenon observed for retinal and renal complications, also applied to new-onset (incident) neuropathy (Martin et al., 2006). The MNSI however, was not designed to evaluate DSPN severity nor was it initially conceived as a tool for longitudinal assessment of neuropathy. Further, it lacked the sensitivity and specificity of the original DCCT evaluations and therefore could not be used as a surrogate for the robust DCCT primary outcome of confirmed clinical neuropathy.

2.1.3 Additional DSPN Measures in EDIC

Vibration perception threshold (VPT) and a neuropathy-specific quality-of-life instrument (NeuroQOL) were included at EDIC year 13/14 (NeuroEDIC) based on recommendations of post-DCCT era consensus panels that recommended quantitative sensory testing and quality-of-life measures as key components of neuropathy outcome measurement in clinical trials (Boulton, Malik, Arezzo, & Sosenko, 2004; Boulton et al., 2005). VPT was measured using a forced-choice algorithm of decreasing vibration intensity at the dominant index finger and great toe (Martin et al., 2010). Abnormal VPT was defined as a threshold value more than 2.5 standard deviations above the age-adjusted mean value obtained from nondiabetic referents (Martin et al., 2010). Quality of life was assessed using the NeuroQOL, an 18-item survey that addresses six specific domains (pain, lost/reduced feeling, sensory/motor symptoms, functional, social, and emotional experiences), as well as an assessment of overall quality of life (Vileikyte et al., 2003).

2.1.4 Assessment of CAN in DCCT

CAN was assessed during DCCT at baseline, every 2 years and at DCCT-end using standard cardiovascular reflex testing (CART) that included: R-R variation to paced breathing (RRV), R-R response to Valsalva maneuver (VR), and postural changes in blood pressure (DCCT, 1998; Martin et al., 2014; Pop-Busui et al., 2010). These tests are objective, standardized, simple to use, and highly reproducible (Low et al., 1997; Pop-Busui et al., 2009). All subjects were uniformly prepared for CAN testing by instruction that included fasting, abstaining from caffeine, tobacco, and medications the morning of testing, avoidance of vigorous physical activity, and alcohol consumption for 48 h prior to testing and absence of hypoglycemia prior to, and during, testing (Martin et al., 2014). Subjects who experienced hypoglycemia after midnight (defined as a blood glucose ≤50 mg/dL/2.775 mmol/L and/or signs/symptoms of hypoglycemia) or subjects with acute illness on the day of testing were excluded (Pop-Busui et al., 2009). Subjects with active proliferative retinopathy, history of laser therapy or vitrectomy, suspected (unconfirmed) proliferative retinopathy, and/or no eye examination in the last 4 years were excluded from performing the Valsalva maneuver (Pop-Busui et al., 2009). The primary CAN outcome was defined as any of the following criteria: abnormal R-R variation (R-R <15); R-R variation of <20 plus a Valsalva ratio ≤1.5, or a decrease of >10 mmHg in diastolic blood pressure at any point during 10 min of standing after a period of 30 min of supine rest (postural hypotension; DCCT, 1998; Pop-Busui et al., 2010, 2009). All CAN studies were reviewed for quality control purposes by a masked investigator at the reading center who decided if the technical quality of the recording and conditions of the test met study criteria (Pop-Busui et al., 2009). In the DCCT, there was a provision to confirm postural hypotension by repeated testing with catecholamine assessment but it was rarely required (DCCT, 1998).

2.1.5 Assessment of CAN in EDIC

All CAN evaluations described above were repeated on two occasions in EDIC; first during EDIC years 13/14, and again during EDIC years 16/17. The catecholamine confirmation of postural hypotension was rarely required during DCCT and it was not included in the EDIC CAN protocol (Pop-Busui et al., 2009). Testing was performed with the Hokanson ANS 2000 device (Hokanson Inc., Bellevue, WA) and results were analyzed by a single reader at a central reading center. During EDIC, the reproducibility of the CAN measurements was evaluated at the reading center where a random sample of 10% of tests were re-read, demonstrating a coefficient of reliability between the primary and the repeat readings of 0.999 (Pop-Busui et al., 2009). Test–retest reproducibility was also evaluated on randomly selected subjects during EDIC with good reproducibility noted for R-R (89%) and Valsalva ratios (91%) (Martin et al., 2014). The same DCCT CAN outcomes were analyzed in EDIC (Pop-Busui et al., 2009). However, considering that since DCCT-end new evidence has emerged regarding some limitations of the categorical CAN outcomes definitions, especially related to an age-related and possibly gender-related effect, new secondary CAN outcomes in EDIC included the age- and gender-adjusted continuous R-R variation and Valsalva ratio (Martin et al., 2014; Pop-Busui et al., 2010, 2009).

The Autonomic Symptom Profile (ASP) (Low et al., 2004), a self-administered survey of autonomic symptoms that collects information on symptoms of postural hypotension, gastroparesis, diarrhea, colonic atony, genitourinary dysfunction, and hypoglycemic unawareness was added to the battery of CAN assessments in EDIC. The ASP survey was added to better describe the participant burden of a variety of autonomic complications (Low et al., 2004; Pop-Busui et al., 2009).

3. COMPLEMENTARY ASSESSMENTS IN EDIC

3.1 Evaluation of Urologic Complications

The UroEDIC ancillary study was conducted at EDIC years 10 and 16 as cross-sectional evaluations to describe the prevalence of diabetic genitourinary complications that included erectile dysfunction (ED) and lower urinary tract symptoms (LUTS) in men, and female sexual dysfunction (FSD) and urinary incontinence (UI) in women, and to investigate relationships between these and other diabetes complications (Martin et al., 2014; Sarma et al., 2009; Wessells et al., 2011). ED was defined as an erectile function domain score of <20 using the International Index of Erectile Function (IIEF) and by response to a single questionnaire item regarding "confidence to get and keep an erection." LUTS were assessed using the American Urological Association Symptom Index (AUASI) with moderate-to-severe LUTS defined by scores >7 (Martin et al., 2014).

3.2 Gastroparesis

The DCCT did not assess gastroparesis, arguably one of the most debilitating autonomic complications of diabetes and certainly one of the most frustrating from a clinical management perspective. In EDIC, the assessment of gastroparesis in the entire cohort is entirely based on self-report of symptoms obtained at annual EDIC visits. However, a formal assessment of gastric emptying as a measure of gastroparesis was performed in 78 DCCT/EDIC participants at EDIC year 20 using the 13C-Spirulina gastric emptying breath test (GEBT, 223 kcal with 19.2 g carbohydrates, 12 g protein, and 10.9 g fat; AB Diagnostics, Brentwood, TN) under an investigator-initiated investigational new drug (IND) from the Food and Drug Administration (FDA) as described (Bharucha et al., 2015).

3.3 Other Evaluations in EDIC

Several self-report measures were also used in EDIC to evaluate the prevalence of hypoglycemia unawareness and use of medications to treat painful neuropathic symptoms (Martin et al., 2014).

4. DCCT/EDIC FINDINGS

4.1 DSPN and CAN Outcomes in the DCCT and EDIC Study

DSPN (confirmed clinical neuropathy) and CAN were uncommon at DCCT start, in part owing to the intentional exclusion of persons with symptomatic neuropathy (DCCT, 1993; Martin et al., 2014; Pop-Busui et al., 2010). The prevalence of confirmed clinical neuropathy increased substantially in the conventional subjects (from 5% to 17%, $P < 0.001$) and only slightly among the intensive group participants (from 7% to 9%; DCCT, 1993, 1995; Martin et al., 2014; Pop-Busui et al., 2010). Adjusting for the presence of confirmed DSPN at baseline, the risk reduction for incident DSPN with intensive glucose control during DCCT was 64% (95% CI, 45–76) (DCCT, 1993, 1995; Martin et al., 2014).The prevalence of CAN almost doubled in the conventional group by DCCT-end, while remaining static in the intensive group (DCCT, 1993, 1998). The risk reduction in incident CAN with intensive therapy during DCCT was 45% (DCCT, 1993, 1998; Martin et al., 2014; Pop-Busui et al., 2010).

During EDIC, the HbA1c separation between former intensive and conventional treatment groups observed through the DCCT quickly waned and by the fifth year of EDIC follow-up, the HbA1c separation between treatment groups completely disappeared (Martin et al., 2014; Pop-Busui et al., 2009; Writing Team for DCCT/EDIC, 2003). Prevalence of DSPN increased during EDIC in both groups and, despite no measureable difference in glucose control from EDIC year 5 through EDIC year 13/14, a 30% reduction in the risk of developing confirmed DSPN by EDIC year 13/14 with prior intensive glucose control persisted (Martin et al., 2014; Pop-Busui et al., 2010). Similar trends were observed on several NCS measures (Albers et al., 2010; Martin et al., 2014; Pop-Busui et al., 2010). After adjusting for NCS results at DCCT closeout, the risk reduction attenuated to 17% (not significant), suggesting that subclinical DSPN (as documented by differences in NCS at DCCT-end, but not meeting criteria for confirmed clinical neuropathy) contributed to the finding of a persistent benefit (Albers et al., 2010, 2007; Martin et al., 2014). Logistic regression analyses to evaluate the effects of glycemic control on prevalent and incident DSPN demonstrated that the odds of all DSPN-related outcomes each increased per unit (%) increase in DCCT and EDIC mean HbA1c (Albers et al., 2010; Ang et al., 2014; Martin et al., 2014; Pop-Busui et al., 2010).

CAN prevalence also increased by EDIC year 13–14 (29% in the intensive vs 35% in the conventional group), with treatment-group differences remaining significant (Martin et al., 2014; Pop-Busui et al., 2009). After adjusting for important covariates including age, sex, cohort assignment, and the level of RRV at DCCT closeout, intensive glucose control during DCCT reduced the risk of incident CAN during EDIC by 31% and of abnormal RRV by 30% (Pop-Busui et al., 2009). The adjusted RRV was significantly higher in the former intensive group as compared to the conventional group (Pop-Busui et al., 2009). The persistent beneficial effects of intensive glucose therapy on CAN after 13–14 years of follow-up in EDIC were explained by the differences in the DCCT and EDIC mean HbA1c levels between groups (Martin et al., 2014; Pop-Busui et al., 2010, 2009). Although approximately 1/3 of DCCT/EDIC participants had CAN by the 13th–14th year of EDIC follow-up, very few participants reported autonomic symptoms (Martin et al., 2014; Pop-Busui et al., 2009). No group differences in the prevalence of autonomic symptoms reported were observed, supporting the notion that autonomic neuropathies may remain largely asymptomatic, even after many years of T1D.

Among the 635 T1D men with both urologic (ED and LUTS) and CAN evaluations at EDIC year 16/17, ED only was reported by 193 (30%), LUTS only by 61 (10%), and ED and LUTS by 97 (15%) participants (Pop-Busui et al., 2015). Men who developed ED and/or LUTS during EDIC had significantly lower R-R variation and Valsalva ratio at DCCT closeout and EDIC year 16/17 compared to those without ED or LUTS. In adjusted analysis, participants with CAN had 2.65 greater odds of ED and LUTS (95% CI, 1.47, 4.79) (Pop-Busui et al., 2015). Among the 580 T1D women DCCT/EDIC participants with both urologic (FSD UI) and CAN evaluations at EDIC year 16/17, 26% and 30% of women presented with FSD and UI, respectively, and there was a significant association between measures of CAN with FSD and UI (Hotaling et al., 2013). Additionally, in cross-sectional analyses, both ED and LUTS in men were also associated with DPN (Pop-Busui et al., 2014), owing perhaps to shared or overlapping mechanisms of neuronal damage.

Epigastric fullness, used as a surrogate marker of gastroparesis in EDIC, was reported by 8% of both INT and CONV subjects (Pop-Busui et al., 2009). In the small pilot study that formally assessed gastric emptying, only 50% presented with normal emptying, while in 47% it was delayed and in 3% it was rapid (Bharucha et al., 2015). The DCCT baseline HbA1c (odds ratio [OR], 1.6; 95% confidence interval [CI], 1.1–2.3), duration of diabetes (OR, 1.2; 95% CI, 1.01–1.3) before DCCT entry, and the mean HbA1c during DCCT-EDIC (OR, 2.2; 95% CI, 1.04–4.5) were associated independently with delayed gastric emptying (Bharucha et al., 2015). However, these data have not yet been confirmed in the entire cohort.

Hypoglycemia unawareness (decreased adrenergic symptoms reported in response to hypoglycemia) was reported by 20% of INT subjects and 25% of CONV subjects in EDIC (p=NS) (Pop-Busui et al., 2009). There were no clinically relevant treatment-group differences in the frequency of reported medication use to reduce neuropathic pain during EDIC (Martin et al., 2014).

5. DISCUSSION

The DCCT/EDIC has furthered the understanding of the role of glucose control on development and progression of DSPN and CAN (Martin et al., 2014; Pop-Busui et al., 2010). Differences in the incidence and prevalence of both DSPN and CAN reflect differences in glucose control, favoring HbA1c levels that are closer to nondiabetic levels.

Confirmed clinical neuropathy increased in both intensive and conventional participants by EDIC year 13–14 but treatment-group differences were eliminated by adjusting for continuous nerve conduction variables at the end of the DCCT (Martin et al., 2014). Among subjects without confirmed clinical neuropathy, former conventional subjects had worse (less normal) electrophysiologic results at the end of the DCCT (Albers et al., 2010, 2007; Martin et al., 2014). These subjects were already closer to the "tipping point" for meeting confirmed clinical neuropathy criteria by DCCT-end and this could partially explain the findings of continued difference in the development of DSPN in the intensive and conventional groups during EDIC (Albers et al., 2010, 2007; Martin et al., 2014). Whether an additional influence of early intensive glucose control might have been apparent earlier during EDIC is unknown, but a persistent metabolic effect was not required to explain the durable beneficial effects on confirmed DSPN at EDIC year 13–14 (Albers et al., 2010, 2007; Martin et al., 2014).

In contrast, the durable benefit of intensive glucose control first observed for retinopathy and nephropathy was also observed for CAN at EDIC year

13–14 and modeling that adjusted for R-R variation at DCCT closeout did not negate the INT-associated risk reduction for development of CAN (Martin et al., 2014; Pop-Busui et al., 2009). These findings parallel the reported long-term benefits of prior intensive glycemic control on retinopathy, nephropathy (DCCT, 2002; The Diabetes Control and Complications Trial/Epidemiology of Diabetes Interventions and Complications Research Group, 2000; Writing Team for DCCT/EDIC, 2003), and cardiovascular disease (Nathan et al., 2005), and are consistent with the phenomenon termed "metabolic memory" (Martin et al., 2014; Pop-Busui et al., 2009; Writing Team for DCCT/EDIC, 2003). This discordance of the impact of DCCT treatment-group effect on longer term outcomes for DSPN and CAN may reflect differences in susceptibility of small and large nerve fibers to glycemic exposure (Martin et al., 2014; Pop-Busui et al., 2010).

Although both DCCT and EDIC confirmed that glycemic control is a significant and robust predictor of neuropathy in T1D, they also show that for most patients with T1D, current strategies for optimizing glucose control are insufficient to fully prevent or delay the development of neuropathic complications, as 25% of subjects in the former intensive treatment group and 35% of subjects in the former conventional treatment group had confirmed DSPN at EDIC year 13–14.

Genitourinary problems associated with diabetes, with the likely exception of ED, are frequently overlooked in clinical practice and are rarely considered in the context of diabetic neuropathies. Including evaluations for the urologic complications both in men and women as part of the UroEDIC study thus afforded an opportunity to define the extent of genitourinary complications of diabetes and to explore any relationships to well-defined complications, including neuropathy (Pop-Busui et al., 2015). Our recent findings reporting strong association between CAN, ED, and LUTS in a large cohort of more than 600 male DCCT/EDIC participants with T1DM demonstrate a link between CAN and more generalized autonomic neuropathy and also suggest that CAN may predict the development of ED and LUTS in men with long-standing T1DM (Pop-Busui et al., 2015). Although prior experimental evidence suggested that peripheral and autonomic dysfunction may be an important factor in the etiology of ED and possibly LUTS, these findings are the first to systematically demonstrate a link between CAN and DSPN, and ED/LUTS in a large cohort of well-characterized men with T1DM, owing perhaps to shared or overlapping mechanisms of neuronal damage (Pop-Busui et al., 2015).

5.1 DCCT/EDIC and Contemporary Neuropathy Trials Design

In multicenter trials, the need to systematically train, certify, and reassess competency of staff performing tests is often overlooked, contributing to inter- and intrasite variability over time. Yet, as demonstrated by the DCCT and EDIC, well-standardized and validated measures of DSPN and CAN can be successfully implemented in large, multicenter, multiyear trials (Albers et al., 2010; DCCT, 1995, 1998; Martin et al., 2014; Pop-Busui et al., 2010, 2009).

The DCCT and EDIC follow-up continue to add to the understanding of diabetes-related neurologic complications. There are also, without doubt, shortcomings and important questions that remain unanswered. Improvements in technology and changes in methods for measurement of complicated longitudinal assessments of well-defined or established outcomes are required. For example, throughout DCCT and for the first 15 years of EDIC, fundus photographs were captured on film, the film developed as slides, and the slides then sent to a central reading center for grading. The emergence of digital photography led to the gradual obsolescence of the cameras which had been used throughout DCCT and EDIC (and at some centers the cameras themselves, or parts to repair them, became unavailable). To ensure continued photographic quality, reproducibility, and comparability, a detailed and time-consuming study was performed to compare same day, same subject film vs digital images prior to adopting digital imaging for future EDIC studies. With regard to CAN assessments, while the testing protocol (eg, subject preparation, 6-min supine R-R during deep breathing, followed by 10 min of standing, followed by at least two Valsalva attempts with standard intervals between test components and centralized analysis), and the Hokanson, Inc. (Bellevue, WA), ANS capture devices used were identical in both DCCT and EDIC, the methods for data capture were not identical. Specifically, during the DCCT the changes in the R-R in response to the cardiovascular reflex tests was recorded onto cassette tape (sent via surface mail to the reading center), and test timing was performed manually (stop-watch), while during EDIC a computer was used for test timing and data capture, and files transmitted digitally to the reading center. Indices of heart rate variability were not part of the CAN evaluations in DCCT or in EDIC, while in a contemporary study design these measures would likely be included.

Another area of continued interest is the role of CAN in the risk of arrhythmia and sudden death, likely mediated through hypoglycemia.

While strong evidence established CAN as an independent predictor of cardiovascular mortality in both T1D and T2D, limitations of the available technologies during DCCT and at the initial EDIC studies, direct evaluations to document an association between CAN, hypoglycemia, and arrhythmia were not possible. Advances in the available technologies for ambulatory cardiac and continuous glucose monitoring will permit the EDIC study to evaluate how changes in blood glucose levels and hypoglycemia may affect heart rate variability and arrhythmia risk in people with and without CAN.

Advances or changes in how clinical disease states are understood and defined may also complicate longitudinal studies such as DCCT/EDIC. For example, the DCCT definition of confirmed clinical neuropathy, at the time was generally well accepted as an appropriate definition for DSPN. This outcome has received recent criticism for its lack of a comprehensive measure of small-fiber dysfunction although, at the time the DCCT was designed, the role and importance of small-fiber dysfunction in the progression of diabetic neuropathy was poorly understood and small-fiber assessments were largely lacking from most neuropathy trials. Arguably, measures of small-fiber neuropathy are most likely to unveil subclinical stages of neuropathy, and are potentially most amenable to interventions (Malik et al., 2011; Smith & Singleton, 2012). Given the natural history of DSPN and CAN in T1D in the current era of standard of care for hyperglycemia and other risk factors, the design of a contemporary trial evaluating neuropathy outcomes would certainly include measures that more directly evaluate small fibers such as intraepidermal nerve fiber density and function, or corneal confocal microscopy. These techniques might also allow tailored trial designs with shorter duration, and/or smaller sample size. Although assessments for other forms of autonomic neuropathy were not performed in DCCT, during EDIC information on urologic complications and some data on gastroparesis were also collected. Yet, to better understand the effects of various risk factors and/or interventions on autonomic neuropathy, adding validated and sensitive measures testing the urogenital, gastrointestinal, or sudomotor autonomic function is necessary.

Integrating patient-specific and meaningful outcomes (for example, quality-of-life measures, pain, sleep, and other functional assessments) that better assess the impact of neuropathy on daily function, burden, and quality of life should be incorporated into modern trial designs. For example, DSPN may compromise balance in daily activities due to progressive loss of proprioception and later weakness, superimposed on the effect of aging, leading to imbalance and unsteadiness in gait, with increased likelihood of a fall.

Lastly, at the time the DCCT was conducted, the limitations of exogenous insulin treatment, namely the significantly greater risk of hypoglycemia associated with INT treatment during DCCT, resulted in overall levels of glucose control that were still not optimal. It may be that, with the use of continuous glucose monitoring and emerging closed-loop insulin delivery systems, normoglycemia without hypoglycemia may be reached. This would indeed offer the opportunity to not only improve our understanding of the impact of glycemia on neurologic outcomes, but to also examine the effects of other risk factors including lipids and obesity.

In summary, the DSPN and CAN measures selected for the DCCT and EDIC represent a carefully considered balance between the aims of the study, and test validity, reproducibility, tolerability, safety, availability, and cost. The DCCT/EDIC participants have been unbelievably generous in their commitment to these studies, now with over 30 years of follow-up. The EDIC study continues to evaluate the incidence, long-term outcomes, and risk factors for neuropathic complications in a large number of well-characterized patients with T1D and continues to demonstrate the value of optimizing glucose control as early as possible in the course of the disease to ameliorate the long-term effects of hyperglycemia.

REFERENCES

Albers, J. W., Herman, W. H., Pop-Busui, R., Feldman, E. L., Martin, C. L., Cleary, P. A., et al. (2010). Effect of prior intensive insulin treatment during the Diabetes Control and Complications Trial (DCCT) on peripheral neuropathy in type 1 diabetes during the Epidemiology of Diabetes Interventions and Complications (EDIC) Study. *Diabetes Care, 33*, 1090–1096.

Albers, J. W., Herman, W. H., Pop-Busui, R., Martin, C. L., Cleary, P., & Waberski, B. (2007). Subclinical neuropathy among diabetes control and complications trial participants without diagnosable neuropathy at trial completion: Possible predictors of incident neuropathy? *Diabetes Care, 30*, 2613–2618.

American Diabetes Association. (2015). Standards of medical care in diabetes-2015: Summary of revisions. *Diabetes Care, 38*(Suppl.), S4.

Ang, L., Jaiswal, M., Martin, C., & Pop-Busui, R. (2014). Glucose control and diabetic neuropathy: Lessons from recent large clinical trials. *Current Diabetes Reports, 14*, 528.

Bharucha, A. E., Batey-Schaefer, B., Cleary, P. A., Murray, J. A., Cowie, C., Lorenzi, G., et al. (2015). Delayed gastric emptying is associated with early and long-term hyperglycemia in type 1 diabetes mellitus. *Gastroenterology, 149*, 330–339.

Boulton, A. J., Malik, R. A., Arezzo, J. C., & Sosenko, J. M. (2004). Diabetic somatic neuropathies. *Diabetes Care, 27*, 1458–1486.

Boulton, A. J., Vinik, A. I., Arezzo, J. C., Bril, V., Feldman, E. L., Freeman, R., et al. (2005). Diabetic neuropathies: A statement by the American diabetes association. *Diabetes Care, 28*, 956–962.

DCCT. (1993). The effect of intensive treatment of diabetes on the development and progression of long-term complications in insulin-dependent diabetes mellitus.

The Diabetes Control and Complications Trial Research Group. *The New England Journal of Medicine*, *329*, 977–986.
DCCT. (1995). Effect of intensive diabetes treatment on nerve conduction in the Diabetes Control and Complications Trial. *Annals of Neurology*, *38*, 869–880.
DCCT. (1998). The effect of intensive diabetes therapy on measures of autonomic nervous system function in the Diabetes Control and Complications Trial (DCCT). *Diabetologia*, *41*, 416–423.
DCCT. (2002). Effect of intensive therapy on the microvascular complications of type 1 diabetes mellitus. *JAMA*, *287*, 2563–2569.
Epidemiology of Diabetes Interventions and Complications (EDIC). (1999). Design, implementation, and preliminary results of a long-term follow-up of the Diabetes Control and Complications Trial cohort. *Diabetes Care*, *22*, 99–111.
Feldman, E. L., Stevens, M. J., Thomas, P. K., Brown, M. B., Canal, N., & Greene, D. A. (1994). A practical two-step quantitative clinical and electrophysiological assessment for the diagnosis and staging of diabetic neuropathy. *Diabetes Care*, *17*, 1281–1289.
Herman, W. H., Pop-Busui, R., Braffett, B. H., Martin, C. L., Cleary, P. A., Albers, J. W., et al. (2012). Use of the Michigan Neuropathy Screening Instrument as a measure of distal symmetrical peripheral neuropathy in type 1 diabetes: Results from the Diabetes Control and Complications Trial/Epidemiology of Diabetes Interventions and Complications. *Diabetic Medicine*, *29*, 937–944.
Hotaling, J., Sarma, A. V., Pop-Busui, R., Cleary, P., Dunn, R. L., & Wessels, H. (2013). Autonomic neuropathy, sexual dysfunction and urinary incontinence in women with type 1 diabetes: Findings from the DCCT/EDIC. *Diabetologia*, *56*, S1–S566.
Low, P. A., Benrud-Larson, L. M., Sletten, D. M., Opfer-Gehrking, T. L., Weigand, S. D., O'Brien, P. C., et al. (2004). Autonomic symptoms and diabetic neuropathy: A population-based study. *Diabetes Care*, *27*, 2942–2947.
Low, P. A., Denq, J. C., Opfer-Gehrking, T. L., Dyck, P. J., O'Brien, P. C., & Slezak, J. M. (1997). Effect of age and gender on sudomotor and cardiovagal function and blood pressure response to tilt in normal subjects. *Muscle & Nerve*, *20*, 1561–1568.
Malik, R., Veves, A., Tesfaye, S., Smith, G., Cameron, N., Zochodne, D., et al. (2011). Small fiber neuropathy: Role in the diagnosis of diabetic sensorimotor polyneuropathy. *Diabetes/Metabolism Research and Reviews*, *27*, 678–684.
Martin, C. L., Albers, J., Herman, W. H., Cleary, P., Waberski, B., Greene, D. A., et al. (2006). Neuropathy among the diabetes control and complications trial cohort 8 years after trial completion. *Diabetes Care*, *29*, 340–344.
Martin, C. L., Albers, J. W., & Pop-Busui, R. (2014). Neuropathy and related findings in the diabetes control and complications trial/epidemiology of diabetes interventions and complications study. *Diabetes Care*, *37*, 31–38.
Martin, C. L., Waberski, B. H., Pop-Busui, R., Cleary, P. A., Catton, S., Albers, J. W., et al. (2010). Vibration perception threshold as a measure of distal symmetrical peripheral neuropathy in type 1 diabetes: Results from the DCCT/EDIC study. *Diabetes Care*, *33*, 2635–2641.
Nathan, D. M., Cleary, P. A., Backlund, J. Y., Genuth, S. M., Lachin, J. M., Orchard, T. J., et al. (2005). Intensive diabetes treatment and cardiovascular disease in patients with type 1 diabetes. *The New England Journal of Medicine*, *353*, 2643–2653.
Pop-Busui, R., Braffett, B. H., Hotaling, J., Cleary, P., Martin, C. L., Dunn, R. L., et al. (2014). Diabetic neuropathy and urologic complications in men with type 1 diabetes in the Diabetes Control and Complications Trial/Epidemiology of Diabetes Intervention and Complications Study (DCCT/EDIC). *Diabetes*, *63*, A149–A580.
Pop-Busui, R., Herman, W. H., Feldman, E. L., Low, P. A., Martin, C. L., Cleary, P. A., et al. (2010). DCCT and EDIC studies in type 1 diabetes: Lessons for diabetic neuropathy regarding metabolic memory and natural history. *Current Diabetes Reports*, *10*, 276–282.

Pop-Busui, R., Hotaling, J., Braffett, B. H., Cleary, P. A., Dunn, R. L., Martin, C. L., et al. (2015). Cardiovascular autonomic neuropathy, erectile dysfunction and lower urinary tract symptoms in men with type 1 diabetes: Findings from the DCCT/EDIC. *The Journal of Urology, 193*, 2045–2051.

Pop-Busui, R., Low, P. A., Waberski, B. H., Martin, C. L., Albers, J. W., Feldman, E. L., et al. (2009). Effects of prior intensive insulin therapy on cardiac autonomic nervous system function in type 1 diabetes mellitus: The Diabetes Control and Complications Trial/Epidemiology of Diabetes Interventions and Complications study (DCCT/EDIC). *Circulation, 119*, 2886–2893.

Sarma, A. V., Kanaya, A. M., Nyberg, L. M., Kusek, J. W., Vittinghoff, E., Rutledge, B., et al. (2009). Urinary incontinence among women with type 1 diabetes—How common is it? *The Journal of Urology, 181*, 1224–1230. discussion 1230.

Smith, A. G., & Singleton, J. R. (2012). Diabetic neuropathy. *Continuum (Minneapolis, Minn), 18*, 60–84.

Tesfaye, S., Boulton, A. J., Dyck, P. J., Freeman, R., Horowitz, M., Kempler, P., et al. (2010). Diabetic neuropathies: Update on definitions, diagnostic criteria, estimation of severity, and treatments. *Diabetes Care, 33*, 2285–2293.

The Diabetes Control and Complications Trial/Epidemiology of Diabetes Interventions and Complications Research Group. (2000). Retinopathy and nephropathy in patients with type 1 diabetes four years after a trial of intensive therapy. *The New England Journal of Medicine, 342*, 381–389.

Vileikyte, L., Peyrot, M., Bundy, C., Rubin, R. R., Leventhal, H., Mora, P., et al. (2003). The development and validation of a neuropathy- and foot ulcer-specific quality of life instrument. *Diabetes Care, 26*, 2549–2555.

Wessells, H., Penson, D. F., Cleary, P., Rutledge, B. N., Lachin, J. M., McVary, K. T., et al. (2011). Effect of intensive glycemic therapy on erectile function in men with type 1 diabetes. *The Journal of Urology, 185*, 1828–1834.

Writing Team for DCCT/EDIC. (2003). Sustained effect of intensive treatment of type 1 diabetes mellitus on development and progression of diabetic nephropathy: The Epidemiology of Diabetes Interventions and Complications (EDIC) study. *JAMA, 290*, 2159–2167.

CHAPTER THREE

The Perfect Clinical Trial

V. Bril[1]
Ellen and Martin Prosserman Centre for Neuromuscular Diseases, University Health Network, University of Toronto, Toronto, ON, Canada
[1]Corresponding author: e-mail address: vera.bril@utoronto.ca

Contents

Abstract

Multiple phase III clinical trials have failed to show disease-modifying benefits for diabetic sensorimotor polyneuropathy (DSP) and this may be due to the design of the clinical trials. The perfect clinical trial in DSP would enroll sufficiently large numbers of patients having early or minimal disease, as demonstrated by nerve conduction studies (NCS). These patients would be treated with an intervention given at an effective and well-tolerated dose for a sufficient duration of time to show change in the end points selected. For objective or surrogate measures such as NCS and for some small fiber measures, the duration needed to show positive change may be as brief as 6–12 months, but subsequently, trials lasting 5–8 years will be required to demonstrate clinical benefits.

1. INTRODUCTION

Multiple phase III clinical trials have lacked success in identifying disease-modifying therapy for diabetic sensorimotor polyneuropathy (DSP). Although numerous pharmacological agents such as neurotrophic factors, aldose reductase inhibitors, protein kinase C inhibitors, and antioxidant

International Review of Neurobiology, Volume 127
ISSN 0074-7742
http://dx.doi.org/10.1016/bs.irn.2016.02.011

Table 1 Pharmacological Agents Tested in Phase III Trials for Disease Modification in Diabetic Sensorimotor Polyneuropathy

Aldose reductase inhibitors	Sorbinil, Zenarestat, Epalrestat, ICI, Zopolrestat, Ranirestat
Neurotrophic factors	Nerve growth factor, VEGF
Antioxidants	Alpha-lipoic acid, vitamin E
Others	Protein kinase C inhibitors

agents showed promising results in preclinical development and phase I and II clinical trials, they failed in phase III clinical trials where convincing results are critical for regulatory approval and widespread acceptance of therapeutic benefit (Table 1). Reasons for failure at the phase III clinical trial level are controversial although the most likely may be a lack of efficacy of the single targeted intervention in a complex disorder; in other words, the selected target may not be fundamental in the pathophysiological cascade leading to DSP. Alternatively, deficiencies in clinical trial design may be responsible for the repeated failures. Can we design a perfect study to overcome concerns about trial deficiencies that may have led to previous failures to demonstrate an effective treatment for this chronic, degenerative process?

In designing clinical trials, important elements to consider are the selection of study participants, duration of treatment, outcome measures, and safety of patients in the study including safety of all study procedures and the intervention. In designing the perfect clinical trial, we may ask: do we change the outcome measures (the way we measure disease), do we change study design, do we change expectations, and do we look for novel biomarkers that accurately predict long-term outcomes? How should we proceed in this field littered with the fallen corpses of clinical trials in DSP? This chapter aims to describe what the perfect clinical trial in DSP would be given unlimited resources and access to study participants.

2. STUDY AIMS: PREVENT DSP, PREVENT PROGRESSION, OR REVERSE DSP?

DSP is a chronic, axonal, length-dependent polyneuropathy that is slowly progressive and depends on duration of diabetes, glycemic control, and other factors such has hypertension, BMI, smoking history, and lipid profile (Dunnigan et al., 2013; Tesfaye et al., 2005). It is the most prevalent complication in diabetes patients that is found in 50% of subjects with long-standing diabetes and likely in more if diagnosis were based on sensitive nerve measures (Pirart, 1978). The exact mechanisms leading to disease progression are unknown, but there is more nerve fiber loss in patients with worse control and limited axonal regeneration in diabetes results in potentially irreversible axonal loss (Zochodne, 2007, 2015). Given this knowledge, it is generally considered optimal to choose patients who are early in the course of their DSP for clinical trials in order to prevent onset or progression of neuropathy, as reversal may be unachievable. In Type 1 patients who have diabetes duration that can be well defined, prevention of DSP may be the optimal goal in any given clinical trial. In Type 2 patients who may have diabetes as long as 10 years prior to diagnosis and who often have DSP at the time of diagnosis of diabetes, the goal of treatment may be prevention of progression rather than prevention of DSP or even reversal of disease.

3. SELECTION OF STUDY PARTICIPANTS

The onset of Type 1 diabetes can be determined fairly accurately, and these patients tend to develop DSP after 5 years of diabetes. So, one might recommend limiting clinical trials to Type 1 patients with diabetes for less than 5 years' duration in a prevention study. Of course this approach would limit enrollment in clinical trials mostly to children and would leave out the 90% of patients who have Type 2 diabetes. In Type 2 diabetes patients, the onset of diabetes is unknown and patients may have diabetes for as long as 10 years before diagnosis. Even patients with a prediabetes state such as impaired glucose tolerance can develop neuropathy indistinguishable from DSP as nerve damage develops over the spectrum of glucose dysregulation (Singleton, Smith, & Bromberg, 2001; Sumner, Sheth, Griffin, Cornblath, & Polydefkis, 2003). By the time that diabetes is diagnosed in Type 2 patients, DSP can be present and advanced. Consequently, selection of Type 2 patients with early DSP cannot be determined simply by duration of diabetes.

Patients with mild DSP can be selected on clinical or objective criteria. The symptoms and signs of peripheral nerve disease can be summarized in scores such as the Toronto Clinical Neuropathy Score that are related to the degree of axonal loss (Bril & Perkins, 2002). A limitation of this approach is the degree of variability in clinical assessment of neuropathy even when neuromuscular experts perform the examination (Dyck et al., 2010) although the variability when using clinical scales can be less marked (Bril, Tomioka, Buchanan, Perkins, & mTCNS Study Group, 2009; Perkins, Olaleye, Zinman, & Bril, 2001). The presence of clinical neuropathy based on neuropathic symptoms, physical signs, or both, helps satisfy regulatory and diabetes community preferences for investigation of disease that is considered to be relevant to patients. Simple clinical measures, such as the TCNS or mTCNS (Bril & Perkins, 2002; Bril, Tomioka, et al., 2009) or the Michigan Neuropathy Score (Herman et al., 2012) can provide a summary method of compiling the clinical features of disease and help identify the presence of neuropathy reliably (Zilliox, Ruby, Singh, Zhan, & Russell, 2015). Assessment of progression may be improved with scales that are more granular and likely more sensitive such as the mTCNS (Bril, Tomioka, et al., 2009).

Objective criteria, such as nerve conduction studies (NCS), have higher reliability and are preferable in clinical trials (Bril et al., 1998; Dyck et al., 2010). As progressive nerve fiber loss is the hallmark of DSP, loss of the lower-limb sensory nerve responses shows advanced disease and selecting patients who have a recordable sural sensory nerve response ensures that they will have less severe neuropathy and may respond to interventions (Vinik et al., 2005). So, all patients in clinical trials should have NCS to help select patients and to show response to therapy when large fiber function is expected to improve. However, DSP may be too far advanced even with NCS in the normal range, as the healthy values for NCS parameters in any given patient are unknown. The reference range is broad and the results in any given patient may have declined significantly but remain in the lower range of normal. Given unlimited resources, one might suggest that everyone in the population have baseline NCS, so that when diabetes is diagnosed and the NCS are repeated, one could determine the magnitude of changes from baseline and then document the presence and severity of DSP at an earlier stage. Naturally, a cohort of healthy volunteers would need to have matched NCS at time points similar to those in diabetes patients to determine the natural history of NCS in a broad population and show the differences in diabetes subjects. Given that this approach is unrealistic to achieve, additional means of accurately identifying disease presence and status are required.

Identification of small fiber impairments are thought to precede large nerve fiber loss and can help select patients for clinical trials. These methods include evaluation of intraepidermal nerve fiber density (IENFD) (Lauria et al., 1999), quantitative thermal thresholds (QTT) (Arezzo, Schaumburg, & Laudadio, 1986; Yarnitsky, 1997), laser Doppler flare area (Krishnan & Rayman, 2004), and corneal confocal nerve fiber microscopy (CCM) (Ahmed et al., 2012; Malik et al., 2003). Tests of sudomotor function can help supplement these studies (Casellini, Parson, Richardson,

Nevoret, & Vinik, 2013). A combination of abnormal tests can reliably identify the presence of DSP even in the setting of normal NCS (Breiner, Lovblom, Perkins, & Bril, 2014) and these small fiber tests, particularly IENFD, QTT, and CCM should be used for selection of subjects in clinical trials (Breiner et al., 2014). Their use as central end points is limited by the variability of testing, need for combination of tests for reliability, and also the need to demonstrate that changes in these fibers will predict improvement in all nerve fibers.

The population studied should be broad and include all racial groups and multiple countries so that the effects of any intervention can be understood across all people (Hotta, Kawamori, Fukuda, Shigeta, & Aldose Reductase Inhibitor-Diabetes Complications Trial Study Group, 2012). A 60-site study done in 20 countries or more is not unreasonable to consider for an intervention study (Bril et al., 1998).

In summary, it is necessary to select patients with clinical evidence of neuropathy, using NCS to document severity and including small fiber tests for objective evidence of DSP, irrespective of whether NCS are normal or abnormal.

4. TRIAL DURATION

The most successful clinical trial on neuropathy in diabetes to date was the DCCT (The effect of intensive diabetes therapy on the development and progression of neuropathy. The Diabetes Control and Complications Trial Research Group, 1995). In this study of 1441 subjects with Type 1 diabetes, intensive glycemic control for 5 years reduced the prevalence of neuropathy as measured by clinical, electrophysiological, and cardiac autonomic tests, by 60% (The effect of intensive diabetes therapy on the development and progression of neuropathy. The Diabetes Control and Complications Trial Research Group, 1995). This study showed that the end points of clinical evaluation, NCS, and autonomic system testing could reflect changes in DSP and are thus valid for use in clinical trials. Also, this costly study was funded by US governmental agencies, was planned for a duration of 7 years, and had a very large number of subjects who were evaluated comprehensively with clinical examination by neurologists, autonomic system testing, and also NCS. The DCCT showed at 5 years the benefits on DSP of intensive compared to conventional glycemic control in Type 1 patients, so the minimum duration of any study in Type 1 patients should be 5 years when this array of tests are the key clinical endpoints. Assessments at intermediate time-points were not available and so shorter durations of studies cannot be recommended. Large-scale epidemiological studies or intervention studies without similar comprehensive evaluations of neuropathy did not provide strong evidence of benefits with respect to DSP, partly because the neuropathy was not evaluated in a comprehensive manner or because the separation of treatment groups of regular and improved control was suboptimal (U. P. D. S. Group, 1998). Nonetheless, a separation of vibration perception was noted at 8 years in the UKPDS study in Type 2 patients, even with these caveats. Reduction in autonomic neuropathy has been observed with multifactorial interventions (Gaede, Vedel, Parving, & Pedersen, 1999, Greene, Arezzo, Brown, & Zenarestat Study Group, 1999). The experience with Type 2 patients may have differed if different study protocols were employed. However, these studies suggest that durations in this population need to be lengthy and perhaps as long as 8 years or more.

The longest study funded by the pharmacology industry was a 4-year treatment trial of an oral medication (the NATHAN study) that was not effective and did not show benefits in the main end points of the study (Ziegler et al., 2011). The end points were similar to those in the DCCT. This outcome is challenging to understand because evaluations in the DCCT were done at baseline and at 5 years, but not at 4 years. We cannot know whether the reduction in prevalence of DSP in the DCCT was linear over 5 years, or not. If the neuropathy assessments had been done at 4 years, then the outcome of the NATHAN study could be compared directly to the DCCT results, but that is not the case. Also, the study populations were different in that the NATHAN study had a predominance of patients with Type 2 diabetes. When factors that might have influenced the outcome were evaluated, those with a normal BMI, normal blood pressure, and higher burden related to cardiovascular disease, diabetes, and neuropathy were more responsive to treatment (Ziegler, Low, Freeman, Tritschler, & Vinik, 2016), suggesting that a multifactorial approach when treating DSP will be required. Nonetheless, it is likely that the DCCT would have shown some trend to less neuropathy in Type 1 diabetes subjects at 4 years.

In summary, given the fact that DSP is a chronic, insidiously progressive axonal neuropathy, and the results of the DCCT and UKPDS, the perfect clinical trial would last at least 5 years in Type 1 patients and 8 years or more in Type 2 patients.

5. STUDY END POINTS

In assessing benefits of any intervention for DSP, clinical improvement in disease status would be the most convincing evidence of efficacy. The reduction in, or elimination of, symptoms and decreased signs of disease indicate a positive response. Certain quantitative or summary scales can provide evidence of clinical benefit. The modified TCNS, the TCNS, the MNI, and other scales can be used to show changes in DSP although some may be more sensitive than others (Bril, Tomioka, et al., 2009; Herman et al., 2012). A reduction in symptoms with improving objective measures of DSP indicates improvement. However, clinical evaluation can be highly variable necessitating the requirement for large numbers of patients who would require careful assessment for benefit. Deficits in patients with mild disease are mainly sensory in nature and sensory testing is the most variable on clinical examination, so the noise level can be very high (Perkins et al., 2001).

The natural history of nerve conduction slowing in diabetes is a gradual decline in values over many years (Gibbons et al., 2013; Partanen et al., 1995). NCS are the best measure of peripheral nerve function and should show separation of two treatment arms at 5 years of treatment based on the DCCT results (D. C. A. C. T. D. R. Group, 1995). Many phase II studies have shown small changes in shorter treatment duration intervals that have not been replicated in phase III studies and have questionable meaning. Small changes in glycemic control or lipid profile might also have a longer effect on NCS parameters than expected (Perkins, Dholasania, Buchanan, & Bril, 2010). NCS should be used as an adjunct measure to confirm lack of deterioration of nerve function if positive symptoms such as pain disappear, to exclude the possibility that progressive disease or therapeutic interventions have alleviated pain by ablating nerve function. In studies where large nerve fiber function appears to improve by some measures such as vibration perception threshold (VPT) or clinical examination, but NCS do not improve (Ziegler et al., 2011), questions about the validity of the results need to be addressed.

While large fiber function can be tracked iteratively across studies, efforts to assess large fiber structure in clinical trials have been largely limited to single cross-sectional studies, although the evaluation of pre- and posttreatment sural nerve biopsy was at one time thought to provide a means to shorten the clinical trial duration (Greene et al., 1999). The procedure is invasive and not as objective as might be thought (Perry & Bril, 1994) without standardized training of those evaluating the biopsy (Kalichman, Chalk, & Mizisin, 1999), so that large nerve fiber biopsy is not advocated at this time.

End points should include evaluation of small nerve fibers as these can be demonstrated to be affected before NCS falls into the abnormal range (Breiner et al., 2014; Smith, Ramachandran, Tripp, & Singleton, 2001). The use of IENFD and other measures of epidermal and dermal nerve fibers (such as axonal swellings and varicosities) can provide direct structural evidence of nerve status (Mellgren, Nolano, & Sommer, 2013). Nerve regeneration with the capsaicin IENFD system can be another way to assess nerve recovery (Polydefkis et al., 2004). So, clinical trials should include IENFD at baseline and then at end of treatment. Sampling from the distal lower leg and distal thigh are suggested. The procedure is invasive, but minimally so with minor potential side effects of infection and bleeding.

Noninvasive measures of small nerve fiber function are preferable to invasive ones, but should also supplement the invasive measure. Corneal confocal microscopy to directly visualize small terminal trigeminal nerve fibers in Bowman's layer of the cornea may be useful to assess nerve regeneration (Ahmed et al., 2012; Malik et al., 2003). Nerve fiber length on CCM is a surrogate measure for the longest nerve fibers, correlates with longer nerve fiber measures (IENFD, NCS, QST), and may show changes earlier than would be apparent in longer fibers (Hertz et al., 2011; Sivaskandarajah et al., 2013). This end point should be measured at baseline and then every 6 months during a study lasting for a minimum of 5 years (Mehra et al., 2007). If no change in CCM parameters is observed at earlier time points, then the study can be terminated earlier, but if changes are apparent, then the study should be continued for 5 years to demonstrate corresponding changes in the longest nerve fibers and provide a convincing study result.

Quantitative sensory threshold measurements provide a numerical evaluation of sensory function, with thermal thresholds (QTT) indicating small nerve fiber function and VPTs reflecting large nerve fiber function. In QTT, cold detection thresholds (CDT) are more reliable than warming thresholds, and so are preferable (Zinman, Bril, & Perkins, 2004). These measures should be done in clinical trials to help evaluate how sensory nerves are functioning in those with DSP. There is great variation between assessments, but VPT and CDT are valuable end points and should be done at study start and end (Young, Every, & Boulton, 1993; Zinman et al., 2004).

Laser Doppler flare area imaging in response to heat-mediated vasodilation can provide a measure of cutaneous nerve fiber function (Krishnan &

Rayman, 2004). This end point may help provide supplementary evidence of small nerve fiber function (Krishnan, Quattrini, Jeziorska, Malik, & Rayman, 2009).

Autonomic function tests such as sudomotor function, heart rate variability, and others can provide separate evidence of small nerve fiber responsiveness.

Generally, all end points at baseline and end of study should have triplicate measures to improve the accuracy of the results. The median value is used so that one unusual value does not skew the "real" value of the parameter.

6. CORE LABS AND TRAINING

A major concern in any study is standardization of all procedures from the clinical examination to the measurement of nerve function such as in NCS. In order to maximize the chances of observing positive change, the testing in the study needs to be done the same way in all patients at all sites in the study. The degree of variability needs to be reduced to the minimum possible so that changes are detectable, as was done for sural NCV over 5 years in DCCT (The effect of intensive diabetes therapy on the development and progression of neuropathy. The Diabetes Control and Complications Trial Research Group, 1995). A core lab is essential to help control study procedures, particularly those with large technical components that confound inexpert CROs, study monitors, and pharmaceutical staff (Bril et al., 1998; Bril, Katzberg, et al., 2009). Such core lab functions are essential for NCS (Bril et al., 1998), CCM, and LDI. Standardized training sessions are necessary for the clinical examination, clinical scales, and all other study procedures.

7. INTERVENTION

Only therapies that are effective in appropriate preclinical models at doses that can sensibly be used in humans, allowing for interspecies allometric scaling (Freireich, Gehan, Rall, Schmidt, & Skipper, 1966; Huang & Riviere, 2014) should be advanced to human trials. In many unsuccessful clinical trials, a study drug has been tested and found effective in preclinical and phase II studies and then failed in phase III. This has sometimes been related to the use of doses in the phase III study that were lower than the phase II and subsequently proved to be less effective, as discussed for trials of NGF in the previous edition of this volume (Apfel, 2002). The use of interventions at effective doses is essential—the study should not be done at all if there is a safety issue at the effective dose. The interventions tested to date have targeted a part of the presumed pathophysiology of DSP. Multi-level interventions may be required to successfully intervene in the cascade that leads to DSP. Proponents of any particular disease hypothesis need to acknowledge and accept other factors as being important in the pathway to disease so that they can work together to bring about effective interventions. In a perfect clinical trial, multiple interventions are likely to be used together to produce a positive effect, for example, a therapy that combines ARI, antioxidant, and axonal regenerative properties.

In addition to the specific intervention, any long-term clinical trial would need to control factors that might contribute to DSP such as hyperglycemia, hypertension, hyperlipidemia, smoking, and exercise patterns (Gaede et al., 1999; Perkins et al., 2010; Tesfaye et al., 2007). A stable regimen addressing these factors needs to be established prior to the study or for the first 3 months of the study, and then patients could be randomized to a specific intervention.

8. STUDY CONDUCT

For a 5-year or longer clinical trial, retention of study subjects becomes a major concern. The DCCT was successful in large part due to the motivated population under study and the initiatives put in place by the study coordinators: morale and team building items and activities that engaged study participants proved key to the success of this study. For any long duration study, such activities are necessary and should be allowed despite the current ethical climate that focuses on perceived "coercion" of study subjects. It is more unethical to run a study that cannot be completed with the requisite number of subjects without constant positive reinforcement and encouragement by the study staff. Small study perks (mugs, t-shirts, baseball caps, and other items) might convince study participants to stay the course. Adequate remuneration for study subjects' time, coverage of all patient study-related costs, and proper funding for sufficient numbers of study staff so that they can spend the time necessary for subject retention are essential. The ability to enroll subjects who live far from a study center should be enabled by proper budgeting of participant expenses related to the study. Newer communication tools such as videoconferencing with each study participant using personal electronic devices might help with long-term retention. All of the items identified in this section are part of a study done with unlimited resources, but need to be stated explicitly to ensure that they are funded in the current research climate where all costs tend to be minimized as much as possible, if not eliminated entirely, to the detriment of any study.

9. STREAMLINED ETHICS AND CONTRACTS PROCESS

Increasingly, it has become more and more difficult to carry out clinical research studies in the developed world due to an ever-increasing bureaucracy in health care research. Motivated by the view that "we only want to protect patients," governmental agencies, ethics, and contracts personnel raise more and more barriers to clinical research. There may be a conflict of interest at work here as increasing the roles of these groups leads to more employment for them. Wrangles over minor wording changes in contracts, the desire to have "zero" liability in any study, and the current environment of competitive enrollment make clinical research a herculean endeavor. We are depending more on clinical research that is done in China, India, and various locales in South America where the requirements appear to differ. It will be interesting to see whether the results obtained in these countries are applicable to different patient populations. We may have treatments that do not work in local population and how can we check, if we cannot do the clinical research in the first place? How ethical is it to take results from clinical trials that we would not allow to take place in our own country? In a world with perfect clinical trials, it would be sufficient for a central ethics board to approve all study sites, and for all institutions to accept that approval and allow the research to proceed. A standard research contract across all academic hospitals (at least in a single country) and agreed to by all sponsors be they governmental agencies, pharmaceutical sponsors, or private sponsors would enable speedier clinical research and provide results and improved care for patients.

10. SUMMARY

The perfect clinical trial in DSP would enroll sufficiently large numbers of patients having early or minimal disease, as demonstrated by NCS. These patients would be treated with an intervention at an effective and well-tolerated dose for a sufficient duration of time to show change in the end points selected. For objective or surrogate measures such as NCS and for some small fiber measure, this duration can be as brief as 6–12 months, but eventually longer duration trials of 5–8 years will be required to show clinical improvement.

REFERENCES

Ahmed, A., Bril, V., Orszag, A., Paulson, J., Yeung, E., Ngo, M., et al. (2012). Detection of diabetic sensorimotor polyneuropathy by corneal confocal microscopy in type 1 diabetes: A concurrent validity study. *Diabetes Care, 35*(4), 821–828. http://doi.org/10.2337/dc11-1396.

Apfel, S. C. (2002). Nerve growth factor for the treatment of diabetic neuropathy: What went wrong, what went right, and what does the future hold? *International Review of Neurobiology, 50*, 393–413.

Arezzo, J. C., Schaumburg, H. H., & Laudadio, C. (1986). Thermal sensitivity tester. Device for quantitative assessment of thermal sense in diabetic neuropathy. *Diabetes, 35*(5), 590–592.

Breiner, A., Lovblom, L. E., Perkins, B. A., & Bril, V. (2014). Does the prevailing hypothesis that small-fiber dysfunction precedes large-fiber dysfunction apply to type 1 diabetic patients? *Diabetes Care, 37*(5), 1418–1424. http://doi.org/10.2337/dc13-2005.

Bril, V., Ellison, R., Ngo, M., Bergstrom, B., Raynard, D., & Gin, H. (1998). Electrophysiological monitoring in clinical trials. Roche Neuropathy Study Group. *Muscle & Nerve, 21*(11), 1368–1373.

Bril, V., Katzberg, H., Donofrio, P., Banach, M., Dalakas, M. C., Deng, C., et al. (2009). Electrophysiology in chronic inflammatory demyelinating polyneuropathy with IGIV. *Muscle & Nerve, 39*(4), 448–455. http://doi.org/10.1002/mus.21236.

Bril, V., & Perkins, B. A. (2002). Validation of the Toronto Clinical Scoring System for diabetic polyneuropathy. *Diabetes Care, 25*(11), 2048–2052.

Bril, V., Tomioka, S., Buchanan, R. A., Perkins, B. A., & mTCNS Study Group (2009). Reliability and validity of the modified Toronto Clinical Neuropathy Score in diabetic sensorimotor polyneuropathy. *Diabetic Medicine: A Journal of the British Diabetic Association, 26*(3), 240–246. http://doi.org/10.1111/j.1464-5491.2009.02667.x.

Casellini, C. M., Parson, H. K., Richardson, M. S., Nevoret, M. L., & Vinik, A. I. (2013). Sudoscan, a noninvasive tool for detecting diabetic small fiber neuropathy and autonomic dysfunction. *Diabetes Technology & Therapeutics, 15*(11), 948–953. http://doi.org/10.1089/dia.2013.0129.

D. C. A. C. T. D. R. Group (1995). Effect of intensive diabetes treatment on nerve conduction in the diabetes control and complications trial. *Annals of Neurology, 38*(6), 869–880.

Dunnigan, S. K., Ebadi, H., Breiner, A., Katzberg, H. D., Lovblom, L. E., Perkins, B. A., et al. (2013). Conduction slowing in diabetic sensorimotor polyneuropathy. *Diabetes Care, 36*(11), 3684–3690. http://doi.org/10.2337/dc13-0746.

Dyck, P. J., Overland, C. J., Low, P. A., Litchy, W. J., Davies, J. L., Dyck, P. J. B., et al. (2010). Signs and symptoms versus nerve conduction studies to diagnose diabetic sensorimotor polyneuropathy: Cl vs. NPhys trial. *Muscle & Nerve, 42*(2), 157–164. http://doi.org/10.1002/mus.21661.

Freireich, E. J., Gehan, E. A., Rall, D. P., Schmidt, L. H., & Skipper, H. E. (1966). Quantitative comparison of toxicity of anticancer agents in mouse, rat, hamster, dog, monkey, and man. *Cancer Chemotherapy Reports Part 1, 50*(4), 219–244.

Gaede, P., Vedel, P., Parving, H. H., & Pedersen, O. (1999). Intensified multifactorial intervention in patients with type 2 diabetes mellitus and microalbuminuria: The Steno type 2 randomised study. *Lancet, 353*(9153), 617–622. http://doi.org/10.1016/S0140-6736(98)07368-1.

Gibbons, C. H., Freeman, R., Tecilazich, F., Dinh, T., Lyons, T. E., Gnardellis, C., et al. (2013). The evolving natural history of neurophysiologic function in patients with well-controlled diabetes. *Journal of the Peripheral Nervous System: JPNS, 18*(2), 153–161. http://doi.org/10.1111/jns5.12021.

Greene, D. A., Arezzo, J. C., Brown, M. B., & Zenarestat Study Group (1999). Effect of aldose reductase inhibition on nerve conduction and morphometry in diabetic neuropathy. Zenarestat Study Group. *Neurology*, *53*(3), 580–591.

Herman, W. H., Pop-Busui, R., Braffett, B. H., Martin, C. L., Cleary, P. A., Albers, J. W., et al. (2012). Use of the Michigan neuropathy screening instrument as a measure of distal symmetrical peripheral neuropathy in type 1 diabetes: Results from the diabetes control and complications trial/epidemiology of diabetes interventions and complications. *Diabetic Medicine: A Journal of the British Diabetic Association*, *29*(7), 937–944. http://doi.org/10.1111/j.1464-5491.2012.03644.x.

Hertz, P., Bril, V., Orszag, A., Ahmed, A., Ng, E., Nwe, P., et al. (2011). Reproducibility of in vivo corneal confocal microscopy as a novel screening test for early diabetic sensorimotor polyneuropathy. *Diabetic Medicine: A Journal of the British Diabetic Association*, *28*(10), 1253–1260. http://doi.org/10.1111/j.1464-5491.2011.03299.x.

Hotta, N., Kawamori, R., Fukuda, M., Shigeta, Y., & Aldose Reductase Inhibitor-Diabetes Complications Trial Study Group (2012). Long-term clinical effects of epalrestat, an aldose reductase inhibitor, on progression of diabetic neuropathy and other microvascular complications: Multivariate epidemiological analysis based on patient background factors and severity of diabetic neuropathy. *Diabetic Medicine: A Journal of the British Diabetic Association*, *29*(12), 1529–1533. http://doi.org/10.1111/j.1464-5491.2012.03684.x.

Huang, Q., & Riviere, J. E. (2014). The application of allometric scaling principles to predict pharmacokinetic parameters across species. *Expert Opinion on Drug Metabolism & Toxicology*, *10*(9), 1241–1253. http://doi.org/10.1517/17425255.2014.934671.

Kalichman, M. W., Chalk, C. H., & Mizisin, A. P. (1999). Classification of teased nerve fibers for multicenter clinical trials. *Journal of the Peripheral Nervous System: JPNS*, *4*(3–4), 233–244.

Krishnan, S. T. M., Quattrini, C., Jeziorska, M., Malik, R. A., & Rayman, G. (2009). Abnormal LDIflare but normal quantitative sensory testing and dermal nerve fiber density in patients with painful diabetic neuropathy. *Diabetes Care*, *32*(3), 451–455. http://doi.org/10.2337/dc08-1453.

Krishnan, S. T. M., & Rayman, G. (2004). The LDIflare: A novel test of C-fiber function demonstrates early neuropathy in type 2 diabetes. *Diabetes Care*, *27*(12), 2930–2935.

Lauria, G., Holland, N., Hauer, P., Cornblath, D. R., Griffin, J. W., & McArthur, J. C. (1999). Epidermal innervation: Changes with aging, topographic location, and in sensory neuropathy. *Journal of the Neurological Sciences*, *164*(2), 172–178.

Malik, R. A., Kallinikos, P., Abbott, C. A., van Schie, C. H. M., Morgan, P., Efron, N., et al. (2003). Corneal confocal microscopy: A non-invasive surrogate of nerve fibre damage and repair in diabetic patients. . *Diabetologia*, *46*(5), 683–688. http://doi.org/10.1007/s00125-003-1086-8.

Mehra, S., Tavakoli, M., Kallinikos, P. A., Efron, N., Boulton, A. J. M., Augustine, T., et al. (2007). Corneal confocal microscopy detects early nerve regeneration after pancreas transplantation in patients with type 1 diabetes. *Diabetes Care*, *30*(10), 2608–2612. http://doi.org/10.2337/dc07-0870.

Mellgren, S. I., Nolano, M., & Sommer, C. (2013). The cutaneous nerve biopsy: Technical aspects, indications, and contribution. *Handbook of Clinical Neurology*, *115*, 171–188. http://doi.org/10.1016/B978-0-444-52902-2.00010-2.

Partanen, J., Niskanen, L., Lehtinen, J., Mervaala, E., Siitonen, O., & Uusitupa, M. (1995). Natural history of peripheral neuropathy in patients with non-insulin-dependent diabetes mellitus. *The New England Journal of Medicine*, *333*(2), 89–94. http://doi.org/10.1056/NEJM199507133330203.

Perkins, B. A., Dholasania, A., Buchanan, R. A., & Bril, V. (2010). Short-term metabolic change is associated with improvement in measures of diabetic neuropathy: A 1-year

placebo cohort analysis. *Diabetic Medicine: A Journal of the British Diabetic Association, 27*(11), 1271–1279. http://doi.org/10.1111/j.1464-5491.2010.03110.x.
Perkins, B. A., Olaleye, D., Zinman, B., & Bril, V. (2001). Simple screening tests for peripheral neuropathy in the diabetes clinic. *Diabetes Care, 24*(2), 250–256.
Perry, J. R., & Bril, V. (1994). Complications of sural nerve biopsy in diabetic versus non-diabetic patients. *The Canadian Journal of Neurological Sciences Le Journal Canadien Des Sciences Neurologiques, 21*(1), 34–37.
Pirart, J. (1978). Diabetes mellitus and its degenerative complications: A prospective study of 4,400 patients observed between 1947 and 1973. *Diabetes Care, 1*(168–188), 252–263.
Polydefkis, M., Hauer, P., Sheth, S., Sirdofsky, M., Griffin, J. W., & McArthur, J. C. (2004). The time course of epidermal nerve fibre regeneration: Studies in normal controls and in people with diabetes, with and without neuropathy. *Brain: A Journal of Neurology, 127*(Pt. 7), 1606–1615. http://doi.org/10.1093/brain/awh175.
Singleton, J. R., Smith, A. G., & Bromberg, M. B. (2001). Increased prevalence of impaired glucose tolerance in patients with painful sensory neuropathy. *Diabetes Care, 24*(8), 1448–1453.
Sivaskandarajah, G. A., Halpern, E. M., Lovblom, L. E., Weisman, A., Orlov, S., Bril, V., et al. (2013). Structure-function relationship between corneal nerves and conventional small-fiber tests in type 1 diabetes. *Diabetes Care, 36*(9), 2748–2755. http://doi.org/10.2337/dc12-2075.
Smith, A. G., Ramachandran, P., Tripp, S., & Singleton, J. R. (2001). Epidermal nerve innervation in impaired glucose tolerance and diabetes-associated neuropathy. *Neurology, 57*, 1701–1704.
Sumner, C. J., Sheth, S., Griffin, J. W., Cornblath, D. R., & Polydefkis, M. (2003). The spectrum of neuropathy in diabetes and impaired glucose tolerance. *Neurology, 60*(1), 108–111.
Tesfaye, S., Chaturvedi, N., Eaton, S. E. M., Ward, J. D., Manes, C., Ionescu-Tirgoviste, C., et al. (2005). Vascular risk factors and diabetic neuropathy. *The New England Journal of Medicine, 352*(4), 341–350. http://doi.org/10.1056/NEJMoa032782.
Tesfaye, S., Tandan, R., Bastyr, E. J., Kles, K. A., Skljarevski, V., Price, K. L., et al. (2007). Factors that impact symptomatic diabetic peripheral neuropathy in placebo-administered patients from two 1-year clinical trials. *Diabetes Care, 30*(10), 2626–2632. http://doi.org/10.2337/dc07-0608.
The effect of intensive diabetes therapy on the development and progression of neuropathy. The Diabetes Control and Complications Trial Research Group. (1995). *Annals of Internal Medicine, 122*(8), 561–568.
U. P. D. S. Group (1998). Tight blood pressure control and risk of macrovascular and microvascular complications in type 2 diabetes: UKPDS 38. UK Prospective Diabetes Study Group. *BMJ, 317*(7160), 703–713.
Vinik, A. I., Bril, V., Litchy, W. J., Price, K. L., Bastyr, E. J., & MBBQ Study Group (2005). Sural sensory action potential identifies diabetic peripheral neuropathy responders to therapy. *Muscle & Nerve, 32*(5), 619–625. http://doi.org/10.1002/mus.20423.
Yarnitsky, D. (1997). Quantitative sensory testing. *Muscle & Nerve, 20*(2), 198–204.
Young, M. J., Every, N., & Boulton, A. J. (1993). A comparison of the neurothesiometer and biothesiometer for measuring vibration perception in diabetic patients. *Diabetes Research and Clinical Practice, 20*(2), 129–131.
Ziegler, D., Low, P. A., Freeman, R., Tritschler, H., & Vinik, A. I. (2016). Predictors of improvement and progression of diabetic polyneuropathy following treatment with α-lipoic acid for 4 years in the NATHAN 1 trial. *Journal of Diabetes and its Complications, 30*(2), 350–356. http://dx.doi.org/10.1016/j.jdiacomp.2015.10.018.
Ziegler, D., Low, P. A., Litchy, W. J., Boulton, A. J. M., Vinik, A. I., Freeman, R., et al. (2011). Efficacy and safety of antioxidant treatment with α-lipoic acid over 4 years in

diabetic polyneuropathy: The NATHAN 1 trial. *Diabetes Care, 34*(9), 2054–2060. http://doi.org/10.2337/dc11-0503.

Zilliox, L. A., Ruby, S. K., Singh, S., Zhan, M., & Russell, J. W. (2015). Clinical neuropathy scales in neuropathy associated with impaired glucose tolerance. *Journal of Diabetes and its Complications*, *29*(3), 372–377. http://doi.org/10.1016/j.jdiacomp.2015.01.011.

Zinman, L. H., Bril, V., & Perkins, B. A. (2004). Cooling detection thresholds in the assessment of diabetic sensory polyneuropathy: Comparison of CASE IV and Medoc instruments. *Diabetes Care, 27*(7), 1674–1679.

Zochodne, D. W. (2007). Diabetes mellitus and the peripheral nervous system: Manifestations and mechanisms. *Muscle & Nerve, 36*(2), 144–166. http://doi.org/10.1002/mus.20785.

Zochodne, D. W. (2015). Diabetes and the plasticity of sensory neurons. *Neuroscience Letters, 596*, 60–65. http://doi.org/10.1016/j.neulet.2014.11.017.

SECTION II

New Models of Diabetic Neuropathy

CHAPTER FOUR

An Introduction to the History and Controversies of Animal Models of Diabetic Neuropathy

N.A. Calcutt*[,1], P. Fernyhough[†,‡]
*University of California, San Diego, La Jolla, CA, United States
[†]University of Manitoba, Winnipeg, MB, Canada
[‡]St. Boniface Hospital Research Centre, Winnipeg, MB, Canada
[1]Corresponding author: e-mail address: ncalcutt@ucsd.edu

Contents

1. WHY USE ANIMAL MODELS?

The prime value of animal models of diabetic neuropathy to biomedical research has been in identifying pathogenic mechanisms that link features of diabetes such as hyperglycemia, insulinopenia, insulin resistance, or dyslipidemia with nerve dysfunction and damage. Diabetic animals, and rodents in particular, are also widely used for establishing efficacy of therapeutic interventions prior to testing in humans. The extent to which animal models of diabetic neuropathy replicate the human condition has been contentious almost since the advent of such models, and remains so. Retrospective condemnation of animal models used in preclinical drug development programs as not being predictive of drug efficacy in humans is a not uncommon reflex response to the failure of a clinical trial that can prevent closer examination of other flaws in the drug development process. There are indeed many reasons why animal models of diabetic neuropathy cannot be faithful replicates of the human condition—not least of

International Review of Neurobiology, Volume 127
ISSN 0074-7742
http://dx.doi.org/10.1016/bs.irn.2016.03.011

which are their short life span (and thus exposure time to diabetes) and short hairy legs (and thus short distance from cell body to nerve terminal). Whether there are interspecies differences in physiological and cellular responses to diabetes that invalidate the common animal models is more debatable. Support for the pertinence of animal models of diabetic neuropathy comes from recognition that diabetic animals develop many disorders in common with diabetic patients, including nerve conduction slowing, resistance to ischemic conduction block, allodynia, loss of thermal sensation, and reduced small sensory fiber density in the skin and cornea (Biessels et al., 2014). Additional practical benefits of using animal models include the ability to test hypotheses by methods not permissible or practical in humans, the lower cost of preclinical research, and the accelerated temporal progression of neuropathy that allows longitudinal studies over months rather than years or decades. Unfortunately, as with many model systems, the literature is cluttered with the inappropriate use of specialized models or overly exuberant interpretation of positive findings. The fidelity of animal models of diabetic neuropathy remains an important issue that requires frequent consideration, revision, and refinement.

2. WHAT SPECIES?

Diabetes is not an exclusively human disease. There are reports of spontaneous diabetes in many long-lived species that live in close contact with humans, including nonhuman primates (Kuhar, Fuller, & Dennis, 2013), elephants (van der Kolk et al., 2011), camelids (Cebra, 2009), dogs (Fleeman, Rand, & Morton, 2009), and cats (O'Brien, 2002). In many cases, the provision of unnatural diets to wild animals kept in captivity and to species domesticated for farming or companionship is suspected as causing diabetes. Indeed, the rise of type 2 diabetes in the human population is being paralleled by a similar epidemic in the domestic cat population, which is also exposed to poor diet and limited exercise. Diabetic cats are particularly prone to developing a peripheral neuropathy that is strikingly similar to the human condition with nerve conduction slowing accompanied by structural damage to blood vessels, Schwann cells, and axons (Estrella et al., 2008; Mizisin et al., 2007). These observations support use of the diabetic cat population as a bridge between traditional preclinical drug efficacy studies and clinical trials so that drug efficacy can be evaluated in nerves exhibiting the full spectrum of structural pathology. Recruitment of diabetic cats and screening for neuropathy by veterinarians is as viable as similar recruitment exercises for clinical trials but, in our experience, also requires the same attention to recruitment practices such as investigator training and incentives for both investigator and the subject (or their owner). Given the huge cost of clinical trials, efficacy testing in diabetic cats may be useful as a cost-effective final preclinical evaluation point in the drug development pipeline.

Some species show normal physiological features that would indicate diabetes in humans. For example, plasma glucose levels of birds, while varying between species, are much higher (10–40 mM range) than in most mammals. This is not a permanent disease condition so much as a reflection of different physiological goals, with glucagon, not insulin, being the dominant pancreatic hormone in chickens (Braun & Sweazea, 2008). While hyperglycemia and relative insulin deficiency are the hallmarks of type 1 diabetes and implicated in most current pathogenic hypotheses, they do not appear to adversely impact the nervous system of birds. It has been argued that such interspecies variations may prove useful for studying how organs resist damage by hyperglycemia and/or insulinopenia (Singer, 2011). The first foray by one of us into medical research was an attempt to make chickens diabetic in order to study the effects of hyperglycemia on the enticingly long vagus nerve (Calcutt & Tomlinson, 1985). The approach, along with many chickens, was ultimately doomed not only because birds are inherently hyperglycemic, but also because it became clear that the chicken pancreas is relatively resistant to the diabetogenic drug streptozotocin (STZ) (Modak, Datar, Bhonde, & Ghaskadbi, 2007; Simon & Dubois, 1980). A lifetime quest for a diabetic giraffe continues.

Experimental requirements for an adequate number of observations to move beyond the anecdotal have restricted research in larger species to occasional observations of the effects of diabetes on nerve from primates (Pare et al., 2007), pigs (Juranek et al., 2010), and dogs (Walker et al., 2001), all of which are expensive to maintain and have yet to reveal tangible benefits as models. By far the vast majority of studies of experimental diabetic neuropathy have therefore been performed in rats and mice. Rodents can be bred to consistently develop spontaneous type 1 or type 2 diabetes or else can have diabetes induced by dietary manipulation or chemicals that ablate pancreatic beta cells. For many years, rats were the most widely studied species. This was in part due to their size, which allows surgical manipulations and yields enough tissue for crude biochemical assays. Rats were also preferred because the dominant pathogenic mechanism under evaluation during the 1960s to 1990s was glucose flux through the polyol pathway and polyol pathway metabolites accumulated in the nerve of diabetic rats. Mice were not widely studied as models for diabetic neuropathy during this period, other than basic characterization of the db/db and ob/ob mouse models of spontaneous type 2 diabetes, largely because it was erroneously thought that they lacked polyol pathway activity in their nerves. Recognition that lack of accumulation of polyol pathway intermediates in nerve was due to distinctive enzyme kinetics that makes mice similar to humans in this respect (Calcutt, Willars, & Tomlinson, 1988; Tomlinson, 1989) and that polyol pathway flux can drive nerve dysfunction in mice (Calcutt, Tomlinson, & Biswas, 1990; Ho et al., 2006; Ng et al., 1998), along with the emergence of new technology that favors use of mice for genetic manipulations, has caused the use of mouse models of both type 1 and type 2 diabetes to greatly expand over the last 10–20 years.

3. WHAT DIABETOGENIC INSULT?

The early use of pancreatectomy to produce diabetic animals for the study of insulin secretion and diabetes was initially superseded by alloxan, which reduces production and secretion of insulin and glucagon by damaging pancreatic beta cells, but does not impede secretion of other pancreatic products. Alloxan is a prooxidant that, while having a marked effect on pancreatic beta cells by virtue of cellular uptake by glucose transporters (Elsner, Tiedge, Guldbakke, Munday, & Lenzen, 2002), can also damage other organs involved in detoxification such as the liver and kidney. The first description of neuropathy in a diabetic rodent, as implied by nerve conduction slowing, was made in alloxan-diabetic rats (Eliasson, 1964). However, difficulties in dose titration to ensure the induction of type 1 diabetes without producing off-target effects or killing most of a cohort, make alloxan-induced diabetes a notoriously difficult model. Alloxan is now only infrequently used to produce diabetes in rats or mice, although use has persisted where diabetic rabbits are required, such as in studies of skin wound healing (Pradhan Nabzdyk et al., 2013). Alloxan has been almost completely replaced as a diabetogenic agent by STZ, a product of the soil bacteria *Streptomyces achromogenes* that was originally developed as a therapy for pancreatic cancer. Like alloxan, STZ enters pancreatic beta cells via the GLUT-2 glucose transporter (Elsner, Guldbakke, Tiedge, Munday, & Lenzen, 2000), where it causes cell death by a variety of mechanisms, as reviewed elsewhere (Eleazu, Eleazu, Chukwuma, & Essien, 2013). STZ replaced alloxan because it is relatively easy to dose titrate so that insulin deficient diabetes is consistently induced without significant damage to other organs or high mortality rates (Mansford & Opie, 1968).

Sliding scales of STZ dose have been developed for rats of different ages and weights (Calcutt, 2004). In mice, the severity of insulinopenia and its practical consequences such as conversion rate, survival duration, and frequency of reversal of diabetes arising from beta cell regeneration can be tempered by varying both STZ dose and dose frequency. A variety of protocols have been developed that employ between 2 and 5 low doses of STZ given on consecutive days (Biessels et al., 2014). As STZ is inexpensive and relatively easy to use when following established protocols, the STZ-diabetic rat and mouse have become the dominant models of diabetic neuropathy.

4. STZ TOXICITY

Concerns are intermittently raised that the cachexia that can occur in rodents treated with high doses of STZ makes them poor models, particularly for behavioral studies such as those using nociceptive testing to assess sensory function. Such concerns can be mitigated by appropriate dose titration of STZ that produces hyperglycemia without the extreme insulinopenia that underlies cachexia, or by trace insulin replacement (Calcutt, 2004), along with reporting of performance in rotarod or other tests. High doses of STZ can also produce direct and acute nephrotoxicity in mice (Breyer et al., 2005) and direct STZ neurotoxicity has been evoked to explain tactile allodynia in STZ-injected animals that did not develop overt hyperglycemia. However, STZ-induced insulinopenia that was not sufficient to produce overt hyperglycemia but contributes to the pathogenesis of allodynia is a more likely explanation (Romanovsky, Cruz, Dienel, & Dobretsov, 2006). A number of other observations also appear to argue against meaningful neurotoxic effects of STZ:

- STZ enters cells via the GLUT-2 transporter, which is not found in peripheral nerve.
- STZ is rapidly cleared from the body, disassociating it from the indices of neuropathy that can take weeks to emerge.
- Insulin replacement therapy to normalize blood glucose levels and started days after STZ delivery prevents onset of indices of neuropathy.
- Coinjection of STZ and 6-O-methyl glucose, which competitively inhibits STZ-uptake by GLUT-2 transporters, produces neither diabetes nor neuropathy as measured by NCV slowing and loss of heat-induced pain sensation (Davidson et al., 2009).
- Indices of neuropathy in STZ-injected animals are replicated in other models of type 1 and type 2 diabetes.

The wealth of reference data already collected means that STZ-diabetic rodents are likely to remain the dominant model of diabetic neuropathy. However, all data obtained in STZ-injected rodents should be interpreted with appropriate caution and the provenance of newly identified disorders validated using groups cotreated with 6-*O*-methyl glucose or insulin. Further, given the possibility that the pathogenesis of diabetic neuropathy may differ in type 1 and type 2 diabetes, STZ-diabetic rodents may best be deployed in combination with models of type 2 diabetes to evaluate whether new disorders and therapies are pertinent to both forms of the disease.

5. NOVEL MODELS

The neuropathy phenotype of commonly used rodent models of diabetes has recently been evaluated under the aegis of Neurodiab and the NIH. The resulting consensus statement is comprehensive and may be viewed elsewhere (Biessels et al., 2014) so need not be replicated in this volume. We have therefore taken the opportunity to invite colleagues to instead consider alternative approaches to modeling diabetic neuropathy. A number of such models, particularly models of type 2 diabetes and metabolic syndrome, are currently undergoing characterization. The review "Alternatives to the STZ-Diabetic Rodent" by Yorek is therefore a particularly opportune contribution. Moreover, the pertinence of neuronal and glial cell cultures to model the toxic impact of hyperglycemia, insulinopenia, or other diabetes-associated insults is perhaps even more controversial than use of animal models of diabetic neuropathy, with the advantage of selective manipulation of the local environment offset by the disadvantage of lack of intercellular and physiological interactions. The chapter "Can Diabetic Neuropathy Be Modeled In Vitro?" by Gardiner eloquently addresses these trade-offs to make a case that, within reason, we can indeed "do it in a dish."

REFERENCES

Biessels, G. J., Bril, V., Calcutt, N. A., Cameron, N. E., Cotter, M. A., Dobrowsky, R., et al. (2014). Phenotyping animal models of diabetic neuropathy: A consensus statement of the diabetic neuropathy study group of the EASD (neurodiab). *Journal of the Peripheral Nervous System*, *19*, 77–87.

Braun, E. J., & Sweazea, K. L. (2008). Glucose regulation in birds. *Comparative Biochemistry and Physiology. Part B, Biochemistry & Molecular Biology*, *151*, 1–9.

Breyer, M. D., Bottinger, E., Brosius, F. C., 3rd, Coffman, T. M., Harris, R. C., Heilig, C. W., et al. (2005). Mouse models of diabetic nephropathy. *Journal of the American Society of Nephrology*, *16*, 27–45.

Calcutt, N. A. (2004). Modeling diabetic sensory neuropathy in rats. *Methods in Molecular Medicine, 99*, 55–65.
Calcutt, N. A., & Tomlinson, D. R. (1985). A novel preparation for the study of axonal transport in vitro: The chick nodose ganglion/vagus nerve preparation. *Journal of Physiology, 361*, 11P.
Calcutt, N. A., Tomlinson, D. R., & Biswas, S. (1990). Coexistence of nerve conduction deficit with increased Na(+)-K(+)-ATPase activity in galactose-fed mice. Implications for polyol pathway and diabetic neuropathy. *Diabetes, 39*, 663–666.
Calcutt, N. A., Willars, G. B., & Tomlinson, D. R. (1988). Statil-sensitive polyol formation in nerve of galactose-fed mice. *Metabolism, 37*, 450–453.
Cebra, C. K. (2009). Disorders of carbohydrate or lipid metabolism in camelids. *The Veterinary Clinics of North America. Food Animal Practice, 25*, 339–352.
Davidson, E., Coppey, L., Lu, B., Arballo, V., Calcutt, N. A., Gerard, C., et al. (2009). The roles of streptozotocin neurotoxicity and neutral endopeptidase in murine experimental diabetic neuropathy. *Experimental Diabetes Research, 2009*, 431980.
Eleazu, C. O., Eleazu, K. C., Chukwuma, S., & Essien, U. N. (2013). Review of the mechanism of cell death resulting from streptozotocin challenge in experimental animals, its practical use and potential risk to humans. *Journal of Diabetes and Metabolic Disorders, 12*, 60.
Eliasson, S. G. (1964). Nerve conduction changes in experimental diabetes. *The Journal of Clinical Investigation, 43*, 2353–2358.
Elsner, M., Guldbakke, B., Tiedge, M., Munday, R., & Lenzen, S. (2000). Relative importance of transport and alkylation for pancreatic beta-cell toxicity of streptozotocin. *Diabetologia, 43*, 1528–1533.
Elsner, M., Tiedge, M., Guldbakke, B., Munday, R., & Lenzen, S. (2002). Importance of the GLUT2 glucose transporter for pancreatic beta cell toxicity of alloxan. *Diabetologia, 45*, 1542–1549.
Estrella, J. S., Nelson, R. N., Sturges, B. K., Vernau, K. M., Williams, D. C., LeCouteur, R. A., et al. (2008). Endoneurial microvascular pathology in feline diabetic neuropathy. *Microvascular Research, 75*, 403–410.
Fleeman, L. M., Rand, J. S., & Morton, J. M. (2009). Pharmacokinetics and pharmacodynamics of porcine insulin zinc suspension in eight diabetic dogs. *The Veterinary Record, 164*, 232–237.
Ho, E. C., Lam, K. S., Chen, Y. S., Yip, J. C., Arvindakshan, M., Yamagishi, S., et al. (2006). Aldose reductase-deficient mice are protected from delayed motor nerve conduction velocity, increased c-Jun NH2-terminal kinase activation, depletion of reduced glutathione, increased superoxide accumulation, and DNA damage. *Diabetes, 55*, 1946–1953.
Juranek, J. K., Aleshin, A., Rattigan, E. M., Johnson, L., Qu, W., Song, F., et al. (2010). Morphological changes and immunohistochemical expression of RAGE and its ligands in the sciatic nerve of hyperglycemic pig (Sus scrofa). *Biochemistry Insights, 2010*, 47–59.
Kuhar, C. W., Fuller, G. A., & Dennis, P. M. (2013). A survey of diabetes prevalence in zoo-housed primates. *Zoo Biology, 32*, 63–69.
Mansford, K. R., & Opie, L. (1968). Comparison of metabolic abnormalities in diabetes mellitus induced by streptozotocin or by alloxan. *Lancet, 1*, 670–671.
Mizisin, A. P., Nelson, R. W., Sturges, B. K., Vernau, K. M., Lecouteur, R. A., Williams, D. C., et al. (2007). Comparable myelinated nerve pathology in feline and human diabetes mellitus. *Acta Neuropathologica, 113*, 431–442.
Modak, M. A., Datar, S. P., Bhonde, R. R., & Ghaskadbi, S. S. (2007). Differential susceptibility of chick and mouse islets to streptozotocin and its co-relation with islet antioxidant status. *Journal of Comparative Physiology B, 177*, 247–257.
Ng, T. F., Lee, F. K., Song, Z. T., Calcutt, N. A., Lee, A. Y., Chung, S. S., et al. (1998). Effects of sorbitol dehydrogenase deficiency on nerve conduction in experimental diabetic mice. *Diabetes, 47*, 961–966.

O'Brien, T. D. (2002). Pathogenesis of feline diabetes mellitus. *Molecular and Cellular Endocrinology, 197*, 213–219.
Pare, M., Albrecht, P. J., Noto, C. J., Bodkin, N. L., Pittenger, G. L., Schreyer, D. J., et al. (2007). Differential hypertrophy and atrophy among all types of cutaneous innervation in the glabrous skin of the monkey hand during aging and naturally occurring type 2 diabetes. *The Journal of Comparative Neurology, 501*, 543–567.
Pradhan Nabzdyk, L., Kuchibhotla, S., Guthrie, P., Chun, M., Auster, M. E., Nabzdyk, C., et al. (2013). Expression of neuropeptides and cytokines in a rabbit model of diabetic neuroischemic wound healing. *Journal of Vascular Surgery, 58*, 766–775. e712.
Romanovsky, D., Cruz, N. F., Dienel, G. A., & Dobretsov, M. (2006). Mechanical hyperalgesia correlates with insulin deficiency in normoglycemic streptozotocin-treated rats. *Neurobiology of Disease, 24*, 384–394.
Simon, J., & Dubois, M. P. (1980). Failure of a sublethal streptozotocin dose to induce diabetes in the chicken. *Hormone and Metabolic Research, 12*, 631–632.
Singer, M. A. (2011). Insights into biomedicine from animal adaptations. *Comprehensive Physiology, 1*, 2063–2081.
Tomlinson, D. R. (1989). Polyols and myo-inositol in diabetic neuropathy—Of mice and men. *Mayo Clinic Proceedings, 64*, 1030–1033.
van der Kolk, J. H., Hoyer, M. J., Verstappen, F. A., Wolters, M. S., Treskes, M., Grinwis, G. C., et al. (2011). Diabetes mellitus in a 50-year-old captive Asian elephant (Elaphas maximus) bull. *The Veterinary Quarterly, 31*, 99–101.
Walker, D., Siddique, I., Anderson, H., Gardiner, T. A., Archer, D. B., Boulton, A. J. M., et al. (2001). Nerve pathology in the type 1 diabetic dog: Effects of treatment with sulindac. *Journal of the Peripheral Nervous System, 6*, 219–226.

CHAPTER FIVE

Can Diabetic Neuropathy Be Modeled In Vitro?

N.J. Gardiner[1], O.J. Freeman
University of Manchester, Manchester, United Kingdom
[1]Corresponding author: e-mail address: natalie.gardiner@manchester.ac.uk

Contents

Abstract

Diabetic neuropathy is a common secondary complication of diabetes that impacts on patient's health and well-being. Distal axon degeneration is a key feature of diabetic neuropathy, but the pathological changes which underlie axonal die-back are incompletely understood; despite decades of research a treatment has not yet been identified. Basic research must focus on understanding the complex mechanisms underlying changes that occur in the nervous system during diabetes. To this end, tissue culture techniques are invaluable as they enable researchers to examine the intricate mechanistic responses of cells to high glucose or other factors in order to better understand the pathogenesis of nerve dysfunction. This chapter describes the use of in vitro models to study a wide range of specific cellular effects pertaining to diabetic neuropathy including apoptosis, neurite outgrowth, neurodegeneration, activity, and bioenergetics. We consider problems associated with in vitro modeling and future refinement such as use of induced pluripotent stem cells and microfluidic technology.

International Review of Neurobiology, Volume 127
ISSN 0074-7742
http://dx.doi.org/10.1016/bs.irn.2016.02.004

1. INTRODUCTION

Diabetic neuropathy affects approximately half of all patients with diabetes mellitus (Dyck et al., 1993). It can be a debilitating condition that impacts on patient's health and quality of life and presents a substantial expense for health-care providers (Hex, Bartlett, Wright, Taylor, & Varley, 2012). Patients typically present with sensory symptoms arising in a "glove and stocking" distribution, with feet commonly affected before hands. Sensory symptoms are divided into the "positive": neuropathic pain, paraesthesia, and dysesthesia; and the "negative": loss of sensation, decreased nerve conduction velocity, and nerve dysfunction. Autonomic and motor neuropathies also occur, and increasing evidence suggests that this secondary complication of diabetes extends to the central nervous system (CNS). Neurons show a distal–proximal pattern of injury, with the longest axons showing the worst pathology.

Diabetic neuropathy is multifactorial in etiology and many pathogenic mechanisms have been described (Fernyhough & Calcutt, 2010; Fernyhough, Roy Chowdhury, & Schmidt, 2010; Freeman et al., 2016; Obrosova, 2002, 2003; Zochodne, 2014). Despite this, treatment options are currently limited to glycemic control and/or symptomatic treatment and are inadequate. Experimental rodent models of diabetes have long been utilized to assess behavioral, electrophysiological, and morphological deficits in diabetic neuropathy and to test the efficacy of novel therapeutics. These models will be considered elsewhere in this volume. The sad fact stands that, despite many decades of research, a treatment for diabetic neuropathy that significantly reverses or ameliorates the progression of the disease has not been identified. The search for efficacious drugs to better treat neuropathic pain, prevent axon degeneration, and/or promote nerve regeneration is therefore an urgent clinical need.

Basic research in diabetic neuropathy must focus on understanding the complex mechanisms underlying the changes that occur in the nervous system. To this end, tissue and cell culture techniques are valuable as they enable researchers to dissect out the intricate mechanistic responses of cells to high glucose or other exogenous factors. The crucial benefit afforded by an in vitro over an in vivo approach is the ability to precisely control and manipulate the environment (Table 1), thereby allowing study of a wide range of specific cellular events ranging from survival to death; neurite outgrowth to neurodegeneration; excitation to inhibition;

Table 1 Advantages and Disadvantages to Consider When Selecting an In Vitro Modeling Approach for Diabetic Neuropathy

	In Vitro Techniques
Advantages	Tight control of the extracellular environment
	Ability to study specific cell responses to individual exogenous stimuli
	Allows real-time imaging and recording of complex intracellular pathways and events
	Ability for high-throughput screening of drugs
	Fewer ethical issues than using in vivo models
Disadvantages	Simplistic
	Fails to recapitulate the complex in vivo environment
	Variable purity of primary cultures
	Acutely dissociated primary cells will be in a regenerative rather than homeostatic mode
	Inability to test therapeutics on functional changes
	Translation—may not mimic the clinical condition as well as experimental models of diabetes

subcellular bioenergetics; and axonal transport. While in vitro systems are by their nature simplistic, and findings should only be extrapolated to the clinical condition following validation in other more complex models, data can be used to better understand the pathogenesis of nerve dysfunction and to identify potential therapeutic mechanisms to inform in vivo studies.

Pathogenic mechanisms associated with diabetes have been investigated using a range of in vitro techniques such as dissociated neuronal and/or glial cell culture, cocultures, and organotypic explants. This chapter will guide the reader through the choices of cell types and culture system to consider and will also highlight phenotypic responses of cells to hyperglycemia and other stressors to emphasize the diversity and usefulness of in vitro models to study diabetic neuropathy. We have focussed on work that can only currently be done in vitro but will also consider the important caveats for in vitro modeling approaches (Table 1) and the exciting new technologies and techniques that may allow better modeling of a "nerve in a dish."

2. DIABETIC NEUROPATHY

The first symptoms of diabetic neuropathy typically arise in the somatosensory aspect of the peripheral nervous system (PNS). The skin is innervated by peptidergic and nonpeptidergic unmyelinated afferents that form free axon terminals in the epidermis, and myelinated afferents that terminate at specialized mechanoreceptors (Fig. 1) (Hilliges, Wang, & Johansson, 1995; Owens & Lumpkin, 2014). Distal processes of sensory nerves degenerate in both clinical and experimental diabetic neuropathy (Kennedy & Zochodne, 2000, 2005; Tomlinson & Gardiner, 2008; Yasuda et al., 2003), and degeneration occurs at a faster rate than reinnervation, leading to loss of sensation. This degeneration, along with other structural changes in the nerve such as axonal dystrophy, Schwann cell pathology, paranodal demyelination, and endoneurial microangiopathy, coincides with electrophysiological changes and sensory dysfunction (Tomlinson & Gardiner, 2008). Autonomic and motor neuropathies are less well characterized than sensory neuropathies, possibly reflective of the lower incidence of motor and visceral autonomic neuropathy in patients with diabetes (Dyck et al., 1993; Spallone et al., 2011) and the comparative ease of studying the somatosensory system in both humans and experimental models of diabetes. Since the interpretation of in vitro experiments requires knowledge of the context in which different cell populations function, this chapter will open with a brief overview of the development, maturation, and structure of the PNS.

3. THE SOMATOSENSORY NERVOUS SYSTEM

The PNS is divided into the somatic (sensory and motor neurons) and autonomic (sympathetic, parasympathetic, and enteric neurons) nervous systems and links peripheral tissues with the brain and spinal cord components of the CNS. PNS efferents include motor neurons and autonomic neurons, whose cell bodies in the ventral horn of the spinal cord or autonomic ganglia extend axons to neuromuscular junctions in skeletal muscle and internal organs, respectively. The cell bodies of sensory neurons reside within the dorsal root ganglia (DRG) in the intervertebral foramen adjacent to the vertebral column, and in the trigeminal ganglia. Sensory neurons are pseudo-unipolar neurons and extend an axon which bifurcates, with one branch projecting to the CNS and the other to peripheral target tissue.

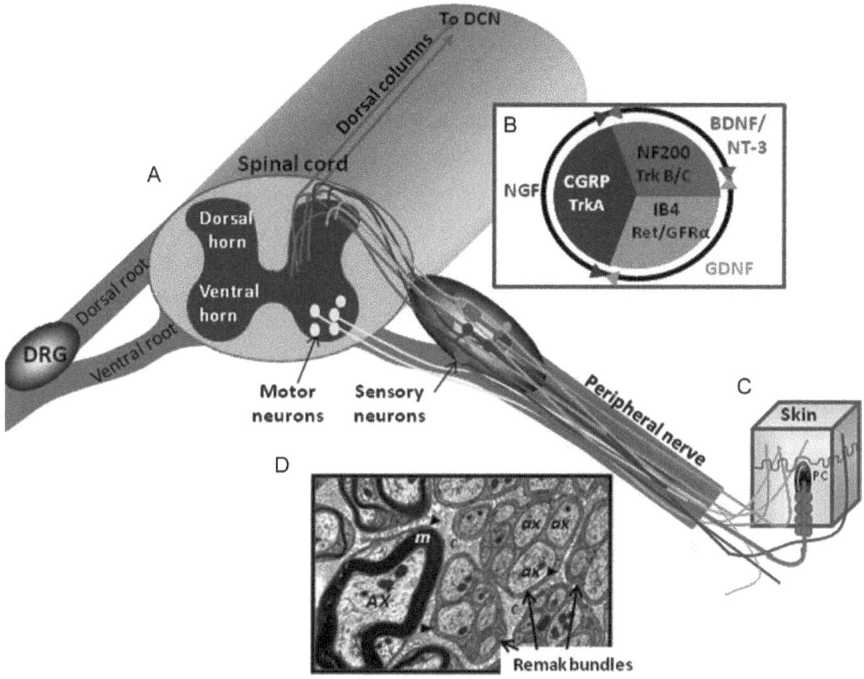

Fig. 1 The peripheral nervous system. (A) The cell bodies of most peripheral sensory neurons are located in the dorsal root ganglia (DRG). The motor neuron cell bodies within the ventral horn of the spinal cord extend efferents via the ventral roots to join the peripheral nerve and terminate in specialized neuromuscular junctions. (B) Sensory neurons are a heterogeneous population, express different receptors and phenotypic markers, and respond to different neurotrophic factors. Peripheral sensory afferents terminate as free nerve endings or in specialized sensory receptors such as Pacinian corpuscles (PC) in the skin (C). Central branches are directed to the spinal cord and through collateral branches to the dorsal column nuclei (DCN). The electron micrograph (D; kindly provided by Prof. David Tomlinson, University of Manchester) shows the internal organization of a peripheral nerve. Large-diameter sensory and motor axons (*AX*) are surrounded by compacted myelin (*m*), while groups of small-diameter unmyelinated sensory axons (*ax*) are grouped in Remak bundles (*arrows*). Endoneurial ECM, collagen fibrils (*c*) surround Schwann cell basement membranes (*arrowheads*) of both Remak bundles and myelinated axon/Schwann cell units. *Reprinted from Gardiner, N. J. (2011). Integrins and the extracellular matrix: Key mediators of development and regeneration of the sensory nervous system.* Developmental Neurobiology, 71, *1054–1072, with permission from John Wiley and Sons © 2011.*

Sensory afferents are capable of transducing a wide variety of modalities, such as touch, temperature, pain, and itch, and transmit this information to the CNS. They are highly diverse in somal size, axon diameter, and neurochemistry. Schwann cells are the principal glial cell of the PNS and play a central role in the homeostasis of the endoneurial environment (Mirsky et al., 2002). Axons in the peripheral nerve and spinal roots are tightly associated with myelinating or nonmyelinating Schwann cells and enclosed within discrete basement membranes (Fig. 1).

Both Schwann cells and sensory neurons arise from a migratory population of neural crest cells (NCCs) derived from the dorsal surface of the neural tube (Harris & Erickson, 2007). These NCCs migrate ventrally to regions proximal to the neural tube and there differentiate into the DRG and cells of the PNS (Bronner-Fraser, Stern, & Fraser, 1991; Sanes, 1983). The two main intermediate cells involved in the generation of Schwann cells from NCCs are the Schwann cell precursor, typically found in rat nerves at embryonic (E) day 14/15 (mouse: E12/13), and the immature Schwann cell, present from E17 (mouse E15; Jessen & Mirsky, 1997, 2005). These associate with developing axons to form a cooperative relationship with each cell promoting survival and differentiation of the other (Jessen & Mirsky, 1997). Immature Schwann cells integrate signals derived from the axon and the basement membrane to differentiate along a myelinating or nonmyelinating fate. Axonal-derived β1 neuregulin (NRG1) binds to the glial Erbb3 receptor to regulate development of Schwann cells. Assignment of the myelinating/nonmyelinating Schwann cell–axon phenotype is completed within the early postnatal period when Schwann cells send cytoplasmic processes into axon bundles, a process known as radial sorting (Chen & Strickland, 2003; Jessen & Mirsky, 1997; Mirsky et al., 2002). Large-diameter axons produce high levels of NRG1 type III, which induces Schwann cell precursors to differentiate into myelinating Schwann cells, while small-diameter axons produce lower levels and associated Schwann cell precursors therefore differentiate into nonmyelinating Schwann cells (Taveggia et al., 2005).

There are two distinct periods of sensory neuron generation, starting around E10–13 in mouse and E9–14 in rat (Fig. 2). Large-diameter TrkC-positive sensory neurons fated to become proprioceptors or low-threshold mechanoreceptors are formed first, followed by a second period in which small-diameter TrkA-expressing neurons are generated (Lawson & Biscoe, 1979; Marmigere & Ernfors, 2007; Marmigere et al., 2006). The factors underlying the initial differentiation of sensory neurons into subpopulations are coupled to the discrete spatial and temporal expression of extracellular matrix (ECM) and neuronal integrins, a family of ECM receptors (Gardiner, 2011; Guan, Puthenveedu, & Condic, 2003). The brain-specific homeobox/POU domain protein 3A gene (*Brn3*) and Neurogenin 1 and 2

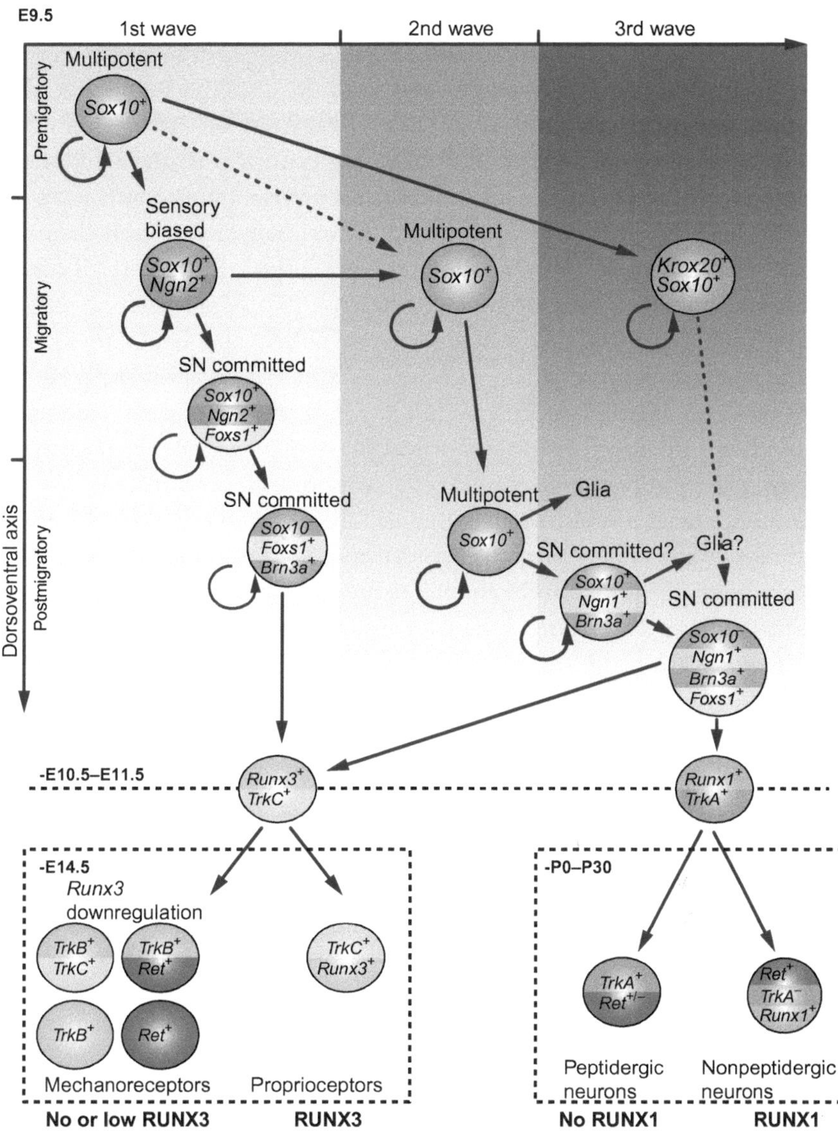

Fig. 2 Genetic control of development of the somatosensory nervous system. In the first wave of neurogenesis, SOX10^{+} cells migrate and begin to express neurogenin 2 (Ngn2). Ngn2^{+} cells subsequently commit to a sensory neuronal (SN) fate and express forkhead transcription factor (Foxs1^{+}) during migration. These postmigratory neurons then begin to express Brn3a and form large sensory neurons (TrkC^{+} and Runx3^{+}) at an early developmental stage. In contrast, cells with little or no Ngn2 remain uncommitted throughout migration. Postmigratory cells in the DRG then start to express Brn3a, Ngn1, and Foxs1. They either produce TrkC^{+} or TrkA^{+} populations of sensory neurons by *(Continued)*

are important factors in controlling neurogenesis (Lallemend & Ernfors, 2012; Marmigere & Ernfors, 2007; Marmigere et al., 2006). The transcription factor Runx1 is expressed by TrkA-positive, but not TrkC-positive neurons at early embryonic stages (Fig. 2; Chen et al., 2006; Kramer et al., 2006; Marmigere et al., 2006). Sensory neurons must make connections with their peripheral target tissues, the source of neurotrophins, to survive (Silos-Santiago et al., 1995). During the early postnatal period, around half of the small-diameter neurons downregulate Runx1 and retain TrkA (Molliver, Radeke, Feinstein, & Snider, 1995; Molliver & Snider, 1997; Molliver et al., 1997). The remaining neurons retain Runx1, lose TrkA, and begin to express Ret, the transmembrane signaling component of the receptor for glial cell line-derived neurotrophic factor (GDNF) family members (Bennett, Averill, Clary, Priestley, & McMahon, 1996; Molliver et al., 1997). Large-diameter trkC-positive neurons, which constitute approximately 30% of the sensory neuron population in the adult DRG, have large-diameter myelinated axons (Aβ fibers) that transmit low-threshold tactile and proprioceptive information to the CNS and respond to neurotrophin (NT)-3. Small- to medium-diameter sensory neurons have either thinly myelinated or unmyelinated axons (Aδ and C-fibers), mechanosensing, thermal and/or nociceptive functions and are divided into either trkA-positive, peptidergic neurons that respond to NGF or nonpeptidergic, Ret receptor-positive neurons that respond to the GDNF family of neurotrophic factors (Averill, McMahon, Clary, Reichardt, & Priestley, 1995; Bennett et al., 1996; Karchewski, Kim, Johnston, McKnight, & Verge, 1999; McMahon, Armanini, Ling, & Phillips, 1994; Wright & Snider, 1995).

Fig. 2—Cont'd expressing Runx3 or Runx1, respectively. A third wave of neurogenesis arises from cells expressing Sox10 and Krox20 which contribute mainly to the RUNX1/TrkA^{+} population of neurons and to glia. Cells from the first wave that maintain Runx3 expression maintain TrkC expression and become proprioceptors, whereas cells that lose or reduce Runx3 expression may express TrkB and TrkC, TrkB alone, RET alone, or RET and TrkB to become mechanoreceptive neurons. The small neurons which maintain Runx1 expression become nonpeptidergic small-diameter neurons (Ret^{+}/TrkA^{-}/Runx1^{+} cells), while downregulation of Runx1 fates neurons toward an NGF-responsive peptidergic phenotype (TrkA^{+}/Ret$^{+/-}$/Runx1^{-}). *E*, embryonic day; *P* postnatal day. *Reprinted from Marmigere, F., & Ernfors, P. (2007). Specification and connectivity of neuronal subtypes in the sensory lineage.* Nature Reviews Neuroscience, *8, 114–127, with permission from Macmillan Publishers Ltd © 2007.*

It has long been acknowledged that there is overlap between phenotypic markers and neurotrophic factor responsiveness. Categorizing adult sensory neurons into three populations (Fig. 1) based on their neurochemistry (expression of specific ion channels neurotransmitters and neurotrophin receptors) or cytoskeletal components (Lawson, 1995; Lawson, Perry, Prabhakar, & McCarthy, 1993; McMahon et al., 1994), while convenient, is simplistic. Recent advances in genomic and single-cell sorting techniques have enabled a much more thorough expression profiling of rodent somatosensory neurons (Thakur et al., 2014; Usoskin et al., 2015). A comprehensive unbiased transcriptome analysis of mouse sensory neurons using single-cell RNA-seq (Usoskin et al., 2015) revealed not three, but eleven distinct subpopulations: three low-threshold mechanoreceptive, two proprioceptive, and six principal types of thermosensitive, itch-sensitive, low-threshold mechanosensitive, and nociceptive neurons (Usoskin et al., 2015). This highlights the diversity and complexity of adult sensory neurons.

Large-diameter axons are ensheathed by a specialized plasma membrane of the myelinating Schwann cells that wraps in concentric rings and compacts around the axon. Small-diameter unmyelinated axons are sorted into Remak bundles, which can contain over 20 axons isolated from each other by processes of a nonmyelinating Schwann cell and enclosed within a basement membrane (Murinson & Griffin, 2004; Murinson, Hoffman, Banihashemi, Meyer, & Griffin, 2005). Schwann cells are important for the proper formation of nodes of Ranvier (Salzer, Brophy, & Peles, 2008) and localization of ion channels within the node (Feinberg et al., 2010). Schwann cells are not purely structural however, and they contribute toward the trophic support of axons. For example, signaling from axonal neuregulin-1 to Schwann cells (Chen et al., 2003) and Schwann cell to axon transfer of ciliary-derived neurotrophic factor (Calcutt, Muir, Powell, & Mizisin, 1992) is critical for healthy axonal function. Schwann cells regulate myelin sheath parameters (Michailov et al., 2004; Taveggia et al., 2005) and axon cytoskeletal components (Kirkpatrick & Brady, 1994). Schwann cells play critical roles in mediating successful axon regeneration. After a peripheral nerve injury, dedifferentiated Schwann cells remove distal myelin and axon debris, form bands of Büngner that provide ECM-rich pathways for regenerating axons, neurotrophic support, and then remyelinate regenerated axons (Allodi, Udina, & Navarro, 2012). There is also a developing literature describing the role of Schwann cells in the metabolic support of peripheral axons. For example, it has been demonstrated that direct impairment of Schwann cell mitochondrial function leads to neuropathy in mice (Viader et al., 2011, 2013). In addition, Schwann cells have the ability to transfer ribosomes and mRNA to axons (Court, Hendriks, MacGillavry, Alvarez, & van Minnen, 2008), implying that Schwann cells and axons have extensive functional coupling. The PNS is closely associated with a wide range of other cell and tissue types: keratinocytes, immune cells, and endothelial cells of blood vessels to name but a few. PNS structure and function also depends upon association with a complex architecture of ECM proteins, modification of which in disease can all impact upon neuronal phenotype and nerve function.

4. CAN WE MODEL DIABETIC NEUROPATHY IN VITRO?

Tissue culture paradigms have long been used alongside animal models of disease or nerve injury to study physiological and pathological mechanisms underlying neuronal phenotype. The use of cell culture significantly reduces the legitimate ethical problems associated with using in vivo models of nerve injury or disease. Russell and Burch's ethical framework for the humane use of animals in scientific research (reviewed in Tannenbaum & Bennett, 2015) describes "The Three R's" (Replacement, Reduction, and Refinement). Culture techniques using cells or tissues derived from animals are regarded as "relative replacements" and so offer a valuable ethical advantage.

Tissue culture allows researchers the exquisite advantage of a carefully controlled extracellular environment in which to examine phenotypic responses of neurons/glia to exogenous stimuli and to elucidate intracellular mechanisms underlying neurite outgrowth, bioenergetics, and cell death. But can one realistically expect to model such a complex, multifactorial disorder as diabetic neuropathy in a dish? Is simply adding high concentrations of glucose and/or limiting the extracellular insulin concentration of neurons good enough? Is it best to use homogenous well-characterized cell lines, dissociate primary cultures of neurons and glia, or use tissue explants which preserve cellular diversity, interactions, and overall tissue structure? Should we use embryonic or adult tissue? Should we use tissue from control animals or tissue from animals with experimental diabetes? What should our endpoint measures be? These questions represent a significant challenge for the in vitro biologist.

4.1 Choice of Cells

One must first decide whether to utilize primary tissue culture or an immortalized cell line and how to maintain them under optimized conditions that best replicate the in vivo environment.

4.1.1 Immortalized Cell Lines

Tumor-derived cell lines including SH-SY5Y, NTERA-2, and particularly PC12 cells, a line derived from a pheochromocytoma of the rat adrenal medulla, have been used extensively in basic neurophysiological research. These cell lines grow continuously as undifferentiated cells but can be stimulated with compounds such as retinoic acid, phorbol esters, cAMP, or NGF to differentiate into neuronal-like cells that develop long processes expressing neuronal markers such as βIII-tubulin, synaptophysin, microtubule-associated protein-2 (MAP2), and NeuN. Immortalized cell lines have also been generated from sensory neurons of the DRG. For example, the 50B11 rat embryonic DRG cell line can be differentiated by forskolin to extend processes that express nociception-specific receptors (Chen, Mi, Haughey, Oz, & Hoke, 2007) such as trkA or GFRα1/α2, and respond to both NGF and GDNF by extending these processes (Bhattacherjee, Liao, & Smith, 2014). Since cells are generated from DRG isolated from E14.5 rats prior to the in vivo influence of Runx1 (Fig. 2), this coexpression of diverse neurotrophin receptors is not surprising and 50B11 cells most accurately represent immature sensory neurons. Schwann cell lines have also been established from Schwannomas, by transfection of Schwann cells with oncogenes or through spontaneous immortalization. ITMS32 cells from adult ICR mice show distinct proliferative Schwann cell phenotypes but do not myelinate, whereas IFRS1 cells can myelinate sensory neurons in coculture models (Sango, Yanagisawa, Takaku, Kawakami, & Watabe, 2011).

Immortalized cell lines are useful for electrophysiological, molecular, and biochemical studies and for high-throughput screening of drugs, which requires a large number of cells. They have distinct advantages over primary cultures, notably their ability to proliferate and provide an unlimited supply of a homogeneous population of cells and their relative ease of transfection. There is also no need to cull animals and dissect each culture, which represents considerable time saving and ethical advantages. However, since they do not replicate the heterogeneity of adult neuron phenotypes, caution must always be taken when interpreting data, and findings should be validated in primary cell culture before being extrapolated to the in vivo situation.

4.1.2 Primary Tissue Culture

Primary tissue culture includes the isolation and maintenance of either whole tissue organotypic explants/slices or the dissociation of tissue by mechanical and/or enzymatic means to produce a cell monolayer. If cells of the primary culture are proliferative, as are Schwann cells, they can be maintained and passaged as a secondary culture or cell line (Kaewkhaw, Scutt, & Haycock, 2012). Cells can then be maintained in optimized conditions to mimic the in vivo nerve environment. The use of acutely dissociated primary neuron and glial cultures expanded considerably following identification and subsequent widespread availability of specific neurotrophic factors and extracellular components (eg, laminin) for use as a substratum to support survival and enable growth of distinct neuron populations (Lindsay, 1988). While preparation and culture of primary neurons is undoubtedly challenging, DRG, sympathetic ganglia, peripheral nerve, hippocampus, and cortex can all be extracted from chicken, mouse or rat embryos, neonates, or adult animals. Excellent protocols exist in the literature describing methodologies for successfully culturing neuronal and Schwann cells. Dissociated adult rat and mouse sensory neurons have been widely utilized in a range of molecular, physiological, and live-cell imaging studies to investigate mechanisms of growth cone dynamics and neurite outgrowth, neurotrophin responsiveness and signal transduction pathways, programmed cell death, axonal transport, electrophysiology, calcium signaling, and mitochondrial function.

While embryonic and neonatal neurons require serum or exogenous growth factors to survive, adult mammalian neurons can be maintained for long periods in the absence of trophic support (Gavazzi, Kumar, McMahon, & Cohen, 1999; Lindsay, 1988). Adult sensory neuron cultures reflect the broad diversity of neuronal populations and show predicted neuritogenic population responses to exogenous growth factors when plated on laminin (Fig. 3; Gardiner et al., 2007). In contrast, fewer neurons respond to neurotrophin stimulation when plated on fibronectin and the response is significantly less than expected, given the distribution of neurotrophin receptors in DRG (Bennett et al., 1996; Gardiner et al., 2005; Gavazzi et al., 1999; McMahon et al., 1994). This difference perhaps reflects the normal association of axons with ECM components in vivo. In peripheral nerve, laminin is localized to the internal surfaces of the Schwann cell basement membranes

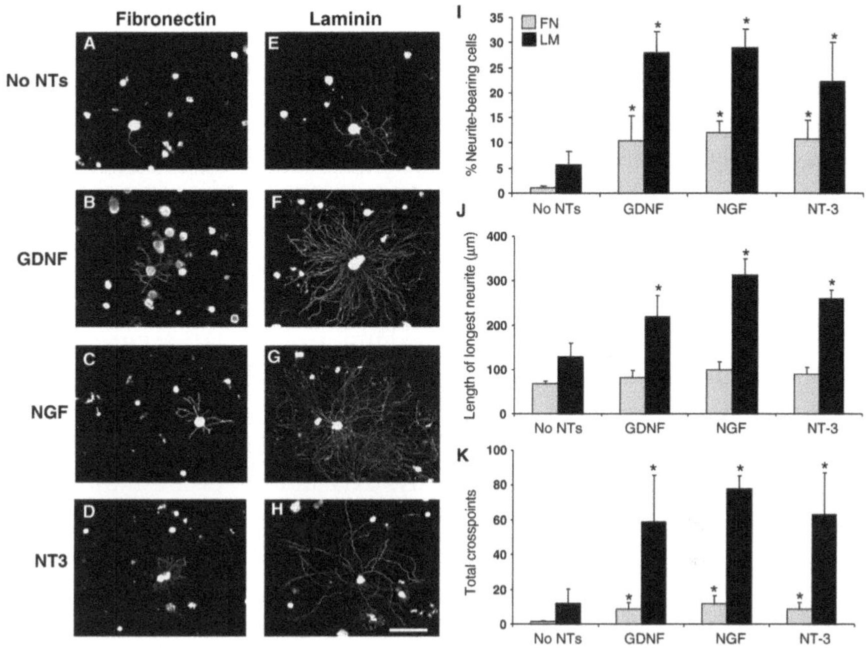

Fig. 3 Neurotrophin-stimulated neurite outgrowth from adult rat dissociated neurons plated on laminin (LM) and fibronectin (FN). Dissociated adult rat sensory neurons were plated on FN or LM treated with GDNF (50 ng/mL, B, F), NGF (10 ng/mL, C, G), and NT-3 (50 ng/mL, D, H), or untreated (A, E). For all parameters of growth assessed, LM was significantly better than FN at supporting neurite outgrowth. Fewer neurons extended neurites in response to neurotrophins when plated on FN (I, mean % neurite-bearing cells ± S.D.). Those that were extended were significantly shorter than those grown on LM (J, mean length of longest neurite ± S.D.) and less dense (total neurite density was quantified by a series of concentric circles overlaid onto an image of a neuron, the number of times neurites crossed each circle was measured; K, mean number of crosspoints per cell ± S.D., $n=4$ independent cultures). *Reprinted from Gardiner, N. J., Moffatt, S., Fernyhough, P., Humphries, M. J., Streuli, C. H., & Tomlinson, D. R. (2007). Preconditioning injury-induced neurite outgrowth of adult rat sensory neurons on fibronectin is mediated by mobilisation of axonal α5 integrin.* Molecular and Cellular Neurosciences, 35, *249–260, © 2007 with permission from Elsevier.*

that ensheath axons. Fibronectin is more diffuse, being localized along external surfaces of endoneurial tubes and the perineurium. This distinction highlights the need to optimize the environment of the culture systems to that of the cell being studied.

Neurons do not naturally exist in simple two-dimensional monolayers, but occupy a complex 3-dimensional space closely associated with multiple cell types and ECM in an environment that is constantly adapting and changing. Axon terminals in vivo constantly undergo dynamic collateral branching with terminal fields being remodeled according to local environmental factors or activity. A primary cell monolayer is therefore a useful but highly simplistic approach to modeling. The limitation of studying dissociated and purified neurons in isolation can be addressed using organotypic cultures obtained from dorsal root or sympathetic ganglia to maintain tissue structure, neuronal–glial interaction, and a three-dimensional environment. For this approach, ganglia are removed from adult or embryonic animals, cut into small pieces, with or without attached nerve stumps, and embedded in collagen or Matrigel™ with defined medium, treated with exogenous factors and cultured for several days. Axons extend from the isolated tissue into the surrounding matrix and Schwann cells proliferate and migrate radially (Fig. 4). This system is not without its limitations, as lack of blood supply and density of tissue mass can lead to hypoxia and cell death in the center of the explant. It is also difficult to use these cultures to investigate specific cellular effects other than measuring Schwann cell migration and axon extension distances. Future development of self-assembled aggregates of cells of the PNS could act as a bridge between traditional two-dimensional PNS-derived cell monolayers and in vivo models. Bioengineering approaches have already been taken with cells of the CNS, which can be maintained in culture for several weeks to enable neural network maturation

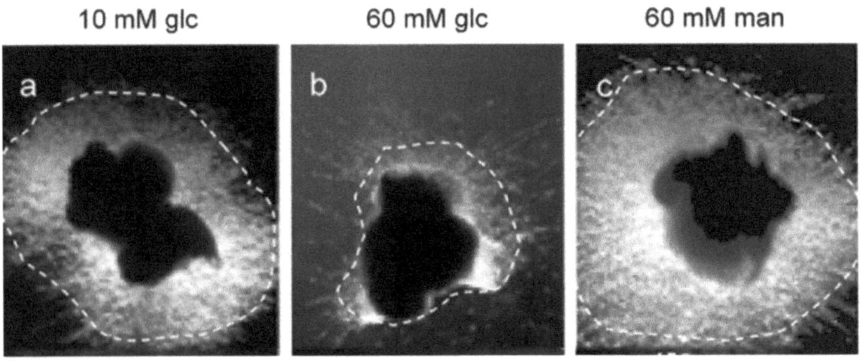

Fig. 4 High glucose inhibits Schwann cell migration. Bright-field micrographs from adult wild-type (WT) mouse DRG explants cultured for 8 days in media containing NGF (20 ng/mL) with (A, B) 10 or 60 mM glucose (glc) or (C) 60 mM mannitol (man) show evidence of reduced Schwann cell migration (demarcated with *dashed line*) in high glucose. *Reprinted from Gumy, L. F., Bampton, E. T., & Tolkovsky, A. M. (2008). Hyperglycaemia inhibits Schwann cell proliferation and migration and restricts regeneration of axons and Schwann cells from adult murine DRG.* Molecular and Cellular Neurosciences, 37, *298–311, © 2008 with permission from Elsevier.*

(Chwalek, Tang-Schomer, Omenetto, & Kaplan, 2015; Hopkins, DeSimone, Chwalek, & Kaplan, 2015). Such cortical spheroids formed from primary postnatal rat cortex contain neurons and glia networks that are electrically active with excitatory and inhibitory synapses (Dingle et al., 2015).

Other disadvantages of primary neuronal culture approaches include the low yield, due to neurons being postmitotic, and the labor-intensive nature of collecting neurons. Purity of cultures can vary, especially of those derived from adult animals, as ganglia consist of both sensory neuron cell bodies and nonneuronal cells (Schwann cells, satellite glia, and fibroblasts). Furthermore, isolation of cell bodies by necessity involves nerve injury (axotomy) so that cultured primary neurons switch from a homeostatic "housekeeping" in vivo phenotype to an injury-response "regenerative" phenotype (Smith & Skene, 1997). However, while these caveats must be accounted for in experimental design and interpretation, the use of primary neurons offers the distinct and valuable advantage that one can isolate DRG to compare phenotypic changes in mature adult sensory neurons obtained from control and diabetic rodents.

4.1.3 Induced Pluripotent Stem Cells

The continued failure of translation of drugs from the bench to the clinic highlights the need for better model systems that predict clinical efficacy. Interspecies differences between rodents commonly used as sources for primary cultures and human patients may be one underlying cause of this. A better in vitro model of neuropathy may in the future be provided through induced pluripotent stem cell (iPSC) technology. iPSCs are particularly useful for the study of hereditary diseases as one can obtain a skin biopsy from human patients, isolate fibroblasts, reprogram them into pluripotent cells via the insertion of four genes (*oct3/4*, *sox2*, *klf4*, *c-myc*; Takahashi et al., 2007), and then differentiate them into many different cell types. Published protocols describe differentiation into sensory neurons (Chambers et al., 2012) and Schwann cells (Liu et al., 2012). It is also possible to compare phenotypic changes between cells from genetically defined individuals to gain insight into pathogenic mechanisms. For example, iPSCs have been used to define transcriptional and functional changes induced in human motor neurons by mutant *SOD1* (Kiskinis et al., 2014). While using iPSC-derived neurons/glia to study disease is not without limitations (Sandoe & Eggan, 2013), this technology has the potential to transform our future understanding of the pathogenesis of diabetic neuropathy.

4.2 Choice of Stimuli

Diabetic neuropathies, which include both focal mononeuropathies and diffuse polyneuropathies, have a multifactorial etiology. Many pathogenic mechanisms have been suggested including hyperglycemia, microangiopathy and subsequent hypoxia, increases in polyol flux through the aldose reductase pathway leading to sorbitol and fructose accumulation, oxidative stress, NAD(P)-redox imbalances, mitochondrial dysfunction, calcium dysregulation, dyslipidemia, glycation of intracellular and extracellular proteins, and also reduced trophic support from insulin and other factors (Fernyhough & Calcutt, 2010; Fernyhough et al., 2010; Freeman et al., 2016; Obrosova, 2002, 2003; Zochodne, 2014). Intersection and synergy between these multiple pathways leads to progressive pathology. Primary cultures of peripheral neurons and glia have been widely used to investigate and compare effects of hyperglycemia, dyslipidemia, or stressors such as loss of insulin signaling and other trophic support.

The majority of studies to date that have used in vitro models of diabetic neuropathy have investigated the effects of raised glucose on cells using concentrations ranging from 10 to 200 mM. Immediate considerations for such a study should include whether to use cells derived only from control rodents, thereby allowing assessment of acute pathogenic mechanisms, or to also include cells from diabetic rodents that may have developed an altered metabolic phenotype in response to prolonged hyperglycemia. Similarly, since insulin has direct effects on sensory neurons (Huang et al., 2003; Huang, Verkhratsky, & Fernyhough, 2005), it is important to consider whether insulin, under its guise as a trophic factor, should be present in culture media and at what concentration. For example, cells derived from a control adult rodent will have matured in an environment of 4–8 mM glucose and fluctuating insulin, whereas those from a type 1 diabetic rodent may have spent months adapting to an environment with little insulin and excess glucose. This can make the choice of what constitutes control and experimental conditions quite complex. It is worth noting that the vast majority of published cell culture studies that used neurons but were unconcerned with studying diabetes employed basal glucose concentrations (20–30 mM in DMEM) that would be considered hyperglycemic by those studying diabetes in vitro and in vivo. Glucose levels of 50 mM+ have been used in some in vitro studies but are unlikely to be of pathophysiological relevance (see below) and 20–50 mM glucose represents more realistic levels that occur in the blood of poorly controlled diabetic rodents. Direct modeling of human diabetes requires external glucose in the 7–20 mM range.

The value of in vitro techniques can be illustrated by studies that investigated the hypothesized role of intracellular oxidative stress as a pathogenic mechanism downstream of hyperglycemia by application of pro-oxidants such as hydrogen peroxide and diethyl maleate, which binds to glutathione, thereby compromising the glutathione redox cycle (Purves et al., 2001). Other potentially toxic products of hyperglycemia, including lipid peroxidation products such as 4-hydroxynonenal (Zherebitskaya, Akude, Smith, & Fernyhough, 2009) and reactive dicarbonyl intermediates, including glyoxal, methylglyoxal, and glycolaldehyde (Radu, Dumitrescu, Mustaciosu, & Radu, 2012), have also been directly applied in vitro. Hyperlipidemia has been modeled using oxidized LDLs (Vincent et al., 2009) or long chain fatty acids (such as palmitate, linoleate, and oleate; Hinder et al., 2014). Another less precise, but perhaps more pertinent, approach has been to use serum or plasma from diabetic patients or rats and nondiabetic controls as a direct challenge to neurons in vitro (Bierhaus et al., 2012; Ristic, Srinivasan, Hall, Sima, & Wiley, 1998; Srinivasan, Stevens, Sheng, Hall, & Wiley, 1998). The complexity of diabetes presents an enormous challenge for the in vitro neurobiologist to select an appropriate stimulus for the cultured cells to mimic the diabetic milieu and this is a significant limitation to in vitro modeling of diabetes.

4.3 Choice of Output Measure

The cellular responses most amenable to study in vitro include cell survival or death, proliferation and migration, and neurite initiation and extension. Live-cell imaging and electrophysiological approaches may also be applied.

4.3.1 Life or Death?

A simple output measure for neurodegeneration in vitro is to assess cell death. Methods to measure death include the trypan blue exclusion assay, the lactate dehydrogenase release assay, and the MTT (3-(4,5-dimethylthiazol-2-yl)-2,5-diphenyltetrazolium bromide) tetrazolium reduction assay. Quantification of apoptosis-related events such as cleavage of caspases and detection of DNA fragmentation by terminal deoxynucleotidyl transferase dUTP nick end labeling can also be used. The response of neurons to high glucose differs depending on their developmental stage (Fig. 2). Studies using cultures of sensory neurons derived from E15 rat DRG and already maintained in 20 mM+ glucose found that exposure to further excess glucose induced apoptosis, oxidative stress, and cell death through a mitochondrial-dependent pathway (Russell et al., 2002; Vincent, Olzmann, Brownlee, Sivitz, & Russell, 2004; Vincent, Russell, Low, & Feldman, 2004). In contrast, studies using cultures of adult sensory neurons found no evidence of glucose-induced apoptosis (Gumy, Bampton, & Tolkovsky, 2008; Huang et al., 2003; Purves et al., 2001; Zherebitskaya et al., 2009). DRG explants or isolated adult sensory neurons cultured in serum-free medium containing NGF, and 10–60 mM glucose showed no evidence of hyperglycemia-induced DNA fragmentation, cyt-*c* release from mitochondria, or activation of caspase-3 (Gumy et al., 2008). Apoptosis of sympathetic neurons from superior cervical ganglion (SCG) in response to high glucose has been described in both embryonic (Russell & Feldman, 1999) and adult neurons (Semra, Smith, & Lincoln, 2004), and decreased viability of adult celiac/superior mesenteric ganglia (CG/SMG) neurons relative to SCG neurons was observed at 100 mM glucose (Semra et al., 2004).

It has been suggested that a reason for these discrepancies is the developmental difference in the neurotrophic requirements of embryonic and adult DRG neurons. During early sensory neuron development all three Trk receptors and p75 receptors are expressed and embryonic DRG neurons depend on neurotrophins for survival. As sensory neurons mature, they develop mechanisms to prevent apoptosis (Vogelbaum, Tong, & Rich, 1998). If neurons are harvested from embryos during the developmental period of programmed cell death, they will require neurotropic support in order to survive in culture (Snider & Silos-Santiago, 1996). The differential apoptotic response of embryonic and adult cells to high glucose may therefore simply reflect developmental stage. Indeed, embryonic sensory neurons cultured for several weeks with neurotrophic support will mature and then do not show high glucose-induced apoptosis (Yu, Rouen, & Dobrowsky, 2008). Adult DRG explants cultured for 8 days in 60 mM glucose also showed no evidence of Schwann cell apoptosis (Gumy et al., 2008), but conflicting evidence exists regarding the neonatal Schwann cell response (Delaney, Russell, Cheng, & Feldman, 2001; Gumy et al., 2008). Since dramatic apoptosis does not occur in the sensory or autonomic ganglia of diabetic rodents (Cheng & Zochodne, 2003; Kamiya, Zhang, & Sima, 2006; Kamiya, Zhangm, & Sima, 2005; Schmidt, 2001) and neuronal cell loss does not appear to underlie the distal axon pathology, studying in vitro apoptotic responses of neurons to high glucose is difficult to relate to clinical pathology.

4.3.2 Altered Bioenergetics and Oxidative Stress

Oxidative stress is a common phenomenon in many theories of the pathogenesis of diabetic neuropathy (Obrosova, 2002). Early studies in cultured endothelial cells identified formation of reactive oxygen species by increased electron donation to the mitochondrial electron transport chain under hyperglycemic conditions, and overload of oxidative defense mechanisms was presented as a pivotal event linking glucotoxic mechanisms of pathogenesis of diabetic complications (Brownlee, 2001, 2005; Nishikawa et al., 2000). Although support of such a unifying mechanism of diabetic complications has waned, metabolic dysfunction and oxidative stress are considered to be key contributors to the pathogenesis of diabetic neuropathy (Chowdhury, Smith, & Fernyhough, 2013; Fernyhough, 2015; Fernyhough & Calcutt, 2010; Fernyhough & McGavock, 2014; Obrosova, 2002). Intracellular calcium flux, mitochondrial dynamics, and oxidative stress are all highly amenable to confocal imaging in vitro, using cell-permeable fluorescent dyes. These dyes can provide real-time measurement of fluctuations of intracellular Ca^{2+} concentration, mitochondrial membrane potential, or increased reactive oxygen species in response to stimuli such as high glucose, or to compare responses of cells obtained from control and diabetic rats. For example, mitochondrial membrane polarization status in sensory neurons from age-matched control or diabetic rats was assessed using tetramethyl rhodamine methyl ester (Zherebitskaya et al., 2009) and rhodamine 123 (Huang et al., 2003) with reactive oxygen species detected using MitoSOX™ red (Akude et al., 2011). Using these dyes, it has been demonstrated that the mitochondrial inner membrane potential is depolarized in adult sensory neurones cultured from STZ-diabetic rats and that this could be prevented by systemic treatment with low doses of insulin or NT-3 (Huang et al., 2003, 2005). Recently, these imaging studies have been complemented by in vitro work using the Seahorse Biosciences XF Analyzer to assess and compare the mitochondrial function of dissociated sensory neurons obtained from control or STZ-diabetic rats (Chowdhury et al., 2013). By measuring oxygen consumption rate in cultured cells in response to specific inhibitors, one can examine various functions of the electron transport chain (Akude et al., 2011; Chowdhury et al., 2013). These studies provide excellent examples of how in vivo and in vitro analyses can work in tandem. Many of the studies described above utilize primary cultures of DRG neurons derived from animals with STZ diabetes and thus benefit from using an in vivo model as the diabetes-stimulus, but utilize in vitro methods to probe the specific mechanistic changes.

Evidence of bioenergetic dysfunction in diabetic neuropathy has been expanded with the use of "-omic" technologies, in which in vitro studies play a central role. Stable isotope labeling of amino acids in cell culture is an in vitro method of proteomic analysis which has been performed on mitochondrial extracts of DRG neurons from STZ-diabetic rats (Akude et al., 2011) and control neonatal Schwann cells cultured in high glucose (Zhang, Yu, et al., 2010). In neurons derived from the DRG of 22-week-STZ-diabetic rats and healthy controls, there was a significant downregulation of proteins related to oxidative phosphorylation and the tricarboxylic acid (TCA) cycle in diabetes (Akude et al., 2011). In contrast, in Schwann cells derived from the sciatic nerve of neonatal rats and cultured in high glucose for 2, 6, or 16 days, there was an upregulation of oxidative phosphorylation and the TCA cycle under hyperglycemia at all time points (Zhang, Yu, et al., 2010). Use of purified cell culture allows detailed examination of differential cellular responses to stimuli, rather than a global mixed-cell tissue response, and therefore enables cell-specific mechanisms, theories of pathogenesis, and therapeutic targets to be identified.

4.3.3 Neuronal Hyperexcitability

Altered expression of ion channels and increased activity of sensory afferents are likely to contribute to neuropathic pain and nerve dysfunction in diabetes. For example, dysregulated expression of genes for sodium channels Na(v)1.3, Na(v)1.6, Na(v)1.8, and Na(v)1.9 has been described in the rat DRG in experimental diabetes (Craner, Klein, Renganathan, Black, & Waxman, 2002). Patch-clamp recordings and electrophysiological studies have been performed on the whole excised DRG, nerve, or dissociated neurons in vitro. Current-clamp recordings from cultured mouse DRG neurons have characterized the effects of exogenous methylglyoxal on sensory neuron excitability (Bierhaus et al., 2012). In another recent study, STZ-induced diabetes was associated with increased TTX-S and TTX-R current densities, while current densities of both transient and sustained components of potassium current in sensory neurons in L4/5 DRG were decreased but could be ameliorated by treating rats with docosahexaenoic acid (Heng, Qi, Yang, & Xu, 2015). These studies and others highlight the increase of sensory neuron excitability in diabetes. Another in vitro modeling approach compared responses of isolated skin–nerve preparations from control or diabetic rats and found an increase in low-frequent ongoing activity of C-fibers from diabetic animals compared to control rats (Fuchs, Birklein, Reeh, & Sauer, 2010). After baseline recordings, the skin was superfused with media aerated with 97.5% N_2/2.5% CO_2 instead of normal carbogen for 30 min to model hypoxia. The response to hypoxia was significantly increased in fibers recorded from diabetic animals compared to control rats in either 2.5 or 25 mM glucose, suggesting that fibers from diabetic rats were more sensitive to hypoxic acidosis (Fuchs et al., 2010). Such detailed electrophysiological studies can only be performed in vitro due to the ability to control the extracellular environment.

4.3.4 Axonal Degeneration and Regeneration

Distal axon retraction or pruning and axonal dystrophy are both hallmark features of clinical and rodent models of diabetic neuropathy (Kennedy & Zochodne, 2000, 2005; Polydefkis et al., 2004; Yasuda et al., 2003). Sensory and autonomic axons degenerate before motor neurons in both clinical and experimental diabetes (Wilson & Wright, 2014; Zochodne, 2014) and neurodegeneration is first evident in the long sensory axons, notably those innervating feet and hands. Reinnervation of distal target tissues can either occur by regeneration of damaged axons or by collateral branching of undamaged axons. However, these processes are impaired in diabetes, and there is a significant reduction in the number of regenerating axons observed in clinical nerve biopsies indicating a reduction in intrinsic growth capacity of sensory neurons in diabetes (Bradley et al., 1995) and/or development of an unsupportive extracellular environment in diabetes. STZ-induced diabetic rats show impaired axonal regeneration following focal nerve injury, and the ability of nerves to regenerate is inversely proportional to the duration of diabetes (Bisby, 1980; Ekstrom & Tomlinson, 1989; Kennedy & Zochodne, 2000). This all culminates in axonal degeneration occurring at a faster rate than regeneration, and permanent loss of end-target innervation. The need to protect against diabetes-associated axon degeneration and promote regeneration are therefore major clinical needs.

A cell culture approach to studying the mechanisms underlying impaired axon regeneration in diabetes is relatively easy to employ as neurons can be stimulated to extend neurites by different agonists/antagonists/stressors and neurite initiation, elongation, and density measures are amenable for quantification in short-term cultures (see Fig. 3). Given the diversity in axonal morphology (particularly in primary cell culture), it is imperative that the researcher conducts analysis in a blinded and randomized fashion to remove bias. One can evaluate changes in neurite morphology, and efficacy of therapeutics, before taking studies forward to in vivo studies. Distal axon degeneration is more problematic to model in vitro, although recent advances in compartmentalized microfluidic systems mean that it is now easier to separate axon terminals from the cell body and these systems have been used in models of chemotherapy-associated neuropathy to study axonal degeneration in similar "die-back" neuropathies (see Fig. 5; Yang, Siddique,

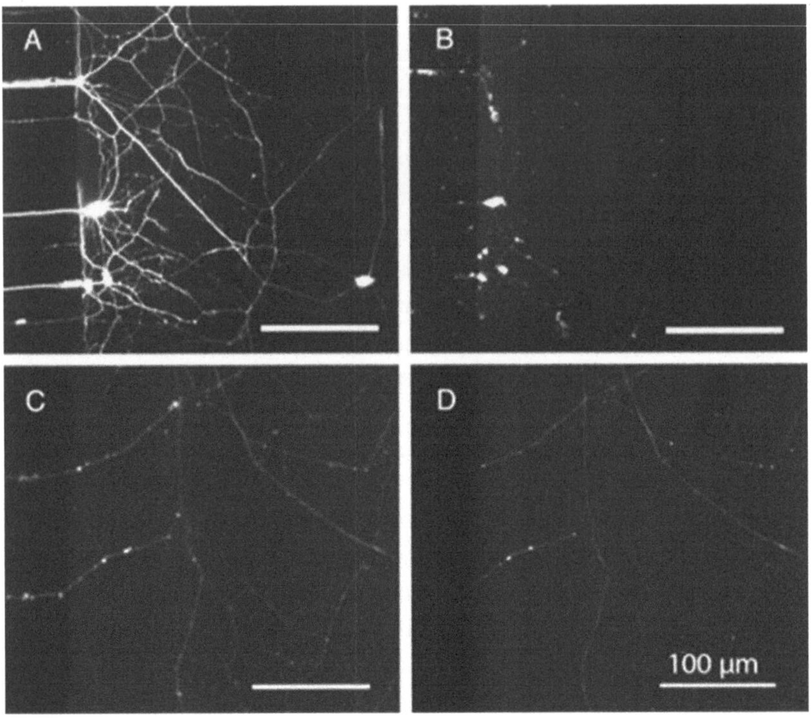

Fig. 5 Paclitaxel-induced axonal degeneration. DRG neurons were added to microfluidic devices and cultured for 5–7 days, for axons to grow through the channels into the other chamber. Paclitaxel (25 ng/mL) was then applied to either the soma or neurite chamber. Axonal degeneration was detected after local axonal application (A = before paclitaxel; B = after paclitaxel) but not when paclitaxel was applied to the cell body compartment (C = before paclitaxel; D = after paclitaxel). *Reprinted from Yang, H., Siddique, R., Hosmane, S., Thakor, N., & Höke, A. (2009). Compartmentalized microfluidic culture platform to study mechanism of paclitaxel-induced axonal degeneration.* Experimental Neurology, 218, *124–128, © 2009 with permission from Elsevier.*

Hosmane, Thakor, & Hoke, 2009). We envisage that this system will soon be employed to generate a "diabetes-induced" model of distal axonopathy (see Section 4.4).

Immortalized cell lines, especially PC12 cells, have long been utilized in regeneration studies and are useful tools to study neurite outgrowth since cells differentiated with NGF extend neurite-like processes that can be easily measured. While neurite outgrowth appears reduced in cells cultured in high glucose conditions, apoptosis is also often described (Koshimura, Tanaka, Murakami, & Kato, 2002; Lelkes, Unsworth, & Lelkes, 2001; Sharifi, Mousavi, Farhadi, & Larijani, 2007), meaning that caution should be exerted in translating results to the in vivo situation. The ease of use of immortalized cell lines has enabled the development of rapid high-throughput assays, primarily exploited by the pharmaceutical industry (Radio, Breier, Shafer, & Mundy, 2008), to quantify the effects of compounds on neuritogenesis. High-throughput assay systems treat cells in multiwell plates with compounds, then automatically capture images using fluorescent or bright-field microscopy, and quantify neurite extension. This unbiased automated approach to assess neurite outgrowth is useful for rapid drug screening and toxicological assessment and may be used to identify compounds with regenerative properties using novel compounds or existing drug libraries (Vincent & Feldman, 2008). However, using immortalized cell lines or embryonic neurons for the study of regenerative pathways is simplistic, due to the heterogeneous nature of adult neurons, as discussed above. Therefore it is preferable to use dissociated adult neurons, with their mature complement of neurotrophic factor receptors, for neurite outgrowth assays.

Following an initial quiescent period, adult sensory neurons in culture exhibit two distinct forms of neurite outgrowth: an early arborizing form that can be enhanced by addition of neurotrophic factors such as NGF, NT-3, or GDNF (Fig. 3), and a later elongating regenerative growth mode which depends on novel gene transcription (Diamond, Foerster, Holmes, & Coughlin, 1992; Gavazzi et al., 1999; Kimpinski, Campenot, & Mearow, 1997; Smith & Skene, 1997). The regenerative responses of cells exposed directly to high glucose and diabetes-associated cell stressors have been assessed using a number of culture paradigms, including comparisons between cells derived from either control or diabetic rodents. Indeed, reduced intrinsic regenerative capacity of neurons may be most evident using tissue derived from diabetic rats. Work from the Fernyhough laboratory clearly illustrates a difference in susceptibility of neurons from STZ-diabetic rats to metabolic stress (Akude et al., 2011; Akude, Zherebitskaya, Roy Chowdhury, Girling, & Fernyhough, 2010; Zherebitskaya et al., 2009, 2012).

While treatment of dissociated sensory neurons from control or diabetic rats with elevated glucose does not impair neurite outgrowth per se, neurites of neurons from diabetic rats displayed abnormal swelling and beading along the axon that immunostained for 4-HNE adducts indicative of lipid peroxidation (Zherebitskaya et al., 2009). This suggests that high glucose induces oxidative stress and dysmorphology in axons of adult sensory neurons, but only those isolated from diabetic rats. Direct exposure of control adult rat sensory neurons to 4-HNE at concentrations as low as 3 μM caused a significant reduction in neurite outgrowth and similar changes to axonal morphology (Akude et al., 2010). Similar studies have been performed using dissociated CG/SMG neurons, which have a slower regenerative response in vitro than SCG neurons and were more sensitive to high glucose than SCG neurons in that exposure to high glucose caused a significant decrease in neurite-bearing cells from CG/SMG but not SCG (Semra et al., 2004). While hyperglycemia can impact neurite initiation and morphology, neurite length is not altered by up to 45 mM glucose in adult mouse DRG explants, although supraphysiological glucose levels (60 mM) had some effect (Gumy et al., 2008). There was also no effect on the density of growth cones at any glucose level tested, indicating that exposure to supraphysiological glucose did not reduce neurite initiation, but impaired rate of neurite extension. These experiments also revealed a striking inhibition of Schwann cell migration from explants in high glucose (Fig. 4) that was confirmed using a Matrigel™ drop assay of isolated neonatal Schwann cells in 30 mM glucose (Gumy et al., 2008). The proliferation of Schwann cells thus appears particularly sensitive to elevated glucose concentrations, which could impact on regenerative capacity in diabetic neuropathy.

Another key aspect to the regenerative deficit in diabetic neuropathy is an unsupportive extracellular environment, both in terms of neurotrophic support from other cells and the ECM (Bradley, King, Muddle, & Thomas, 2000; Calcutt, Jolivalt, & Fernyhough, 2008; Gardiner, 2011; Kennedy & Zochodne, 2005; King, Llewelyn, Thomas, Gilbey, & Watkins, 1989; Yasuda et al., 2003). The ECM of peripheral nerve plays a key role in supporting axonal regeneration, and attachment of neurons to ECM is an important phase for successful neurite outgrowth in culture (Duran-Jimenez et al., 2009). Dissociated sensory neurons from adult diabetic mice show reduced adhesion to laminin, type I and IV collagens, and fibronectin (Sango, Horie, Okamura, Inoue, & Takenaka, 1995), possibly through altered expression of ECM receptors. ECM proteins are glycated in the peripheral nerve in diabetes (Duran-Jimenez et al., 2009), and exogenous glycation of purified ECM proteins using glucose, glyceraldehyde, or methylglyoxal inhibited neurite outgrowth from neuroblastoma cells (Federoff, Lawrence, & Brownlee, 1993), dissociated neonatal (Luo, King, Lewin, & Thomas, 2002) and adult sensory neurons (Duran-Jimenez et al., 2009; Radu et al., 2012), and adult DRG explants (Ozturk, Sekeroglu, Erdogan, & Ozturk, 2006). Reducing AGE formation in the peripheral nerve and target tissues may thereby diminish the inhibitory environment and enable reinnervation of target tissue.

4.4 Looking Forward to a Better In Vitro Model of Diabetic Neuropathy

The ultimate aim of all in vitro models of neuropathy should be to create a system that best mimics the in vivo microenvironment but still affords the advantage of exquisite control of the cellular environment. Peripheral neurons in vivo extend long axons whose terminals may be over a meter away from their soma so that distinct regions of a neuron may face different metabolic demands and extracellular environments. Conventional monolayer or explant cultures cannot effectively model this and do not allow the researcher to control and manipulate different microenvironments. A compartmentalized approach to neuronal cell culture was first developed by Robert Campenot in 1977. His system consisted of three chambers separated by a fluid-impermeable barrier so that the composition of media in each chamber could be individually manipulated. "Campenot Chambers" were first used to study the subcellular actions of NGF, and it was found that neurites from rat sympathetic neurons would preferentially grow into chambers containing NGF (Campenot, 1977). This system has now been widely used to study axonal transport and regeneration as well as to isolate axons from cell bodies for molecular and biochemical analysis. A disadvantage of the Campenot chamber is that creation of the barrier is tricky and leakage between chambers is common. Consequently, this pioneering technique has been developed and modified over the past decades and many elegant devices now exist in which isolated neurons are introduced into one compartment and axons grow into another. One example is the recent development of microfluidic platforms that consist of one or more chambers separated by thin "microgrooves." Microfluidic devices can be bonded to glass slides to create a three-dimensional fluid-filled microenvironment. Neuron cell bodies are plated in one chamber and treated with growth factors so that neurites extend through the microgrooves to the other chamber. Since cell somata cannot pass through the microgrooves and hydrostatic pressure is present between chambers, this system can spatially and fluidically isolate cell bodies from axons. One can thus begin to probe the specific effects of exogenous chemicals/factors on the axon or soma compartments. Studies have used these chambers with live-cell imaging methodologies to study axonal transport of NGF (Zhang, Osakada, et al., 2010), calcium imaging (Tsantoulas et al., 2013), and genesis and conduction of action potentials in sensory axons using patch-clamp methodology (Tsantoulas et al., 2013).

Compartmentalized culture systems have been used with models of HIV- and chemotherapy-associated neuropathy to probe whether distal axon degeneration is due to local axonal damage or originates from the neuronal cell body (Melli, Jack, Lambrinos, Ringkamp, & Hoke, 2006; Melli, Keswani, Fischer, Chen, & Hoke, 2006; Silva, Wang, Wang, Ravula, & Glass, 2006; Yang et al., 2009). The potential of such systems will be of obvious interest to researchers modeling diabetic neuropathy in vitro, particularly if combined with coculture methods. It has long been possible to drive myelination of dissociated sensory neurons by their long-term coculture with Schwann cells and supplementation with ascorbic acid (Eldridge, Bunge, Bunge, & Wood, 1987). In such a system, Schwann cells and neurites can interact and mature, as illustrated by the formation of nodes of Ranvier (Zhang et al., 2012). A microfluidic-based coculture system could be used to investigate the interactions of Schwann cells and other pertinent cells such as keratinocytes in modulating axonal function. Combining such techniques with real-time imaging to monitor biochemical, electrophysiological, and morphological changes (for example, see Fig. 6; Tsantoulas et al., 2013) may move us closer to modeling a "nerve in a dish."

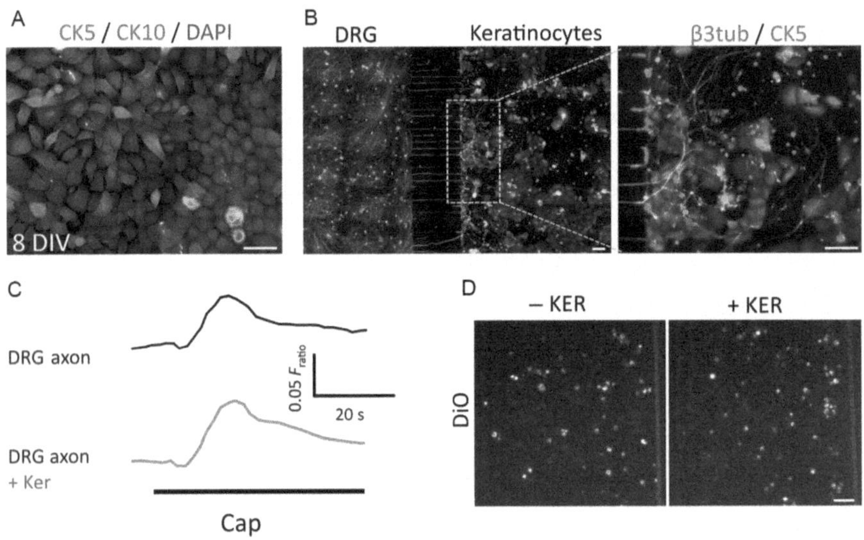

Fig. 6 Sensory neuron–keratinocyte coculture using a microfluidic system. (A) Neonatal rat keratinocytes immunostained with the keratinocyte marker cytokeratin 5 (CK5) and differentiation marker cytokeratin 10 (CK10). (B) Dissociated neonatal rat sensory neurons (immunostained with β3-tubulin$^+$, seeded into left DRG compartment) extend axons into the keratinocytes (CK5$^+$)-containing compartment. (D) Coculture with keratinocytes did not alter regeneration of sensory axons—with no evident differences in DiO tracer uptake between cultures grown with or without keratinocytes. Microfluidics can be used to study a wide variety of phenotypes, including regeneration, axonal transport, and activity. For example, (C) shows Ca^{2+} imaging of the soma in response to capsaicin stimulation of axonal compartment ± keratinocytes. *Reprinted from Tsantoulas, C., Farmer, C., Machado, P., Baba, K., McMahon, S. B., & Raouf, R. (2013). Probing functional properties of nociceptive axons using a microfluidic culture system.* PLoS One, 8, *e80722, © 2013, an open-access article distributed under the terms of the Creative Commons Attribution License.*

5. CONCLUSIONS

Can we model diabetic neuropathy in vitro? The answer is perhaps "no, or at least not in its entirety." However, we can now model many aspects of neuropathy effectively. The in vitro models used to study the pathophysiological responses to hyperglycemia, reduced insulin signaling, hyperlipidemia, and other stressors, while simplistic, do enable us to predict how neurons may behave in more complex in vivo environments. They also enable us to investigate subcellular responses and dissect out mechanistic changes that contribute to the pathogenesis of the disease, test neuroprotective strategies, and develop high-throughput drug discovery screens for drugs to promote axonal regeneration. As such, they will remain a valuable component to the neurobiologists' toolbox. Future refinement of models, including use of iPSCs and the ability to challenge different neuronal compartments using microfluidic devices, will hopefully enable us to discover underlying mechanisms that predispose the distal axon to degeneration in diabetic neuropathy and improve treatment options.

ACKNOWLEDGMENTS

The authors would like to extend thanks to the Juvenile Diabetes Research Foundation, Diabetes UK and the Medical Research Council, UK for funding their research over the past years. We apologize to the many authors whose work we were not able to cite due to space constraints. We thank the editors Paul Fernyhough and Nigel Calcutt for their helpful comments.

REFERENCES

Akude, E., Zherebitskaya, E., Chowdhury, S. K., Smith, D. R., Dobrowsky, R. T., & Fernyhough, P. (2011). Diminished superoxide generation is associated with respiratory chain dysfunction and changes in the mitochondrial proteome of sensory neurons from diabetic rats. *Diabetes*, *60*, 288–297.

Akude, E., Zherebitskaya, E., Roy Chowdhury, S. K., Girling, K., & Fernyhough, P. (2010). 4-Hydroxy-2-nonenal induces mitochondrial dysfunction and aberrant axonal outgrowth in adult sensory neurons that mimics features of diabetic neuropathy. *Neurotoxicity Research*, *17*, 28–38.

Allodi, I., Udina, E., & Navarro, X. (2012). Specificity of peripheral nerve regeneration: Interactions at the axon level. *Progress in Neurobiology*, *98*, 16–37.

Averill, S., McMahon, S. B., Clary, D. O., Reichardt, L. F., & Priestley, J. V. (1995). Immunocytochemical localization of trkA receptors in chemically identified subgroups of adult rat sensory neurons. *The European Journal of Neuroscience*, 7, 1484–1494.

Bennett, D. L., Averill, S., Clary, D. O., Priestley, J. V., & McMahon, S. B. (1996). Postnatal changes in the expression of the trkA high-affinity NGF receptor in primary sensory neurons. *The European Journal of Neuroscience*, *8*, 2204–2208.

Bhattacherjee, A., Liao, Z., & Smith, P. G. (2014). Trophic factor and hormonal regulation of neurite outgrowth in sensory neuron-like 50B11 cells. *Neuroscience Letters*, *558*, 120–125.

Bierhaus, A., Fleming, T., Stoyanov, S., Leffler, A., Babes, A., Neacsu, C., et al. (2012). Methylglyoxal modification of Nav1.8 facilitates nociceptive neuron firing and causes hyperalgesia in diabetic neuropathy. *Nature Medicine*, *18*, 926–933.

Bisby, M. A. (1980). Axonal transport of labeled protein and regeneration rate in nerves of streptozocin-diabetic rats. *Experimental Neurology*, *69*, 74–84.

Bradley, J. L., King, R. H., Muddle, J. R., & Thomas, P. K. (2000). The extracellular matrix of peripheral nerve in diabetic polyneuropathy. *Acta Neuropathologica*, *99*, 539–546.

Bradley, J. L., Thomas, P. K., King, R. H., Muddle, J. R., Ward, J. D., Tesfaye, S., et al. (1995). Myelinated nerve fibre regeneration in diabetic sensory polyneuropathy: Correlation with type of diabetes. *Acta Neuropathologica*, *90*, 403–410.

Bronner-Fraser, M., Stern, C. D., & Fraser, S. (1991). Analysis of neural crest cell lineage and migration. *Journal of Craniofacial Genetics and Developmental Biology*, *11*, 214–222.

Brownlee, M. (2001). Biochemistry and molecular cell biology of diabetic complications. *Nature*, *414*, 813–820.

Brownlee, M. (2005). The pathobiology of diabetic complications: A unifying mechanism. *Diabetes*, *54*, 1615–1625.

Calcutt, N. A., Jolivalt, C. G., & Fernyhough, P. (2008). Growth factors as therapeutics for diabetic neuropathy. *Current Drug Targets*, *9*, 47–59.

Calcutt, N. A., Muir, D., Powell, H. C., & Mizisin, A. P. (1992). Reduced ciliary neurotrophic factor-like activity in nerves from diabetic or galactose-fed rats. *Brain Research*, *575*, 320–324.

Campenot, R. B. (1977). Local control of neurite development by nerve growth factor. *Proceedings of the National Academy of Sciences of the United States of America*, *74*, 4516–4519.

Chambers, S. M., Qi, Y., Mica, Y., Lee, G., Zhang, X. J., Niu, L., et al. (2012). Combined small-molecule inhibition accelerates developmental timing and converts human pluripotent stem cells into nociceptors. *Nature Biotechnology*, *30*, 715–720.

Chen, C. L., Broom, D. C., Liu, Y., de Nooij, J. C., Li, Z., Cen, C., et al. (2006). Runx1 determines nociceptive sensory neuron phenotype and is required for thermal and neuropathic pain. *Neuron*, *49*, 365–377.

Chen, W., Mi, R., Haughey, N., Oz, M., & Hoke, A. (2007). Immortalization and characterization of a nociceptive dorsal root ganglion sensory neuronal line. *Journal of the Peripheral Nervous System: JPNS*, *12*, 121–130.

Chen, S., Rio, C., Ji, R. R., Dikkes, P., Coggeshall, R. E., Woolf, C. J., et al. (2003). Disruption of ErbB receptor signaling in adult non-myelinating Schwann cells causes progressive sensory loss. *Nature Neuroscience*, *6*, 1186–1193.

Chen, Z. L., & Strickland, S. (2003). Laminin gamma1 is critical for Schwann cell differentiation, axon myelination, and regeneration in the peripheral nerve. *The Journal of Cell Biology*, *163*, 889–899.

Cheng, C., & Zochodne, D. W. (2003). Sensory neurons with activated caspase-3 survive long-term experimental diabetes. *Diabetes*, *52*, 2363–2371.

Chowdhury, S. K., Smith, D. R., & Fernyhough, P. (2013). The role of aberrant mitochondrial bioenergetics in diabetic neuropathy. *Neurobiology of Disease*, *51*, 56–65.

Chwalek, K., Tang-Schomer, M. D., Omenetto, F. G., & Kaplan, D. L. (2015). In vitro bioengineered model of cortical brain tissue. *Nature Protocols*, *10*, 1362–1373.

Court, F. A., Hendriks, W. T., MacGillavry, H. D., Alvarez, J., & van Minnen, J. (2008). Schwann cell to axon transfer of ribosomes: Toward a novel understanding of the role of glia in the nervous system. *The Journal of Neuroscience: The Official Journal of the Society for Neuroscience*, *28*, 11024–11029.

Craner, M. J., Klein, J. P., Renganathan, M., Black, J. A., & Waxman, S. G. (2002). Changes of sodium channel expression in experimental painful diabetic neuropathy. *Annals of Neurology, 52*, 786–792.

Delaney, C. L., Russell, J. W., Cheng, H. L., & Feldman, E. L. (2001). Insulin-like growth factor-I and over-expression of Bcl-xL prevent glucose-mediated apoptosis in Schwann cells. *Journal of Neuropathology and Experimental Neurology, 60*, 147–160.

Diamond, J., Foerster, A., Holmes, M., & Coughlin, M. (1992). Sensory nerves in adult rats regenerate and restore sensory function to the skin independently of endogenous NGF. *The Journal of Neuroscience: The Official Journal of the Society for Neuroscience, 12*, 1467–1476.

Dingle, Y. L., Boutin, M. E., Chirila, A. M., Livi, L. L., Labriola, N. R., Jakubek, L. M., et al. (2015). 3D neural spheroid culture: An in vitro model for cortical studies. *Tissue Engineering Part C, Methods, 21*, 1274–1283.

Duran-Jimenez, B., Dobler, D., Moffatt, S., Rabbani, N., Streuli, C. H., Thornalley, P. J., et al. (2009). Advanced glycation end products in extracellular matrix proteins contribute to the failure of sensory nerve regeneration in diabetes. *Diabetes, 58*, 2893–2903.

Dyck, P. J., Kratz, K. M., Karnes, J. L., Litchy, W. J., Klein, R., Pach, J. M., et al. (1993). The prevalence by staged severity of various types of diabetic neuropathy, retinopathy, and nephropathy in a population-based cohort: The Rochester Diabetic Neuropathy Study. *Neurology, 43*, 817–824.

Ekstrom, A. R., & Tomlinson, D. R. (1989). Impaired nerve regeneration in streptozotocin-diabetic rats. Effects of treatment with an aldose reductase inhibitor. *Journal of the Neurological Sciences, 93*, 231–237.

Eldridge, C. F., Bunge, M. B., Bunge, R. P., & Wood, P. M. (1987). Differentiation of axon-related Schwann cells in vitro. I. Ascorbic acid regulates basal lamina assembly and myelin formation. *The Journal of Cell Biology, 105*, 1023–1034.

Federoff, H. J., Lawrence, D., & Brownlee, M. (1993). Nonenzymatic glycosylation of laminin and the laminin peptide CIKVAVS inhibits neurite outgrowth. *Diabetes, 42*, 509–513.

Feinberg, K., Eshed-Eisenbach, Y., Frechter, S., Amor, V., Salomon, D., Sabanay, H., et al. (2010). A glial signal consisting of gliomedin and NrCAM clusters axonal Na + channels during the formation of nodes of Ranvier. *Neuron, 65*, 490–502.

Fernyhough, P. (2015). Mitochondrial dysfunction in diabetic neuropathy: A series of unfortunate metabolic events. *Current Diabetes Reports, 15*, 89.

Fernyhough, P., & Calcutt, N. A. (2010). Abnormal calcium homeostasis in peripheral neuropathies. *Cell Calcium, 47*, 130–139.

Fernyhough, P., & McGavock, J. (2014). Mechanisms of disease: Mitochondrial dysfunction in sensory neuropathy and other complications in diabetes. *Handbook of Clinical Neurology, 126*, 353–377.

Fernyhough, P., Roy Chowdhury, S. K., & Schmidt, R. E. (2010). Mitochondrial stress and the pathogenesis of diabetic neuropathy. *Expert Review of Endocrinology & Metabolism, 5*, 39–49.

Freeman, O. J., Unwin, R. D., Dowsey, A. W., Begley, P., Ali, S., Hollywood, K. A., et al. (2016). Metabolic dysfunction is restricted to the sciatic nerve in experimental diabetic neuropathy. *Diabetes, 65*, 228–238.

Fuchs, D., Birklein, F., Reeh, P. W., & Sauer, S. K. (2010). Sensitized peripheral nociception in experimental diabetes of the rat. *Pain, 151*, 496–505.

Gardiner, N. J. (2011). Integrins and the extracellular matrix: Key mediators of development and regeneration of the sensory nervous system. *Developmental Neurobiology, 71*, 1054–1072.

Gardiner, N. J., Fernyhough, P., Tomlinson, D. R., Mayer, U., von der Mark, H., & Streuli, C. H. (2005). Alpha7 integrin mediates neurite outgrowth of distinct populations of adult sensory neurons. *Molecular and Cellular Neurosciences, 28*, 229–240.

Gardiner, N. J., Moffatt, S., Fernyhough, P., Humphries, M. J., Streuli, C. H., & Tomlinson, D. R. (2007). Preconditioning injury-induced neurite outgrowth of adult rat sensory neurons on fibronectin is mediated by mobilisation of axonal alpha5 integrin. *Molecular and Cellular Neurosciences*, *35*, 249–260.

Gavazzi, I., Kumar, R. D., McMahon, S. B., & Cohen, J. (1999). Growth responses of different subpopulations of adult sensory neurons to neurotrophic factors in vitro. *The European Journal of Neuroscience*, *11*, 3405–3414.

Guan, W., Puthenveedu, M. A., & Condic, M. L. (2003). Sensory neuron subtypes have unique substratum preference and receptor expression before target innervation. *The Journal of Neuroscience: The Official Journal of the Society for Neuroscience*, *23*, 1781–1791.

Gumy, L. F., Bampton, E. T., & Tolkovsky, A. M. (2008). Hyperglycaemia inhibits Schwann cell proliferation and migration and restricts regeneration of axons and Schwann cells from adult murine DRG. *Molecular and Cellular Neurosciences*, *37*, 298–311.

Harris, M. L., & Erickson, C. A. (2007). Lineage specification in neural crest cell pathfinding. *Developmental Dynamics: An Official Publication of the American Association of Anatomists*, *236*, 1–19.

Heng, L. J., Qi, R., Yang, R. H., & Xu, G. Z. (2015). Docosahexaenoic acid inhibits mechanical allodynia and thermal hyperalgesia in diabetic rats by decreasing the excitability of DRG neurons. *Experimental Neurology*, *271*, 291–300.

Hex, N., Bartlett, C., Wright, D., Taylor, M., & Varley, D. (2012). Estimating the current and future costs of Type 1 and Type 2 diabetes in the UK, including direct health costs and indirect societal and productivity costs. *Diabetic Medicine: A Journal of the British Diabetic Association*, *29*, 855–862.

Hilliges, M., Wang, L., & Johansson, O. (1995). Ultrastructural evidence for nerve fibers within all vital layers of the human epidermis. *The Journal of Investigative Dermatology*, *104*, 134–137.

Hinder, L. M., Figueroa-Romero, C., Pacut, C., Hong, Y., Vivekanandan-Giri, A., Pennathur, S., et al. (2014). Long-chain acyl coenzyme A synthetase 1 overexpression in primary cultured Schwann cells prevents long chain fatty acid-induced oxidative stress and mitochondrial dysfunction. *Antioxidants & Redox Signaling*, *21*, 588–600.

Hopkins, A. M., DeSimone, E., Chwalek, K., & Kaplan, D. L. (2015). 3D in vitro modeling of the central nervous system. *Progress in Neurobiology*, *125*, 1–25.

Huang, T. J., Price, S. A., Chilton, L., Calcutt, N. A., Tomlinson, D. R., Verkhratsky, A., et al. (2003). Insulin prevents depolarization of the mitochondrial inner membrane in sensory neurons of type 1 diabetic rats in the presence of sustained hyperglycemia. *Diabetes*, *52*, 2129–2136.

Huang, T. J., Verkhratsky, A., & Fernyhough, P. (2005). Insulin enhances mitochondrial inner membrane potential and increases ATP levels through phosphoinositide 3-kinase in adult sensory neurons. *Molecular and Cellular Neurosciences*, *28*, 42–54.

Jessen, K. R., & Mirsky, R. (1997). Embryonic Schwann cell development: The biology of Schwann cell precursors and early Schwann cells. *Journal of Anatomy*, *191*(Pt. 4), 501–505.

Jessen, K. R., & Mirsky, R. (2005). The origin and development of glial cells in peripheral nerves. *Nature Reviews Neuroscience*, *6*, 671–682.

Kaewkhaw, R., Scutt, A. M., & Haycock, J. W. (2012). Integrated culture and purification of rat Schwann cells from freshly isolated adult tissue. *Nature Protocols*, *7*, 1996–2004.

Kamiya, H., Zhang, W., & Sima, A. A. (2006). Degeneration of the Golgi and neuronal loss in dorsal root ganglia in diabetic BioBreeding/Worcester rats. *Diabetologia*, *49*, 2763–2774.

Kamiya, H., Zhangm, W., & Sima, A. A. (2005). Apoptotic stress is counterbalanced by survival elements preventing programmed cell death of dorsal root ganglions in subacute type 1 diabetic BB/Wor rats. *Diabetes*, *54*, 3288–3295.

Karchewski, L. A., Kim, F. A., Johnston, J., McKnight, R. M., & Verge, V. M. (1999). Anatomical evidence supporting the potential for modulation by multiple neurotrophins in the majority of adult lumbar sensory neurons. *The Journal of Comparative Neurology*, *413*, 327–341.

Kennedy, J. M., & Zochodne, D. W. (2000). The regenerative deficit of peripheral nerves in experimental diabetes: Its extent, timing and possible mechanisms. *Brain: A Journal of Neurology*, *123*(Pt. 10), 2118–2129.

Kennedy, J. M., & Zochodne, D. W. (2005). Impaired peripheral nerve regeneration in diabetes mellitus. *Journal of the Peripheral Nervous System: JPNS*, *10*, 144–157.

Kimpinski, K., Campenot, R. B., & Mearow, K. (1997). Effects of the neurotrophins nerve growth factor, neurotrophin-3, and brain-derived neurotrophic factor (BDNF) on neurite growth from adult sensory neurons in compartmented cultures. *Journal of Neurobiology*, *33*, 395–410.

King, R. H., Llewelyn, J. G., Thomas, P. K., Gilbey, S. G., & Watkins, P. J. (1989). Diabetic neuropathy: Abnormalities of Schwann cell and perineurial basal laminae. Implications for diabetic vasculopathy. *Neuropathology and Applied Neurobiology*, *15*, 339–355.

Kirkpatrick, L. L., & Brady, S. T. (1994). Modulation of the axonal microtubule cytoskeleton by myelinating Schwann cells. *The Journal of Neuroscience: The Official Journal of the Society for Neuroscience*, *14*, 7440–7450.

Kiskinis, E., Sandoe, J., Williams, L. A., Boulting, G. L., Moccia, R., Wainger, B. J., et al. (2014). Pathways disrupted in human ALS motor neurons identified through genetic correction of mutant SOD1. *Cell Stem Cell*, *14*, 781–795.

Koshimura, K., Tanaka, J., Murakami, Y., & Kato, Y. (2002). Involvement of nitric oxide in glucose toxicity on differentiated PC12 cells: Prevention of glucose toxicity by tetrahydrobiopterin, a cofactor for nitric oxide synthase. *Neuroscience Research*, *43*, 31–38.

Kramer, I., Sigrist, M., de Nooij, J. C., Taniuchi, I., Jessell, T. M., & Arber, S. (2006). A role for Runx transcription factor signaling in dorsal root ganglion sensory neuron diversification. *Neuron*, *49*, 379–393.

Lallemend, F., & Ernfors, P. (2012). Molecular interactions underlying the specification of sensory neurons. *Trends in Neurosciences*, *35*, 373–381.

Lawson, S. N. (1995). Neuropeptides in morphologically and functionally identified primary afferent neurons in dorsal root ganglia: Substance P, CGRP and somatostatin. *Progress in Brain Research*, *104*, 161–173.

Lawson, S. N., & Biscoe, T. J. (1979). Development of mouse dorsal root ganglia: An autoradiographic and quantitative study. *Journal of Neurocytology*, *8*, 265–274.

Lawson, S. N., Perry, M. J., Prabhakar, E., & McCarthy, P. W. (1993). Primary sensory neurones: Neurofilament, neuropeptides, and conduction velocity. *Brain Research Bulletin*, *30*, 239–243.

Lelkes, E., Unsworth, B. R., & Lelkes, P. I. (2001). Reactive oxygen species, apoptosis and altered NGF-induced signaling in PC12 pheochromocytoma cells cultured in elevated glucose: An in vitro cellular model for diabetic neuropathy. *Neurotoxicity Research*, *3*, 189–203.

Lindsay, R. M. (1988). Nerve growth factors (NGF, BDNF) enhance axonal regeneration but are not required for survival of adult sensory neurons. *The Journal of Neuroscience: The Official Journal of the Society for Neuroscience*, *8*, 2394–2405.

Liu, Q., Spusta, S. C., Mi, R., Lassiter, R. N., Stark, M. R., Hoke, A., et al. (2012). Human neural crest stem cells derived from human ESCs and induced pluripotent stem cells: Induction, maintenance, and differentiation into functional schwann cells. *Stem Cells Translational Medicine*, *1*, 266–278.

Luo, Z. J., King, R. H., Lewin, J., & Thomas, P. K. (2002). Effects of nonenzymatic glycosylation of extracellular matrix components on cell survival and sensory neurite extension in cell culture. *Journal of Neurology*, *249*, 424–431.

Marmigere, F., & Ernfors, P. (2007). Specification and connectivity of neuronal subtypes in the sensory lineage. *Nature Reviews Neuroscience*, *8*, 114–127.
Marmigere, F., Montelius, A., Wegner, M., Groner, Y., Reichardt, L. F., & Ernfors, P. (2006). The Runx1/AML1 transcription factor selectively regulates development and survival of TrkA nociceptive sensory neurons. *Nature Neuroscience*, *9*, 180–187.
McMahon, S. B., Armanini, M. P., Ling, L. H., & Phillips, H. S. (1994). Expression and coexpression of Trk receptors in subpopulations of adult primary sensory neurons projecting to identified peripheral targets. *Neuron*, *12*, 1161–1171.
Melli, G., Jack, C., Lambrinos, G. L., Ringkamp, M., & Hoke, A. (2006). Erythropoietin protects sensory axons against paclitaxel-induced distal degeneration. *Neurobiology of Disease*, *24*, 525–530.
Melli, G., Keswani, S. C., Fischer, A., Chen, W., & Hoke, A. (2006). Spatially distinct and functionally independent mechanisms of axonal degeneration in a model of HIV-associated sensory neuropathy. *Brain: A Journal of Neurology*, *129*, 1330–1338.
Michailov, G. V., Sereda, M. W., Brinkmann, B. G., Fischer, T. M., Haug, B., Birchmeier, C., et al. (2004). Axonal neuregulin-1 regulates myelin sheath thickness. *Science*, *304*, 700–703.
Mirsky, R., Jessen, K. R., Brennan, A., Parkinson, D., Dong, Z., Meier, C., et al. (2002). Schwann cells as regulators of nerve development. *Journal of Physiology, Paris*, *96*, 17–24.
Molliver, D. C., Radeke, M. J., Feinstein, S. C., & Snider, W. D. (1995). Presence or absence of TrkA protein distinguishes subsets of small sensory neurons with unique cytochemical characteristics and dorsal horn projections. *The Journal of Comparative Neurology*, *361*, 404–416.
Molliver, D. C., & Snider, W. D. (1997). Nerve growth factor receptor TrkA is down-regulated during postnatal development by a subset of dorsal root ganglion neurons. *The Journal of Comparative Neurology*, *381*, 428–438.
Molliver, D. C., Wright, D. E., Leitner, M. L., Parsadanian, A. S., Doster, K., Wen, D., et al. (1997). IB4-binding DRG neurons switch from NGF to GDNF dependence in early postnatal life. *Neuron*, *19*, 849–861.
Murinson, B. B., & Griffin, J. W. (2004). C-fiber structure varies with location in peripheral nerve. *Journal of Neuropathology and Experimental Neurology*, *63*, 246–254.
Murinson, B. B., Hoffman, P. N., Banihashemi, M. R., Meyer, R. A., & Griffin, J. W. (2005). C-fiber (Remak) bundles contain both isolectin B4-binding and calcitonin gene-related peptide-positive axons. *The Journal of Comparative Neurology*, *484*, 392–402.
Nishikawa, T., Edelstein, D., Du, X. L., Yamagishi, S., Matsumura, T., Kaneda, Y., et al. (2000). Normalizing mitochondrial superoxide production blocks three pathways of hyperglycaemic damage. *Nature*, *404*, 787–790.
Obrosova, I. G. (2002). How does glucose generate oxidative stress in peripheral nerve? *International Review of Neurobiology*, *50*, 3–35.
Obrosova, I. G. (2003). Update on the pathogenesis of diabetic neuropathy. *Current Diabetes Reports*, *3*, 439–445.
Owens, D. M., & Lumpkin, E. A. (2014). Diversification and specialization of touch receptors in skin. *Cold Spring Harbor Perspectives in Medicine*, *4*, pii: a013656.
Ozturk, G., Sekeroglu, M. R., Erdogan, E., & Ozturk, M. (2006). The effect of non-enzymatic glycation of extracellular matrix proteins on axonal regeneration in vitro. *Acta Neuropathologica*, *112*, 627–632.
Polydefkis, M., Hauer, P., Sheth, S., Sirdofsky, M., Griffin, J. W., & McArthur, J. C. (2004). The time course of epidermal nerve fibre regeneration: Studies in normal controls and in people with diabetes, with and without neuropathy. *Brain: A Journal of Neurology*, *127*, 1606–1615.
Purves, T., Middlemas, A., Agthong, S., Jude, E. B., Boulton, A. J., Fernyhough, P., et al. (2001). A role for mitogen-activated protein kinases in the etiology of diabetic

neuropathy. *FASEB Journal: Official Publication of the Federation of American Societies for Experimental Biology, 15*, 2508–2514.
Radio, N. M., Breier, J. M., Shafer, T. J., & Mundy, W. R. (2008). Assessment of chemical effects on neurite outgrowth in PC12 cells using high content screening. *Toxicological Sciences: An Official Journal of the Society of Toxicology, 105*, 106–118.
Radu, B. M., Dumitrescu, D. I., Mustaciosu, C. C., & Radu, M. (2012). Dual effect of methylglyoxal on the intracellular Ca2+ signaling and neurite outgrowth in mouse sensory neurons. *Cellular and Molecular Neurobiology, 32*, 1047–1057.
Ristic, H., Srinivasan, S., Hall, K. E., Sima, A. A., & Wiley, J. W. (1998). Serum from diabetic BB/W rats enhances calcium currents in primary sensory neurons. *Journal of Neurophysiology, 80*, 1236–1244.
Russell, J. W., & Feldman, E. L. (1999). Insulin-like growth factor-I prevents apoptosis in sympathetic neurons exposed to high glucose. *Hormone and Metabolic Research = Hormon- und Stoffwechselforschung = Hormones et metabolisme, 31*, 90–96.
Russell, J. W., Golovoy, D., Vincent, A. M., Mahendru, P., Olzmann, J. A., Mentzer, A., et al. (2002). High glucose-induced oxidative stress and mitochondrial dysfunction in neurons. *FASEB Journal: Official Publication of the Federation of American Societies for Experimental Biology, 16*, 1738–1748.
Salzer, J. L., Brophy, P. J., & Peles, E. (2008). Molecular domains of myelinated axons in the peripheral nervous system. *Glia, 56*, 1532–1540.
Sandoe, J., & Eggan, K. (2013). Opportunities and challenges of pluripotent stem cell neurodegenerative disease models. *Nature Neuroscience, 16*, 780–789.
Sanes, J. R. (1983). Roles of extracellular matrix in neural development. *Annual Review of Physiology, 45*, 581–600.
Sango, K., Horie, H., Okamura, A., Inoue, S., & Takenaka, T. (1995). Diabetes impairs DRG neuronal attachment to extracellular matrix proteins in vitro. *Brain Research Bulletin, 37*, 533–537.
Sango, K., Yanagisawa, H., Takaku, S., Kawakami, E., & Watabe, K. (2011). Immortalized adult rodent Schwann cells as in vitro models to study diabetic neuropathy. *Experimental Diabetes Research, 2011*, 374943.
Schmidt, R. E. (2001). Neuronal preservation in the sympathetic ganglia of rats with chronic streptozotocin-induced diabetes. *Brain Research, 921*, 256–259.
Semra, Y. K., Smith, N. C., & Lincoln, J. (2004). Comparative effects of high glucose on different adult sympathetic neurons in culture. *Neuroreport, 15*, 2321–2325.
Sharifi, A. M., Mousavi, S. H., Farhadi, M., & Larijani, B. (2007). Study of high glucose-induced apoptosis in PC12 cells: Role of bax protein. *Journal of Pharmacological Sciences, 104*, 258–262.
Silos-Santiago, I., Molliver, D. C., Ozaki, S., Smeyne, R. J., Fagan, A. M., Barbacid, M., et al. (1995). Non-TrkA-expressing small DRG neurons are lost in TrkA deficient mice. *The Journal of Neuroscience: The Official Journal of the Society for Neuroscience, 15*, 5929–5942.
Silva, A., Wang, Q., Wang, M., Ravula, S. K., & Glass, J. D. (2006). Evidence for direct axonal toxicity in vincristine neuropathy. *Journal of the Peripheral Nervous System: JPNS, 11*, 211–216.
Smith, D. S., & Skene, J. H. (1997). A transcription-dependent switch controls competence of adult neurons for distinct modes of axon growth. *The Journal of Neuroscience: The Official Journal of the Society for Neuroscience, 17*, 646–658.
Snider, W. D., & Silos-Santiago, I. (1996). Dorsal root ganglion neurons require functional neurotrophin receptors for survival during development. *Philosophical Transactions of the Royal Society of London. Series B, Biological Sciences, 351*, 395–403.
Spallone, V., Ziegler, D., Freeman, R., Bernardi, L., Frontoni, S., Pop-Busui, R., et al. (2011). Cardiovascular autonomic neuropathy in diabetes: Clinical impact, assessment, diagnosis, and management. *Diabetes/Metabolism Research and Reviews, 27*, 639–653.

Srinivasan, S., Stevens, M. J., Sheng, H., Hall, K. E., & Wiley, J. W. (1998). Serum from patients with type 2 diabetes with neuropathy induces complement-independent, calcium-dependent apoptosis in cultured neuronal cells. *The Journal of Clinical Investigation, 102*, 1454–1462.

Takahashi, K., Tanabe, K., Ohnuki, M., Narita, M., Ichisaka, T., Tomoda, K., et al. (2007). Induction of pluripotent stem cells from adult human fibroblasts by defined factors. *Cell, 131*, 861–872.

Tannenbaum, J., & Bennett, B. T. (2015). Russell and Burch's 3Rs then and now: The need for clarity in definition and purpose. *Journal of the American Association for Laboratory Animal Science: JAALAS, 54*, 120–132.

Taveggia, C., Zanazzi, G., Petrylak, A., Yano, H., Rosenbluth, J., Einheber, S., et al. (2005). Neuregulin-1 type III determines the ensheathment fate of axons. *Neuron, 47*, 681–694.

Thakur, M., Crow, M., Richards, N., Davey, G. I., Levine, E., Kelleher, J. H., et al. (2014). Defining the nociceptor transcriptome. *Frontiers in Molecular Neuroscience, 7*, 87.

Tomlinson, D. R., & Gardiner, N. J. (2008). Glucose neurotoxicity. *Nature Reviews Neuroscience, 9*, 36–45.

Tsantoulas, C., Farmer, C., Machado, P., Baba, K., McMahon, S. B., & Raouf, R. (2013). Probing functional properties of nociceptive axons using a microfluidic culture system. *PLoS One, 8*, e80722.

Usoskin, D., Furlan, A., Islam, S., Abdo, H., Lonnerberg, P., Lou, D., et al. (2015). Unbiased classification of sensory neuron types by large-scale single-cell RNA sequencing. *Nature Neuroscience, 18*, 145–153.

Viader, A., Golden, J. P., Baloh, R. H., Schmidt, R. E., Hunter, D. A., & Milbrandt, J. (2011). Schwann cell mitochondrial metabolism supports long-term axonal survival and peripheral nerve function. *The Journal of Neuroscience: The Official Journal of the Society for Neuroscience, 31*, 10128–10140.

Viader, A., Sasaki, Y., Kim, S., Strickland, A., Workman, C. S., Yang, K., et al. (2013). Aberrant Schwann cell lipid metabolism linked to mitochondrial deficits leads to axon degeneration and neuropathy. *Neuron, 77*, 886–898.

Vincent, A. M., & Feldman, E. L. (2008). Can drug screening lead to candidate therapies for testing in diabetic neuropathy? *Antioxidants & Redox Signaling, 10*, 387–393.

Vincent, A. M., Hayes, J. M., McLean, L. L., Vivekanandan-Giri, A., Pennathur, S., & Feldman, E. L. (2009). Dyslipidemia-induced neuropathy in mice: The role of oxLDL/LOX-1. *Diabetes, 58*, 2376–2385.

Vincent, A. M., Olzmann, J. A., Brownlee, M., Sivitz, W. I., & Russell, J. W. (2004). Uncoupling proteins prevent glucose-induced neuronal oxidative stress and programmed cell death. *Diabetes, 53*, 726–734.

Vincent, A. M., Russell, J. W., Low, P., & Feldman, E. L. (2004). Oxidative stress in the pathogenesis of diabetic neuropathy. *Endocrine Reviews, 25*, 612–628.

Vogelbaum, M. A., Tong, J. X., & Rich, K. M. (1998). Developmental regulation of apoptosis in dorsal root ganglion neurons. *The Journal of Neuroscience: The Official Journal of the Society for Neuroscience, 18*, 8928–8935.

Wilson, N. M., & Wright, D. E. (2014). Experimental motor neuropathy in diabetes. *Handbook of Clinical Neurology, 126*, 461–467.

Wright, D. E., & Snider, W. D. (1995). Neurotrophin receptor mRNA expression defines distinct populations of neurons in rat dorsal root ganglia. *The Journal of Comparative Neurology, 351*, 329–338.

Yang, I. H., Siddique, R., Hosmane, S., Thakor, N., & Hoke, A. (2009). Compartmentalized microfluidic culture platform to study mechanism of paclitaxel-induced axonal degeneration. *Experimental Neurology, 218*, 124–128.

Yasuda, H., Terada, M., Maeda, K., Kogawa, S., Sanada, M., Haneda, M., et al. (2003). Diabetic neuropathy and nerve regeneration. *Progress in Neurobiology, 69*, 229–285.

Yu, C., Rouen, S., & Dobrowsky, R. T. (2008). Hyperglycemia and downregulation of caveolin-1 enhance neuregulin-induced demyelination. *Glia*, *56*, 877–887.

Zhang, Y., Bekku, Y., Dzhashiashvili, Y., Armenti, S., Meng, X., Sasaki, Y., et al. (2012). Assembly and maintenance of nodes of Ranvier rely on distinct sources of proteins and targeting mechanisms. *Neuron*, *73*, 92–107.

Zhang, K., Osakada, Y., Vrljic, M., Chen, L., Mudrakola, H. V., & Cui, B. (2010). Single-molecule imaging of NGF axonal transport in microfluidic devices. *Lab on a Chip*, *10*, 2566–2573.

Zhang, L., Yu, C., Vasquez, F. E., Galeva, N., Onyango, I., Swerdlow, R. H., et al. (2010). Hyperglycemia alters the schwann cell mitochondrial proteome and decreases coupled respiration in the absence of superoxide production. *Journal of Proteome Research*, *9*, 458–471.

Zherebitskaya, E., Akude, E., Smith, D. R., & Fernyhough, P. (2009). Development of selective axonopathy in adult sensory neurons isolated from diabetic rats: Role of glucose-induced oxidative stress. *Diabetes*, *58*, 1356–1364.

Zherebitskaya, E., Schapansky, J., Akude, E., Smith, D. R., Van der Ploeg, R., Solovyova, N., et al. (2012). Sensory neurons derived from diabetic rats have diminished internal Ca2+ stores linked to impaired re-uptake by the endoplasmic reticulum. *ASN Neuro*, *4*. pii: e00072.

Zochodne, D. W. (2014). Mechanisms of diabetic neuron damage: Molecular pathways. *Handbook of Clinical Neurology*, *126*, 379–399.

CHAPTER SIX

Alternatives to the Streptozotocin-Diabetic Rodent

M.A. Yorek[1]
Iowa City Health Care System, Iowa City, IA, United States
University of Iowa, Iowa City, IA, United States
Fraternal Order of Eagles Diabetes Research Center, University of Iowa, Iowa City, IA, United States
[1]Corresponding author: e-mail address: mark-yorek@uiowa.edu

Contents

Abstract

The study of diabetic neuropathy has relied primarily on the use of streptozotocin-treated rat and mouse models of type 1 diabetes. This chapter will review the creation and use of other rodent models that have been developed in order to investigate the contribution of factors besides insulin deficiency to the development and progression

International Review of Neurobiology, Volume 127
ISSN 0074-7742
http://dx.doi.org/10.1016/bs.irn.2016.03.002

of diabetic neuropathy as it occurs in obesity, type 1 or type 2 diabetes. Diabetic peripheral neuropathy is a complex disorder with multiple mechanisms contributing to its development and progression. Even though many animal models have been developed and investigated, no single model can mimic diabetic peripheral neuropathy as it occurs in humans. Nonetheless, animal models can play an important role in improving our understanding of the etiology of diabetic peripheral neuropathy and in performing preclinical screening of potential new treatments. To date treatments found to be effective for diabetic peripheral neuropathy in rodent models have failed in clinical trials. However, with the identification of new endpoints for the early detection of diabetic peripheral neuropathy and the understanding that a successful treatment may require a combination therapeutic approach there is hope that an effective treatment will be found.

1. INTRODUCTION

For years the standard animal model for the study of diabetic neuropathy has been the streptozotocin-treated rodent. In rats depending on the species, age, and delivery, a single dose of streptozotocin ranging from 40 to 75 mg/kg is usually sufficient to destroy enough β cells to cause an insulin-deficient form of diabetes (Rees & Alcolado, 2005; Tesch & Allen, 2007). Such rats fail to gain weight and often lose weight unless supported with a low-dose insulin treatment regime, which can allow such animals to be maintained for extended periods of time in a hyperglycemic state (Calcutt, 2004). In mice, the dose of streptozotocin required to create a model of type 1 diabetes is generally higher than that used for rats, with single doses ranging from 100 to 200 mg/kg (Rees & Alcolado, 2005; Tesch & Allen, 2007). More recently, multiple low dosing of streptozotocin has become the preferred method to induce an insulin-deficient form of diabetes in mice (O'Brien, Sakowski, & Feldman, 2014; Rees & Alcolado, 2005; Tesch & Allen, 2007). Using the latter approach creates a type 1 diabetic mouse model that is more stable in regard to maintaining their initial weight and will even gain weight compared to mice treated with a single high dose of streptozotocin (personal observation). The potential for high doses of streptozotocin to cause nonspecific effects on nerve and kidney has for many years been a criticism of this type 1 diabetic model, even though studies have shown that neurotoxicity is not the cause of slowing of nerve conduction velocity or changes in thermal nociception in streptozotocin-treated diabetic rats or mice (Davidson et al., 2009; Wiese, Matsushita, Lowe, Stokes, & Yorek, 1996). Nonetheless, investigators seeking the ultimate animal model to mimic the human development and progression of diabetic neuropathy have developed additional models in order to decipher the complex pathogenesis of neuropathy in obesity and types 1 and 2 diabetes. The purpose of this chapter is to review some of the characteristics of these other rodent models of diabetes and diabetic neuropathy.

Peripheral neuropathy is a multifaceted complication of diabetes that frequently leads to foot ulcers and may progress to limb amputations (Kim, Kim, & Yoon, 2012). Even though it is the most common complication of diabetes, the only recommended clinical treatment is good glycemic control which, at best, only delays the onset and slows progression (Callaghan, Little, Feldman, & Hughes, 2012; Figueroa-Romero, Sadidi, & Feldman, 2008). Diabetic peripheral neuropathy has been described by some investigators to be a disease of the vasculature leading to nerve ischemia and altered nerve function (Cameron, Cotter, Archibald, Dines, & Maxfield, 1994; Cameron, Cotter, Dines, et al., 1994; Cameron, Cotter, & Low, 1991; Nukada & Dyck, 1984). Other investigators have proposed that diabetic peripheral neuropathy is caused by a combination of metabolic defects associated with an increased flux of glucose through the aldose reductase pathway leading to a defect in Na^+/K^+-ATPase activity and an alteration of signal transduction pathways in the nerve (Cameron, Cotter, Dines, & Love, 1992; Cotter, Dines, & Cameron, 1993). Additional pathologic contributors to diabetic peripheral neuropathy have been reported to include increased formation of advanced glycation endproducts, reduced neurotrophic support, and increased inflammatory and oxidative stress (Pop-Busui, Sima, & Stevens, 2006; Sima, 2006). These and likely other mechanisms, which are reviewed in detail elsewhere in this volume, cause damage to neurons, Schwann cells, and the vasculature. Ultimately, relentless damage to the nerve complex and surrounding vasculature leads to diabetic peripheral neuropathy. Given the complex etiology of diabetic peripheral neuropathy, creating an animal model that mimics all the pathologies and developmental pattern of neuropathy seen in human diabetic subjects is probably an impossible task. Therefore, investigators have focused on gaining what information they can from the multiple animal models that have been created before determining whether there is some translational value in these data.

2. RODENT MODELS OF OBESITY

2.1 High-Fat Fed Sprague-Dawley Rats and C57Bl6/J Mice

Feeding rats or mice a high fat or Western diet to induce obesity has recently become a more common approach to creating models of prediabetes (Coppey, Davidson, Lu, Gerard, & Yorek, 2011; Davidson, Coppey,

Calcutt, Oltman, & Yorek, 2010; Groover et al., 2013; Nowicki, Kosacka, Serke, Bluher, & Spanel-Borowski, 2012; O'Brien, Sakowski, et al., 2014; Obrosova et al., 2007; Yorek et al., 2015). There have been many different formulations of these diets, but in general they cause excess weight gain, raise circulating lipid levels, and cause insulin resistance. Our studies have shown that rats fed a high-fat (45% kcal) diet rapidly gained weight, became hyperinsulinemic, and developed insulin resistance in 4–6 weeks (unpublished data). However, fasting blood glucose and hemoglobin A_1C levels were not increased (Davidson et al., 2010; Davidson, Coppey, Dake, & Yorek, 2011; Davidson, Coppey, Kardon, & Yorek, 2014). In 12 weeks the rats developed sensory neuropathy, as indicated by slowing of sensory nerve conduction velocity and onset of thermal hypoalgesia (Davidson et al., 2010; Davidson, Coppey, Dake, et al., 2011; Davidson, Coppey, Kardon, et al., 2014). C57Bl6/J mice fed a high-fat diet also develop a prediabetes phenotype including insulin resistance, impaired glucose utilization, fasting hyperglycemia, and sensory neuropathy (Coppey et al., 2011; Yorek et al., 2015). In our studies, diet-induced obese mice, like their rat counterparts, had a normal motor nerve conduction velocity after 12 weeks of a high-fat diet. Others have reported both motor and sensory neuropathy in C57Bl6/J mice fed a high-fat diet for 16 weeks (Obrosova et al., 2007) or longer (Anderson, King, Delbruck, & Jolivalt, 2014). Reasons for these differences could include the type of high-fat diet used or study duration. It has been reported that exercise can alleviate some of the neuropathic deficits associated with diet-induced obesity (Groover et al., 2013). C57Bl6/J mice have been commonly used by investigators. However, it should be noted that other strains of mice are also used and may respond differently to high-fat-containing diets. For example, Swiss Webster mice develop insulin resistance and cognitive impairment but do not gain weight or show indices of peripheral neuropathy (Anderson et al., 2014). Human subjects determined to have impaired glucose tolerance have also been reported to develop a sensory neuropathy (Asghar et al., 2014; Kannan et al., 2014; Papanas, Vinik, & Ziegler, 2011; Papanas & Ziegler, 2012; Smith & Singleton, 2008).

2.2 Zucker Rats

The Zucker (fa/fa) rat is the best-known and most widely used rat model of genetic obesity. The fa mutation was discovered by Zucker and Zucker (Kava, Greenwood, & Johnson, 1990) in crosses between Sherman and Merck stock M rats (13M strain). Animals homozygous for the fa allele become noticeably obese by 3–5 weeks of age, and by 14 weeks of age their body composition is over 40% lipid. Obese Zucker rats do not become hyperglycemic, but are hyperlipemic, hypercholesterolemic, hyperinsulinemic, and develop adipocyte hypertrophy and hyperplasia (Alderson et al., 2003; Bray, 1977; Kurtz, Morris, & Pershadsingh, 1989). The Zucker rat has also occasionally been used in investigations of obesity-associated noninsulin-dependent diabetes mellitus. In our studies with this model, we found that vascular and neural dysfunction developed at a slower rate in obese Zucker rats than in the Zucker diabetic fatty (ZDF) rat model of type 2 diabetes (see later). In both models, vascular impairment preceded slowing of motor nerve conduction velocity, which occurs at 12–14 weeks of age in ZDF rats and 32 weeks of age in obese Zucker rats (Oltman et al., 2005).

3. RODENT MODELS OF TYPE 2 DIABETES

3.1 Zucker Diabetic Fatty Rats

There is a wide variety of models duplicating various pathologies associated with type 2 diabetes. In rats and mice the most commonly used to study neuropathy are those that lack a functional leptin receptor (ZDF rat, spontaneously diabetic Torii (SDT) rat, and db/db mouse) or are leptin deficient (ob/ob mouse) (Sasase, Yokoi, Pezzolesi, & Shinohara, 2015). The ZDF rat is recognized as a standard model for metabolic syndrome and type 2 diabetes and has been extensively used. It develops obesity and hyperglycemia at 8–10 weeks of age due to the leptin-receptor deficit and a second mutation that developed in a colony of obese Zucker rats that was isolated through inbreeding (Peterson, Shaw, Neel, Little, & Eichberg, 1990). The ZDF rat has been used to study pain mechanisms associated with early diabetic neuropathy (Gao & Zheng, 2014; Lirk et al., 2012, 2015; Otto, Wyse, Cabot, & Smith, 2011). It has also been used to examine effect of various treatments on diabetic neuropathic endpoints (Li et al., 2006; Lupachyk, Watcho, Hasanova, Julius, & Obrosova, 2012; Lupachyk, Watcho, Obrosov, Stavniichuk, & Obrosova, 2013; Oltman et al., 2008; Oltman, Davidson, Coppey, Kleinschmidt, & Yorek, 2009). We found that monotherapy treatment of ZDF diabetic fatty rats once complications had developed using different classes of drugs for vascular and neural dysfunction did not achieve expected efficacy levels suggesting a complex etiology that includes multiple mechanisms (Oltman et al., 2008).

3.2 Spontaneously Diabetic Torii Rat

The SDT rat is an inbred strain of Sprague-Dawley rat that has been established as a nonobese model of type 2 diabetes (Sasase et al., 2013). Male SDT rats show high plasma glucose levels by 20 weeks of age with pancreatic islet histopathology. Prior to the onset of diabetes, glucose intolerance with hypoinsulinemia is observed (Sasase et al., 2013). SDT rats develop profound complications including retinopathy, nephropathy, and neuropathy (Sasase et al., 2013). However, another group has reported that the time of onset of glucosuria is different between male and female SDT rats with male rats demonstrating a 100% incidence of diabetes at 40 weeks of age, while it is only 33% for female rats (Shinohara et al., 2000). In order to shorten the period of development of hyperglycemia in the SDT rat, a second SDT type 2 diabetic rat model was created by introducing the fa allele of the Zucker fatty rat into the SDT rat genome. This created a new obese model of type 2 diabetes with both male and female SDT fatty rats showing overt obesity, hyperglycemia, and hyperlipidemia at an early age (Kemmochi et al., 2013). Diabetic peripheral neuropathy including decreased nerve conduction velocity and a decrease in sural nerve fibers has been reported in this model (Yamaguchi et al., 2012).

3.3 Zucker Diabetic Sprague-Dawley Rat

As discussed earlier, masking the effects of leptin by either decreasing the leptin receptor or eliminating leptin production has been used to create models of obesity-related type 2 diabetes. One concern for the applicability of the ZDF rats to humans is the recessive homozygous mutation in the leptin receptor (fa) that causes loss of function and induces severe hyperphagia (Davis, Cain, Banz, & Peterson, 2013; Reinwald, Peterson, Allen, & Burr, 2009). Leptin has been shown to have an array of effects that may influence development of the metabolic syndrome (Beltowski, 2012; Gade, Schmit, Collins, & Gade, 2010; Kaur, 2014; Ricci & Bevilacqua, 2012). Thus, loss of leptin signaling in the ZDF rat makes it a less than ideal model for complications of type 2 diabetes in humans, who generally do not have a leptin-receptor deficit. These concerns led to the creation of the Zucker diabetic Sprague-Dawley (ZDSD) rat (Davis et al., 2013; Peterson et al., 2015; Reinwald et al., 2009). The ZDSD rat was developed by crossbreeding the Charles River Laboratory diet-induced obese rat (Sprague-Dawley-derived) with lean ZDF−/− rats. Selective inbreeding produced animals with a predisposition to obesity and a propensity to develop overt diabetes between 15 and 21 weeks of age with nutritional intervention (Reinwald et al., 2009). Importantly, ZDSD rats have an intact leptin-signaling pathway and more modest accumulation of body fat compared to ZDF rats (Reinwald et al., 2009). We obtained ZDSD rats with the objective of characterizing endpoints associated with diabetic neuropathy in comparison to age matched Sprague-Dawley rats (Davidson, Coppey, Holmes, et al., 2014). We found that chronic hyperglycemia in ZDSD rats caused vascular and neural dysfunction similar to that documented in other diabetic rat models and consistent with the development of diabetic neuropathy (Coppey, Davidson, Dunlap, Lund, & Yorek, 2000, Coppey, Gellett, Davidson, Dunlap, & Yorek, 2002; Davidson et al., 2010; Oltman et al., 2005; Terata et al., 1999). The one exception was that chronic hyperglycemia in ZDSD rat did not cause a decrease in intraepidermal nerve fibers (IENF) in the skin, as has been reported for other rat models of diabetes. There was, however, a decrease in epidermal Langerhans cells. Overall, we found that the ZDSD rat is an easily maintained animal model for type 2 diabetes that spontaneously develops hyperglycemia upon dietary manipulation. This was followed by the development of diabetic neuropathy, including loss of nerves in the subepithelial layer of the cornea and decreased corneal function. Since determination of corneal nerve density and function are noninvasive surrogate markers for diabetic neuropathy (Quattrini et al., 2007), the ZDSD rat with its functional leptin pathway may be a good model for preclinical testing of treatments for diabetic neuropathy.

3.4 Goto–Kakizaki Rat

The Goto–Kakizaki (GK) rat is considered one of the best animal models of nonobese type 2 diabetes (Akash, Rehman, & Chen, 2013) as they exhibit many characteristics in common with diabetic patients. The GK rat spontaneously develops type 2 diabetes due to the complex interaction of multiple mechanisms that include the presence of several susceptibility genes, gestational metabolic impairment inducing an epigenetic programming of the offspring pancreas and the major insulin target tissues, and environmental-induced loss of β-cell differentiation due to chronic exposure to hyperglycemia/hyperlipidemia, inflammation, and oxidative stress (Portha, Giroix, Tourrel-Cuzin, Le-Stunff, & Movassat, 2012). Investigations using GK rats have demonstrated a number of diabetes-related neuronal defects in this model compared to the Wistar rats that are commonly used as controls. GK rats have impaired glucose tolerance and progressive insulinopenia and go on to develop peripheral nerve abnormalities in the absence of overt hyperglycemia (Murakawa et al., 2002). Onset of hyperglycemia worsens the development/progression of peripheral neuropathy in GK rats. Several studies have demonstrated that sustained improvement in hyperglycemia can improve diabetic peripheral neuropathy in GK rats (Ueta et al., 2005; Wada et al., 1999). Recently, Wang, Gao, Yin, and Yu (2012) demonstrated that diabetic neuropathy also occurs in the cornea of type 2 diabetic GK rats. They concluded that defects in the sensory nerve and/or tear film may contribute to diabetic keratopathy and delayed wound healing in diabetic corneas. Their finding that subepithelial corneal nerve loss in GK rats was more prominent in the central cornea than in the limbal region is similar to our own findings using the high-fat fed, low-dose streptozotocin-treated rat model as described later (Davidson, Coppey, Kardon, et al., 2014; Wang et al., 2012).

3.5 BioBreeding Zucker Diabetic Rat

The BioBreeding Zucker diabetic rat (BBZDR)/Wor rat was created by breeding the insulin-resistant Zucker fatty rat into the inbred BioBreeding/Worcester (BB/Wor) strain in order to introduce the defective leptin receptor gene (Tirabassi et al., 2004). BBZDR/Wor rats develop hyperglycemia at about 7 weeks of age and become obese with elevated triglyceride and cholesterol levels. Lipid levels can be further increased by feeding a diet containing high fat and sucrose. These rats are insulin resistant with hyperinsulinemia and develop microvascular- and macrovascular-related complications including retinopathy, neuropathy, nephropathy, and coronary artery disease with hypertension (Gao & Zheng, 2014; Tirabassi et al., 2004). With regard to neuropathy, the diabetic BBZDR/Wor rat shows a slowly progressive nerve conduction defect accompanied by mild myelinated fiber atrophy, mild changes of the node of Ranvier, and significant segmental demyelination and Wallerian degeneration (Sima et al., 2000). These changes were much less dramatic than those observed in the type 1 spontaneously diabetic BB/Wor rat maintained with the same degree of hyperglycemia (Kamiya, Murakawa, Zhang, & Sima, 2005; Sima et al., 2000).

3.6 Otsuka Long-Evans Tokushima Fatty Rat

OLETF rats, when fed a diet with or without sucrose, were hyperglycemic compared to control rats (Kamenov, Higashino, Todorova, Kajimoto, &

Suzuki, 2006). At 10 months of age motor nerve conduction velocity and thermal nociception were significantly decreased, and all parameters deteriorated further when OLETF rats received a sucrose-supplemented diet. Other studies with sucrose-fed OLETF rats have demonstrated a reduction in sciatic nerve blood flow and decreased Na^+/K^+-ATPase activity in the sciatic nerve (Nakamura et al., 2001).

3.7 Ob/ob and db/db Mice

The most common mouse models of type 2 diabetes are those with the ob/ob or db/db mutations. These mice are deficient in leptin (ob/ob) or lack a functional leptin receptor (db/db). They are also obese, insulin resistant and develop spontaneous hyperglycemia. Studies by Grote et al. (2013) demonstrated that insulin resistance occurred in muscle, liver, and sciatic nerve. Characterization of peripheral diabetic neuropathy in ob/ob mice revealed that after 9–13 weeks of hyperglycemia motor and sensory nerve conduction velocity are decreased, and mice have thermal hypoalgesia, tactile allodynia, and a significant loss of IENF (Drel et al., 2006; O'Brien, Hur, et al., 2014). Treating ob/ob mice with an aldose reductase inhibitor or with peroxynitrite decomposition catalysts improved diabetic peripheral neuropathy endpoints (Drel et al., 2006; Vareniuk et al., 2007) suggesting that activation of the aldose reductase pathway as well as increased oxidative stress contribute to the development and progression of diabetic peripheral neuropathy in the ob/ob mouse.

The leptin-receptor deficiency present in db/db mice has been crossed into a number of background strains allowing for studies of multiple pathological conditions (Sullivan et al., 2007). The db/db mouse, like the ZDF rat, is obese and insulin resistant. It develops a severe hyperglycemia and peripheral neuropathy with decreased nerve conduction velocity and morphological changes in peripheral nerves (Cho et al., 2014; Norido, Canella, Zanoni, & Gorio, 1984; Nowicki et al., 2012; Pande et al., 2011; Robertson & Sima, 1980; Shi et al., 2013; Sima & Robertson, 1978). In a study comparing the effect of hyperglycemia in 4-month-old C57BL6 (control), ob/ob, and db/db mice it was found that myelin thickness was significantly reduced in small, medium-sized, and large axons of db/db mice compared with control C57Bl6 mice (Nowicki et al., 2012). In contrast, only large fibers showed a decrease in myelin sheath thickness in ob/ob mice, while the number of nonmyelinated nerve fibers was lower in ob/ob mice than in db/db mice. A thickened basal lamina of Schwann cells also occurred only in ob/ob mice. The basement membrane of endoneural microvessels was thickened in both ob/ob and db/db mice. Thus, fundamental differences in some endpoints of diabetic neuropathology exist between ob/ob and db/db mice.

3.8 Tsumura Suzuki Obese Diabetes Mouse

Other mouse models of obesity and spontaneous type 2 diabetes include the Tsumura Suzuki Obese Diabetes (TSOD) mouse (Lizuka et al., 2005). The male mice exhibit polydipsia and polyuria at about 2 months of age followed by hyperglycemia and hyperinsulinemia. Sensory and motor neuropathy deficits can be observed at 12 and 14 months of age, respectively, with weakness of front and hind paws at about 17 months of age (Lizuka et al., 2005). At this time, light microscopic and electron microscopic examination of the sciatic nerve shows a decrease in the density of nerve fibers as well as degenerative changes of myelinated fibers. Interestingly, a study by Kawada et al. (2010) showed that TSOD mice even after 18 months of age did not develop any form of cardiac dysfunction.

3.9 Combined High-Fat Fed, Low-Dose Streptozotocin Models of Type 2 Diabetes

A rodent model for type 2 diabetes that I have found to be useful and easy to work with is the high-fat fed, low-dose streptozotocin-treated rat, or mouse (Gao & Zheng, 2014; Skovso, 2014). The rationale behind this approach is that the high-fat diet renders the rodent insulin resistant, and the low dose of streptozotocin destroys enough of the β cells to induce a constant state of hyperglycemia.

Reed et al. (2000) were the first to use the high-fat fed streptozotocin-treated rat as a model for type 2 diabetes. Sprague-Dawley rats were fed a high-fat diet for 7 weeks and then treated with streptozotocin (50 mg/kg). These rats were hyperglycemic, insulin resistant, and sensitive to the glucose-lowering effects of metformin and troglitazone. Others have also found this model to be suitable for testing agents for the treatment of type 2 diabetes (Ding et al., 2005; Gaikwad, Viswanad, & Ramarao, 2007; Reed et al., 2000; Srinivasan, Viswanad, Asrat, Kaul, & Ramarao, 2005; Wang et al., 2009; Zhang, Lv, Li, Xu, & Chen, 2008; Zhu, Peng, Liu, Zhang, & Li, 2010). It should be noted that the literature contains some variability in the insults used to create this model. A study by Wang, Li, Liu, Liu, and Sun (2011) examined parameters for creating the high-fat diet streptozotocin–diabetic rat model and found that duration of high-fat diet, dosage of streptozotocin, and the age of rats were the most important factors for establishing this model. Generally the high-fat diet is consistent at 40–50% calories as fat for a period of 4–8 weeks (Gao & Zheng, 2014). However, the dose of streptozotocin used ranges from 20 to 50 mg/kg (Reed et al., 2000; Wang et al., 2011). It has been our experience using rats that were 12 weeks of age at the onset of the study, that after 8 weeks on a high-fat diet, 20 mg/kg streptozotocin resulted in a low success rate for inducing hyperglycemia. When we used 40 mg/kg streptozotocin, our mortality rate was high and rats began dying 2–4 weeks following the injection of streptozotocin. With 30 mg/kg streptozotocin our success rate was over 90%, and the mortality rate was less than 5%. Rats do not require insulin treatment to maintain weight, unlike streptozotocin-induced type 1 diabetic rat models (Davidson, Coppey, Holmes, et al., 2011; Davidson, Kleinschmidt, Oltman, Lund, & Yorek, 2007). An important item to remember is that the potency of streptozotocin can vary widely between vendors and between different lots purchased from the same vendor. We agree with Reed et al. (2000) that the high-fat fed streptozotocin–diabetic rat models late stage type 2 diabetes in patients. The diabetes in these rats is analogous to the development of human type 2 diabetes when the decline in hyperinsulinemia is not able to compensate for insulin resistance and hyperglycemia occurs (Reed et al., 2000).

My laboratory was the first to characterize diabetic neuropathy in the high-fat fed low-dose streptozotocin-treated rat (Davidson, Coppey, Holmes, et al., 2011). In our studies, Sprague-Dawley rats were fed a high-fat diet for 8 weeks before treating with streptozotocin. Our initial study examined vascular and neural endpoints at 16 and 24 weeks after the onset of hyperglycemia and high-fat diet, respectively (Davidson, Coppey, Holmes, et al., 2011). Compared to high-fat fed rats we found that following induction of hyperglycemia glucose utilization was further impaired. At the end of the study period, the diabetic rats weighed about the same as control rats, insulin and leptin levels were near control values, and they were hyperlipidemic. Furthermore, we found that at the end of the study period weight of the epididymal fat pad was significantly less in the high-fat streptozotocin–diabetic rats compared to high-fat fed rats. However, epididymal fat pad weight in the high-fat streptozotocin–diabetic rats was still greater than control rats. Serum insulin and leptin levels were significantly decreased in high-fat streptozotocin–diabetic rats compared to high-fat fed rats but similar to control rats. Two notable differences between high-fat fed rats and high-fat streptozotocin–diabetic rats were that in diabetic rats motor nerve conduction velocity was significantly decreased and superoxide levels and nitrotyrosine staining were increased in epineurial arterioles (Davidson et al., 2010; Davidson, Coppey, Holmes, et al., 2011). This indicates that oxidative stress in the microvasculature is increased by hyperglycemia. In epineurial arterioles from high-fat fed rats superoxide levels and nitrotyrosine staining were not increased and motor nerve conduction velocity not significantly impaired (Davidson et al., 2010). Sensory nerve conduction velocity, thermal nociception, and intraepidermal nerve fiber profiles were impaired in both high-fat fed rats and high-fat streptozotocin–diabetic rats. Interestingly, acetylcholine- and CGRP-mediated vascular relaxations of epineurial arterioles were impaired to a similar degree in high-fat fed rats and high-fat streptozotocin–diabetic rats. It was expected that because of the increase in superoxide and nitrotyrosine staining in epineurial arterioles from high-fat streptozotocin–diabetic rats compared to high-fat fed rats that vascular relaxation to acetylcholine would be impaired to a greater degree in high-fat streptozotocin–diabetic rats. This suggests that mechanisms in addition to oxidative stress contribute significantly to impairment of vascular function in epineurial arterioles from high-fat fed rats and high-fat streptozotocin–diabetic rats. More recent studies have revealed loss of corneal nerves and decrease in corneal nerve sensitivity in diet-induced obese rats and high-fat streptozotocin–diabetic rats (Davidson, Coppey, Kardon, et al., 2014).

We have also used the high-fat diet low-dose streptozotocin approach to create a mouse model of type 2 diabetes (Yorek et al., 2015). Unlike rats, feeding C57Bl6/J mice a high-fat diet causes an elevated level of fasting blood glucose (Coppey et al., 2011; Davidson, Coppey, Holmes, et al., 2011). However, hyperglycemia in the high-fat fed mouse is modest and not associated with an increase in hemoglobin A_1C levels. Moreover, blood glucose was not increased in the fed state. Using a low dose of streptozotocin with a high-fat fed C57Bl/6J mouse creates a higher level of blood glucose, which is present in both the fasted and fed state. This type 2 diabetic mouse model has been previously used to examine pharmacological interventions, vascular biology and atherosclerosis, cardiomyocyte hypertrophy, and β-cell function (Bansal et al., 2012; Lin et al., 2015; Lv et al., 2010; Mali et al., 2014; Ullevig, Zhao, Zamora, & Asmis, 2011; Wang, Hsu, Lin, & Chen, 2014; Xue, Ding, & Liu, 2010; Zhu et al., 2014). In our studies comparing the effect of diet-induced obesity, type 1 and type 2 diabetes on neuropathy, we found the severity of hyperglycemia was significantly different between the types 1 and 2 diabetic mouse models (Yorek et al., 2015). However, motor and sensory nerve conduction velocity, thermal and mechanical sensitivity, and intraepidermal nerve fiber density in the skin were all impacted similarly in diet-induced obesity and types 1 and 2 diabetic mice. Loss of corneal nerves in the subepithelial layer and penetrating the corneal epithelium occurred more rapidly in the diet-induced obesity mice and type 2 diabetic mice compared to type 1 diabetic mouse. This suggests that hyperglycemia is not the only factor contributing to nerve fiber loss in the cornea and that loss of intraepidermal and corneal nerve fibers is mediated by different factors and/or occurs at different rates.

Overall, these studies demonstrate that the high-fat fed low-dose streptozotocin-treated rodents are a good animal model for preclinical studies for discovery and evaluation of new treatments for diabetic neuropathy especially in relation to changes in nerve structure in the skin and cornea.

3.10 Streptozotocin–Nicotinamide Rat

For this model the rat is treated with a dose of streptozotocin generally used to create type 1 diabetes followed by nicotinamide (Sharma et al., 2012; Szkudelski, 2012). Nicotinamide is a poly(ADP-ribose) polymerase inhibitor and has been shown to reverse neurological and neurovascular deficits in streptozotocin–diabetic rats (Negi, Kumar, Kaundal, Gulati, & Sharma, 2010; Stevens et al., 2007). When nicotinamide is used in combination with streptozotocin, it protects β cells from the effects of streptozotocin and the result is a model of type 2 diabetes (Szkudelski, 2012). In this model the severity of diabetes strongly depends on the doses of streptozotocin and nicotinamide given to the animal (Szkudelski, 2012). It has been reported that the streptozotocin–nicotinamide-treated rat develops neuropathy (Sharma et al., 2012).

4. RODENT MODELS OF TYPE 1 DIABETES

4.1 Spontaneously Hypertensive Rat

Use of streptozotocin has been the most often used approach to induce type 1 diabetes in rodents. However, inducing type 1 diabetes in Sprague-Dawley or Wistar rats does not cause hypertension, which is an independent risk factor for neuropathy in diabetic patients. Streptozotocin has therefore been used to induce type 1 diabetes in spontaneously hypertensive rats

(Gregory, Jolivalt, Goor, Mizisin, & Calcutt, 2012; Sanada et al., 2015) to study the combined effects of diabetes and hypertension on peripheral neuropathy.

4.2 BioBreeding/Worcester Rat

There are several genetic rat and mouse models that have been used to replicate type 1 diabetes. The BB/Wor rat has been stated to model human insulin-dependent diabetes mellitus (Whalen, Mordes, & Rossini, 2001). This rat requires daily insulin treatment to survive. It has been reported that motor and sensory nerve conduction velocities are decreased after 4 and 6 weeks of diabetes and continue to decline for up to 9 months (Kamiya, Zhang, & Sima, 2009). Myelinated sural nerve fibers show progressive decreases in fiber number and size. Other deficits include thermal hyperalgesia, loss of content of neuropeptides in dorsal root ganglia, and reduced endoneurial blood flow (Kamiya, Zhang, & Sima, 2004; Stevens, Zhang, Li, & Sima, 2004). Interestingly, Sima and colleagues have demonstrated that many of the neurological deficits caused by diabetes in the BB/Wor rat can be prevented/reversed with C-peptide treatment (Kamiya et al., 2004; Sima et al., 2001; Stevens et al., 2004; Zhang et al., 2001).

4.3 Ins2Akita Mouse

The heterozygous Ins2Akita mouse spontaneously develops insulin-dependent diabetes, including hyperglycemia, hypoinsulinemia, polydipsia, and polyuria, which is more severe in males than females. The Akita mouse has been widely used to introduce a diabetic background into other mouse genetic models. For instance, crossing the Akita mouse with the apoE (−/−) mouse created a mouse model used for studying the effect of diabetes on atherosclerosis (Jun, Ma, & Segar, 2011). The C57BL/6-Akita mouse has been proposed as a nonobese model for type 2 diabetes (Yaguchi, Nagashima, Izumi, & Okamoto, 2003) and has been used to examine the effect of diabetes on the central nervous system. Studies using the Akita mouse for determination of the effect of diabetes on autonomic neuropathy have shown that the Akita mouse develops abnormal cardiac function, with a loss in nerve density (Yang & Chon, 2011). In regard to peripheral neuropathy, 16-week-old Ins2Akita mice develop sensory neuropathy including reduced nerve conduction velocity, thermal and mechanical hypoalgesia, and tactile allodynia (Drel et al., 2011). Motor nerve conduction velocity showed a nonsignificant trend to slowing, whereas others (de Preux Charles et al., 2010) have reported a decrease in motor nerve conduction velocity in Ins2Akita mice at 12 weeks of age.

4.4 Nonobese Diabetic Mouse

The nonobese diabetic (NOD) mouse is susceptible to the development of autoimmune diabetes but also multiple other autoimmune diseases (Bour-Jordan et al., 2013). Over 20 susceptibility loci linked to diabetes have been identified in NOD mice (Bour-Jordan et al., 2013), and the strain has been used extensively to cross the diabetic background onto other mice (Gross et al., 2008). Invasive insulitis, seen in NOD mice, causes early sympathetic islet neuropathy, but it remains to be determine whether this is responsible for the impaired glucagon secretion (Taborsky et al., 2009). In addition to the sympathetic neuropathy developed by the NOD mouse, it has also been reported to develop sensory neuropathy with hypoalgesia and correction by use of a peroxynitrite decomposition catalyst (Obrosova et al., 2005).

5. OTHER ANIMAL MODELS

This chapter has focused primarily on rodent models of obesity and diabetes and peripheral neuropathy. A number of other animal models have also been used successfully in the study of the effect of diabetes on peripheral and autonomic neuropathy. These include guinea pigs, rabbits, cats, dogs, pigs, and monkeys (Cohen, Tesfamariam, Weisbrod, & Zitnay, 1990; Estrella et al., 2008; Islam, 2013; Mesangeau, Laude, & Elghozi, 2000; Mizisin et al., 2007; Morgan, Vite, Radhakrishnan, & Hess, 2008). However, ethical, practical, and financial costs concerns will likely continue to limit the extent to which these other models are used.

6. CONCLUSIONS

In 1997, Dr. Tomlinson coauthored a review article titled "Does neuropathy develop in animal models?" (Hounsom & Tomlinson, 1997). Investigators have been addressing this question for the past 18 years. In the abstract of that paper the authors wrote in reference to diabetic neuropathy, "currently the cornerstone of treatment lies with the maintenance of euglycemia and development of effective treatments for diabetic neuropathy is urgently needed." The same statement could be made today. The only accepted treatment for diabetic neuropathy remains good glycemic control, but it is now clear that this is ineffective, especially in patients with type 2 diabetes (Callaghan et al., 2012). Hounsom and Tomlinson further noted that animal models have been developed to investigate the pathogenesis of diabetic neuropathy and evaluate potential therapeutic agents. However, no model is perfect and no one would suggest that diabetic rodents can replicate the human condition fully. Since this article was written, animal models, including those discussed earlier, continue to be used to investigate and identify potential new treatments for diabetic neuropathy. However, translation of findings to humans with diabetic peripheral neuropathy has continued to fail. Many articles have been written to rationalize these failures. We know that diabetic neuropathy is a complex disease with multiple etiologies. It is my belief that animal models of diabetes and diabetic neuropathy can still play a role in discovery of a treatment. However, this treatment will likely be a combination therapy that will delay progression and induce nerve repair when used in combination with glycemic control. It should also be accepted that the preclinical efficacy of any treatment be tested in several animal models utilizing multiple endpoints. A consensus statement regarding the phenotyping of rodent models for diabetic peripheral neuropathy was published following a joint meeting of the Diabetic Neuropathy Study Group of EASD (Biessels et al., 2014). Following these guidelines will help in standardizing studies coming from multiple laboratories.

REFERENCES

Akash, M. S., Rehman, K., & Chen, S. (2013). Goto-Kakizaki rats: Its suitability as non-obese diabetic animal model for spontaneous type 2 diabetes mellitus. *Current Diabetes Reviews*, *9*, 387–396.

Alderson, N. L., Chachich, M. E., Youssef, N. N., Beattie, R. J., Nachtigal, M., Thorpe, S. R., et al. (2003). The AGE inhibitor pyridoxamine inhibits lipemia and development of renal and vascular disease in Zucker obese rats. *Kidney International*, *63*, 2123–2133.

Anderson, N. J., King, M. R., Delbruck, L., & Jolivalt, C. G. (2014). Role of insulin signaling impairment, adiponectin and dyslipidemia in peripheral and central neuropathy in mice. *Disease Models & Mechanisms*, *7*, 625–633.

Asghar, O., Petropoulos, I. N., Alam, U., Jones, W., Jeziorska, M., Marshall, A., et al. (2014). Corneal confocal microscopy detects neuropathy in subjects with impaired glucose tolerance. *Diabetes Care*, *37*, 2643–2646.

Bansal, P., Paul, P., Mudgal, J., Nayak, P., Pannakal, S., Priyadarsini, K., et al. (2012). Antidiabetic, antihyperlipidemic and antioxidant effects of the flavonoid rich fraction of Pilea microphylla (L.) in high fat diet/streptozotocin-induced diabetes in mice. *Experimental and Toxicologic Pathology*, *64*, 651–658.

Beltowski, J. (2012). Leptin and the regulation of endothelial function in physiological and pathological conditions. *Clinical and Experimental Pharmacology and Physiology*, *39*, 168–178.

Biessels, G. J., Bril, V., Calcutt, N. A., Cameron, N. E., Cotter, M. A., Dobrowsky, R., et al. (2014). Phenotyping animal models of diabetic neuropathy: A consensus statement of the diabetic neuropathy study group of the EASD (Neurodiab). *Journal of the Peripheral Nervous System*, *19*, 77–87.

Bour-Jordan, H., Thompson, H. L., Giampaolo, J. R., Davini, D., Rosenthal, W., & Bluestone, J. A. (2013). Distinct genetic control of autoimmune neuropathy and diabetes in the non-obese diabetic background. *Journal of Autoimmunity*, *45*, 58–67.

Bray, G. A. (1977). The Zucker-fatty rat: A review. *Federation Proceedings*, *36*, 148–153.

Calcutt, N. A. (2004). Modeling diabetic sensory neuropathy in rats. *Methods in Molecular Medicine*, *99*, 55–65.

Callaghan, B. C., Little, A. A., Feldman, E. L., & Hughes, R. A. (2012). Enhanced glucose control for preventing and treating diabetic neuropathy. *Cochrane Database of Systematic Reviews*, *6*, CD007543.

Cameron, N. E., Cotter, M. A., Archibald, V., Dines, K. C., & Maxfield, E. K. (1994a). Anti-oxidant and pro-oxidant effects on nerve conduction velocity, endoneurial blood flow and oxygen tension in non-diabetic and streptozotocin-diabetic rats. *Diabetologia*, *37*, 449–459.

Cameron, N. E., Cotter, M. A., Dines, K. C., & Love, A. (1992). Effects of aminoguanidine on peripheral nerve function and polyol pathway metabolites in streptozotocin-diabetic rats. *Diabetologia*, *35*, 946–950.

Cameron, N. E., Cotter, M. A., Dines, K. C., Maxfield, E. K., Carey, F., & Mirrlees, D. J. (1994b). Aldose reductase inhibition, nerve perfusion, oxygenation and function in streptozotocin-diabetic rats: Dose-response considerations and independence from a myo-inositol mechanism. *Diabetologia*, *37*, 651–663.

Cameron, N. E., Cotter, M. A., & Low, P. A. (1991). Nerve blood flow in early experimental diabetes in rats: Relation to conduction deficits. *American Journal of Physiology. Endocrinology and Metabolism*, *261*, E1–E8.

Cho, Y. R., Lim, J. H., Kim, M. Y., Kim, T. W., Hong, B. Y., Kim, Y. S., et al. (2014). Therapeutic effects of fenofibrate on diabetic peripheral neuropathy by improving endothelial and neural survival in db/db mice. *PLoS One*, *9*, e83204.

Cohen, R. A., Tesfamariam, B., Weisbrod, R. M., & Zitnay, K. M. (1990). Adrenergic denervation in rabbits with diabetes mellitus. *American Journal of Physiology. Heart and Circulatory Physiology*, *259*, H55–H61.

Coppey, L. J., Davidson, E. P., Dunlap, J. A., Lund, D. D., & Yorek, M. A. (2000). Slowing of motor nerve conduction velocity in streptozotocin-induced diabetic rats is preceded by impaired vasodilation in arterioles that overlie the sciatic nerve. *International Journal of Experimental Diabetes Research*, *1*, 131–143.

Coppey, L., Davidson, E., Lu, B., Gerard, C., & Yorek, M. (2011). Vasopeptidase inhibitor ilepatril (AVE7688) prevents obesity- and diabetes-induced neuropathy in C57Bl/6J mice. *Neuropharmacology*, *60*, 259–266.

Coppey, L. J., Gellett, J. S., Davidson, E. P., Dunlap, J. A., & Yorek, M. A. (2002). Changes in endoneurial blood flow, motor nerve conduction velocity and vascular relaxation of epineurial arterioles of the sciatic nerve in ZDF-obese diabetic rats. *Diabetes/Metabolism Research and Reviews*, *18*, 49–56.

Cotter, M. A., Dines, K. C., & Cameron, N. E. (1993). Prevention and reversal of motor and sensory peripheral nerve conduction abnormalities in streptozotocin-diabetic rats by the prostacyclin analogue iloprost. *Naunyn Schmiedeberg's Archives of Pharmacology*, *347*, 534–540.

Davidson, E. P., Coppey, L. J., Calcutt, N. A., Oltman, C. L., & Yorek, M. A. (2010). Diet-induced obesity in Sprague-Dawley rats causes microvascular and neural dysfunction. *Diabetes/Metabolism Research and Reviews, 26*, 306–318.

Davidson, E. P., Coppey, L. J., Dake, B., & Yorek, M. A. (2011a). Effect of treatment of Sprague Dawley rats with AVE7688, enalapril, or candoxatril on diet-induced obesity. *Journal of Obesity, 2011*. 686952.

Davidson, E. P., Coppey, L. J., Holmes, A., Dake, B., & Yorek, M. A. (2011b). Effect of treatment of high fat fed/low dose streptozotocin-diabetic rats with Ilepatril on vascular and neural complications. *European Journal of Pharmacology, 668*, 497–506.

Davidson, E. P., Coppey, L. J., Holmes, A., Lupachyk, S., Dake, B. L., Oltman, C. L., et al. (2014a). Characterization of diabetic neuropathy in the Zucker diabetic Sprague-Dawley rat: A new animal model for type 2 diabetes. *Journal of Diabetes Research, 2014*. 714273.

Davidson, E. P., Coppey, L. J., Kardon, R. H., & Yorek, M. A. (2014b). Differences and similarities in development of corneal nerve damage and peripheral neuropathy and in diet-induced obesity and type 2 diabetic rats. *Investigative Ophthalmology and Visual Science, 55*, 1222–1230.

Davidson, E., Coppey, L., Lu, B., Arballo, V., Calcutt, N. A., Gerard, C., et al. (2009). The roles of streptozotocin neurotoxicity and neural endopeptidase in murine experimental diabetic neuropathy. *Experimental Diabetes Research, 2009*. 431980.

Davidson, E. P., Kleinschmidt, T. L., Oltman, C. L., Lund, D. D., & Yorek, M. A. (2007). Treatment of streptozotocin-induced diabetic rats with AVE7688, a vasopeptidase inhibitor: Effect on vascular and neural disease. *Diabetes, 56*, 355–362.

Davis, J. E., Cain, J., Banz, W. J., & Peterson, R. G. (2013). Age-related differences in response to high-fat feeding on adipose tissue and metabolic profile in ZDSD rats. *ISRN Obesity, 2013*. 584547.

de Preux Charles, A. S., Verdier, V., Zenker, J., Peter, B., Medard, J. J., Kuntzer, T., et al. (2010). Global transcriptional programs in peripheral nerve endoneurium and DRG are resistant to the onset of type 1 diabetic neuropathy in Ins2 mice. *PLoS One, 5*, e10832.

Ding, S. Y., Shen, Z. F., Chen, Y. T., Sun, S. J., Liu, Q., & Xie, M. Z. (2005). Pioglitazone can ameliorate insulin resistance in low-dose streptozotocin and high sucrose-fat diet induced obese rats. *Acta Pharmacologica Sinica, 26*, 575–580.

Drel, V. R., Mashtalir, N., Ilnytska, O., Shin, J., Li, F., Lyzogubov, V. V., et al. (2006). The leptin-deficient (ob/ob) mouse: A new animal model of peripheral neuropathy of type 2 diabetes and obesity. *Diabetes, 55*, 3335–3343.

Drel, V. R., Pacher, P., Stavniichuk, R., Xu, W., Zhang, J., Kuchmerovska, T. M., et al. (2011). Poly(ADP-ribose)polymerase inhibition counteracts renal hypertrophy and multiple manifestations of peripheral neuropathy in diabetic Akita mice. *International Journal of Molecular Medicine, 28*, 629–635.

Estrella, J. S., Nelson, R. N., Sturges, B. K., Vernau, K. M., Williams, D. C., LeCouteur, R. A., et al. (2008). Endoneurial microvascular pathology in feline diabetic neuropathy. *Microvascular Research, 75*, 403–410.

Figueroa-Romero, C., Sadidi, M., & Feldman, E. L. (2008). Mechanisms of disease: The oxidative stress theory of diabetic neuropathy. *Review in Endocrine and Metabolic Disorders, 9*, 301–314.

Gade, W., Schmit, J., Collins, M., & Gade, J. (2010). Beyond obesity: The diagnosis and pathophysiology of metabolic syndrome. *Clinical Laboratory Science, 23*, 51–61.

Gaikwad, A. B., Viswanad, B., & Ramarao, P. (2007). PPAR gamma agonists partially restores hyperglycemia induced aggravation of vascular dysfunction to angiotensin II in thoracic aorta isolated from rats with insulin resistance. *Pharmacological Research, 55*, 400–407.

Gao, F., & Zheng, Z. M. (2014). Animal models of diabetic neuropathic pain. *Experimental and Clinical Endocrinology and Diabetes, 122*, 100–106.

Gregory, J. A., Jolivalt, C. G., Goor, J., Mizisin, A. P., & Calcutt, N. A. (2012). Hypertension-induced peripheral neuropathy and the combined effects of hypertension and diabetes on nerve structure and function in rats. *Acta Neuropathologica, 124*, 561–573.

Groover, A. L., Ryals, J. M., Guilford, B. L., Wilson, N. M., Christianson, J. A., & Wright, D. E. (2013). Exercise-mediated improvements in painful neuropathy associated with prediabetes in mice. *Pain, 154*, 2658–2667.

Gross, V., Tank, J., Partke, H. J., Plehm, R., Diedrich, A., da Costa Goncalves, A. C., et al. (2008). Cardiovascular autonomic regulation in non-obese diabetic (NOD) mice. *Autonomic Neuroscience: Basic and Clinical, 138*, 108–113.

Grote, C. W., Groover, A. L., Ryals, J. M., Geiger, P. C., Feldman, E. L., & Wright, D. E. (2013). Peripheral nervous system insulin resistance in ob/ob mice. *Acta Neuropathologica Communications, 1*, 15. http://dx.doi.org/10.1186/2051-5960-1-15.

Hounsom, L., & Tomlinson, D. R. (1997). Does neuropathy develop in animal models? *Clinical Neuroscience, 4*, 380–389.

Islam, M. S. (2013). Animal models of diabetic neuropathy: Progress since 1960s. *Journal of Diabetes Research, 2013*. 149452.

Jun, J. Y., Ma, Z., & Segar, L. (2011). Spontaneously diabetic Ins2(+/Akita): ApoE-deficient mice exhibit exaggerated hypercholesterolemia and atherosclerosis. *American Journal of Physiology. Endocrinology and Metabolism, 301*, E145–E154.

Kamenov, Z., Higashino, H., Todorova, M., Kajimoto, N., & Suzuki, A. (2006). Physiological characteristics of diabetic neuropathy in sucrose-fed Otsuka Long-Evans Tokushima fatty rats. *Methods and Findings in Experimental and Clinical Pharmacology, 28*, 13–18.

Kamiya, H., Murakawa, Y., Zhang, W., & Sima, A. A. (2005). Unmyelinated fiber sensory neuropathy differs in type 1 and type 2 diabetes. *Diabetes/Metabolism Research and Reviews, 21*, 448–458.

Kamiya, H., Zhang, W., & Sima, A. A. (2004). C-peptide prevents nociceptive sensory neuropathy in type 1 diabetes. *Annals of Neurology, 56*, 827–835.

Kamiya, H., Zhang, W., & Sima, A. A. (2009). Dynamic changes of neuroskeletal proteins in DRGs underlie impaired axonal maturation and progressive axonal degeneration in type 1 diabetes. *Experimental Diabetes Research, 2009*. 793281.

Kannan, M. A., Sarva, S., Kandadai, R. M., Paturi, V. R., Jabeen, S. A., & Borgohain, R. (2014). Prevalence of neuropathy in patients with impaired glucose tolerance using various electrophysiological tests. *Neurology India, 62*, 656–661.

Kaur, J. (2014). A comprehensive review on metabolic syndrome. *Cardiology Research and Practice, 2014*. 943162.

Kava, R., Greenwood, M. R. C., & Johnson, P. R. (1990). Zucker *(fa/fa)* rat. *ILAR Journal, 32*, 4–8.

Kawada, T., Miyata, S., Shimada, T., Sanzen, Y., Ito, M., Hemmi, C., et al. (2010). A study of cardiovascular function in Tsumura Suzuki obese diabetes, a new model mouse of type 2 diabetes. *Biological and Pharmaceutical Bulletin, 33*, 998–1003.

Kemmochi, Y., Fukui, K., Maki, M., Kimura, S., Ishii, Y., Sasase, T., et al. (2013). Metabolic disorders and diabetic complications in spontaneously diabetic Torii Lepr (fa) rat: A new obese type 2 diabetic model. *Journal of Diabetes Research, 2013*. 948257.

Kim, H., Kim, J. J., & Yoon, Y. S. (2012). Emerging therapy for diabetic neuropathy: Cell therapy targeting vessels and nerves. *Endocrine, Metabolic and Immune Disorders Drug Targets, 12*, 168–178.

Kurtz, T. W., Morris, R. C., & Pershadsingh, H. A. (1989). The Zucker fatty rat as a genetic model of obesity and hypertension. *Hypertension, 13*, 896–901.

Li, F., Abatan, O. I., Kim, H., Burnett, D., Larkin, D., Obrosova, I. G., et al. (2006). Taurine reverses neurological and neurovascular deficits in Zucker diabetic fatty rats. *Neurobiology of Disease, 22*, 669–676.

Lin, Y., Hu, G., Li, K., Zhao, Y., Wei, J., & Zhen, Y. (2015). The protection of Rhein lysinate to liver in diabetic mice induced by high-fat diet and streptozotocin. *Archives of Pharmacal Research, 38*, 885–892.

Lirk, P., Flatz, M., Haller, I., Hausott, B., Blumenthal, S., Stevens, M. F., et al. (2012). In Zucker diabetic fatty rats, subclinical diabetic neuropathy increases in vivo lidocaine block duration but not in vitro neurotoxicity. *Regional Anesthesia and Pain Medicine, 37*, 601–606.

Lirk, P., Verhamme, C., Boeckh, R., Stevens, M. F., ten Hoope, W., Gerner, P., et al. (2015). Effects of early and late diabetic neuropathy on sciatic nerve block duration and neurotoxicity in Zucker diabetic fatty rats. *British Journal of Anesthesia, 114*, 319–326.

Lizuka, S., Suzuki, W., Tabuchi, M., Nagata, M., Imamura, S., Kobayashi, Y., et al. (2005). Diabetic complications in a new animal model (TSOD mouse) of spontaneous NIDDM with obesity. *Experimental Animals, 54*, 71–83.

Lupachyk, S., Watcho, P., Hasanova, N., Julius, U., & Obrosova, I. G. (2012). Triglyceride, nonesterified fatty acids, and prediabetic neuropathy: Role for oxidative-nitrosative stress. *Free Radical Biology and Medicine, 52*, 1255–1263.

Lupachyk, S., Watcho, P., Obrosov, A. A., Stavniichuk, R., & Obrosova, I. G. (2013). Endoplasmic reticulum stress contributes to prediabetic peripheral neuropathy. *Experimental Neurology, 247*, 342–348.

Lv, L., Wu, S., Wang, G., Zhang, J., Pang, J., Liu, Z., et al. (2010). Effect of astragaloside IV on hepatic glucose-regulating enzymes in diabetic mice induced by a high-fat diet and streptozotocin. *Phytotherapy Research, 24*, 219–224.

Mali, V., Ning, R., Chen, J., Yang, X., Xu, J., & Palaniyandi, S. (2014). Impairment of aldehyde dehydrogenase-2 by 4-hydroxy-2-nonenal adduct formation and cardiomyocyte hypertrophy in mice fed a high-fat diet and injected with low-dose streptozotocin. *Experimental Biology and Medicine, 239*, 610–618.

Mesangeau, D., Laude, D., & Elghozi, J. L. (2000). Early detection of cardiovascular autonomic neuropathy in diabetic pigs using blood pressure and heart rate variability. *Cardiovascular Research, 45*, 889–899.

Mizisin, A. P., Nelson, R. W., Sturges, B. K., Vernau, K. M., Lecouteur, R. A., Williams, D. C., et al. (2007). Comparable myelinated nerve pathology in feline and human diabetes mellitus. *Acta Neuropathologica, 113*, 431–442.

Morgan, M. J., Vite, C. H., Radhakrishnan, A., & Hess, R. S. (2008). Clinical peripheral neuropathy associated with diabetes mellitus in 3 dogs. *The Canadian Veterinary Journal, 49*, 583–586.

Murakawa, Y., Zhang, W., Pierson, C. R., Brismar, T., Ostenson, C. G., Efendic, S., et al. (2002). Impaired glucose tolerance and insulinopenia in the GK-rat causes peripheral neuropathy. *Diabetes/Metabolism Research and Reviews, 18*, 473–483.

Nakamura, J., Hamada, Y., Sakakibara, F., Hara, T., Wakao, T., Mori, K., et al. (2001). Physiological and morphometric analyses of neuropathy in sucrose-fed OLETF rats. *Diabetes Research and Clinical Practice, 51*, 9–20.

Negi, G., Kumar, A., Kaundal, R. K., Gulati, A., & Sharma, S. S. (2010). Functional and biochemical evidence indicating beneficial effect of melatonin and nicotinamide alone and in combination in experimental diabetic neuropathy. *Neuropharmacology, 58*, 585–592.

Norido, F., Canella, R., Zanoni, R., & Gorio, A. (1984). Development of diabetic neuropathy in the C57BL/Ks (db/db) mouse and its treatment with gangliosides. *Experimental Neurology, 83*, 221–232.

Nowicki, M., Kosacka, J., Serke, H., Bluher, M., & Spanel-Borowski, K. (2012). Altered sciatic nerve fiber morphology and endoneural microvessels in mouse models relevant for obesity, peripheral diabetic polyneuropathy, and the metabolic syndrome. *Journal of Neuroscience Research, 90*, 122–131.

Nukada, H., & Dyck, P. J. (1984). Microsphere embolization of nerve capillaries and fiber degeneration. *The American Journal of Pathology*, *115*, 275–287.
O'Brien, P. D., Hur, J., Hayes, J. M., Backus, C., Sakowski, S. A., & Feldman, E. L. (2014). BTBR ob/ob mice as a novel diabetic neuropathy model: Neurological characterization and gene expression analyses. *Neurobiology of Disease*, *73C*, 348–355.
O'Brien, P. D., Sakowski, S. A., & Feldman, E. L. (2014). Mouse models of diabetic neuropathy. *ILAR Journal*, *54*, 259–272.
Obrosova, I. G., Ilnytska, O., Lyzogubov, V. V., Pavlov, I. A., Mashtalir, N., Nadler, J. L., et al. (2007). High-fat diet induced neuropathy of pre-diabetes and obesity: Effects of "healthy" diet and aldose reductase inhibition. *Diabetes*, *56*, 2598–2608.
Obrosova, I. G., Mabley, J. G., Zsengeller, Z., Charniauskaya, T., Abatan, O. I., Groves, J. T., et al. (2005). Role for nitrosative stress in diabetic neuropathy: Evidence from studies with a peroxynitrite decomposition catalyst. *FASEB Journal*, *19*, 401–403.
Oltman, C. L., Coppey, L. J., Gellett, J. S., Davidson, E. P., Lund, D. D., & Yorek, M. A. (2005). Progression of vascular and neural dysfunction in sciatic nerves of Zucker diabetic fatty and Zucker rats. *American Journal of Physiology. Endocrinology and Metabolism*, *289*, E113–E122.
Oltman, C. L., Davidson, E. P., Coppey, L. J., Kleinschmidt, T. L., Lund, D. D., Adebara, E. T., et al. (2008). Vascular and neural dysfunction in Zucker diabetic fatty rats: A difficult condition to reverse. *Diabetes, Obesity and Metabolism*, *10*, 64–74.
Oltman, C. L., Davidson, E. P., Coppey, L. J., Kleinschmidt, T. L., & Yorek, M. A. (2009). Treatment of Zucker diabetic fatty rats with AVE7688 improves vascular and neural dysfunction. *Diabetes, Obesity and Metabolism*, *11*, 223–233.
Otto, K. J., Wyse, B. D., Cabot, P. J., & Smith, M. T. (2011). Longitudinal study of painful diabetic neuropathy in the Zucker diabetic fatty rat model of type 2 diabetes: Impaired basal G-protein activity appears to underpin marked morphine hyposensitivity at 6 months. *Pain Medicine*, *12*, 437–450.
Pande, M., Hur, J., Hong, Y., Backus, C., Hayes, J. M., Oh, S. S., et al. (2011). Transcriptional profiling of diabetic neuropathy in the BKS db/db mouse: A model of type 2 diabetes. *Diabetes*, *60*, 1981–1989.
Papanas, N., Vinik, A. I., & Ziegler, D. (2011). Neuropathy in prediabetes: Does the clock start ticking early? *Nature Reviews. Endocrinology*, *7*, 682–690.
Papanas, N., & Ziegler, D. (2012). Prediabetic neuropathy: Does it exist? *Current Diabetes Reports*, *12*, 376–383.
Peterson, R. G., Jackson, C. V., Zimmeran, K., de Winter, W., Huebert, N., & Hansen, M. K. (2015). Characterization of the ZDSD rat: A translational model for the study of metabolic syndrome and type 2 diabetes. *Journal of Diabetes Research*, *2015*. 487816.
Peterson, R. G., Shaw, W. N., Neel, M. A., Little, L. A., & Eichberg, J. (1990). Zuckerdiabetic fatty rat as a model for non-insulin-dependent diabetes mellitus. *ILAR Journal*, *32*, 16–19.
Pop-Busui, R., Sima, A., & Stevens, M. (2006). Diabetic neuropathy and oxidative stress. *Diabetes/Metabolism Research and Reviews*, *22*, 257–273.
Portha, B., Giroix, M. H., Tourrel-Cuzin, C., Le-Stunff, H., & Movassat, J. (2012). The GK rat: A prototype for the study of non-overweight type 2 diabetes. *Methods in Molecular Biology*, *933*, 125–159.
Quattrini, C., Tavakoli, M., Jeziorska, M., Kallinikos, P., Tesfaye, S., Finnigan, J., et al. (2007). Surrogate markers of small fiber damage in human diabetic neuropathy. *Diabetes*, *56*, 2148–2154.
Reed, M. J., Meszaros, K., Entes, L. J., Claypool, M. D., Pinkett, J. G., Gadbois, T. M., et al. (2000). A new rat model of type 2 diabetes: The fat-fed, streptozotocin-treated rat. *Metabolism*, *49*, 1390–1394.

Rees, D. A., & Alcolado, J. C. (2005). Animal models of diabetes mellitus. *Diabetic Medicine*, *22*, 359–370.

Reinwald, S., Peterson, R. G., Allen, M. R., & Burr, D. B. (2009). Skeletal changes associated with the onset of type 2 diabetes in the ZDF and ZDSD rodent models. *American Journal of Physiology. Endocrinology and Metabolism*, *296*, E765–E774.

Ricci, R., & Bevilacqua, F. (2012). The potential role of leptin and adiponectin in obesity: A comparative review. *Veterinary Journal*, *191*, 292–298.

Robertson, D. M., & Sima, A. A. (1980). Diabetic neuropathy in the mutant mouse [C57BL/ks(db/db)]: A morphometric study. *Diabetes*, *29*, 60–67.

Sanada, L. S., Tavares, M. R., Sato, K. L., Ferreira Rda, S., Neubern, M. C., Castania, J. A., et al. (2015). Association of chronic diabetes and hypertension in sural nerve morphometry: An experimental study. *Diabetology and Metabolic Syndrome*, *7*, 9.

Sasase, T., Ohta, T., Masuyama, T., Yokoi, N., Kakehashi, A., & Shinohara, M. (2013). The spontaneously diabetic Torii rat: An animal model of nonobese type 2 diabetes with severe diabetic complications. *Journal of Diabetes Research*, *2013*. 976209.

Sasase, T., Yokoi, N., Pezzolesi, M. G., & Shinohara, M. (2015). Animal models of diabetes and metabolic disease 2014. *Journal of Diabetes Research*, *2015*. 571809.

Sharma, A. K., Sharma, A., Kumari, R., Kishore, K., Sharma, D., Srinivasan, B. P., et al. (2012). Sitagliptin, sitagliptin and metformin, or sitagliptin and amitriptyline attenuate streptozotocin-nicotinamide induced diabetic neuropathy in rats. *Journal of Biomedical Research*, *26*, 200–210.

Shi, T. J., Zhang, M. D., Zeberg, H., Nilsson, J., Grunler, J., Liu, S. X., et al. (2013). Coenzyme Q10 prevents peripheral neuropathy and attenuates neuron loss in the db-/db- mouse, a type 2 diabetes model. *Proceedings of the National Academy of Sciences of the United States of America*, *110*, 690–695.

Shinohara, M., Masuyama, T., Shoda, T., Takahashi, T., Katsuda, Y., Komeda, K., et al. (2000). A new spontaneously diabetic non-obese Torii rat strain with severe ocular complications. *International Journal of Experimental Diabetes Research*, *1*, 89–100.

Sima, A. A. (2006). Pathological mechanisms involved in diabetic neuropathy: Can we slow the process? *Current Opinion in Investigational Drugs*, *7*, 324–337.

Sima, A. A., & Robertson, D. M. (1978). Peripheral neuropathy in mutant diabetic mouse [C57BL/Ks (db/db)]. *Acta Neuropathologica*, *41*, 85–89.

Sima, A. A., Zhang, W., Sugimoto, K., Henry, D., Li, Z., Wahren, J., et al. (2001). C-peptide prevents and improves chronic type 1 diabetic polyneuropathy in the BB/Wor rat. *Diabetologia*, *44*, 889–897.

Sima, A. A., Zhang, W., Xu, G., Sugimoto, K., Guberski, D., & Yorek, M. A. (2000). A comparison of diabetic polyneuropathy in type II diabetic BBZDR/Wor rats and in type I diabetic BB/Wor rats. *Diabetologia*, *43*, 786–793.

Skovso, S. (2014). Modeling type 2 diabetes in rats using high fat diet and streptozotocin. *Journal of Diabetes Investigation*, *5*, 349–358.

Smith, A. G., & Singleton, J. R. (2008). Impaired glucose tolerance and neuropathy. *The Neurologist*, *14*, 23–29.

Srinivasan, K., Viswanad, B., Asrat, L., Kaul, C. L., & Ramarao, P. (2005). Combination of high-fat diet-fed and low-dose streptozotocin-treated rat: A model for type 2 diabetes and pharmacological screening. *Pharmacological Research*, *52*, 313–320.

Stevens, M. J., Li, F., Drel, V. R., Abatan, O. I., Kim, H., Burnett, D., et al. (2007). Nicotinamide reverses neurological and neurovascular deficits in streptozotocin diabetic rats. *The Journal of Pharmacology and Experimental Therapeutics*, *320*, 458–464.

Stevens, M. J., Zhang, W., Li, F., & Sima, A. A. (2004). C-peptide corrects endoneurial blood flow but not oxidative stress in type 1 BB/Wor rats. *American Journal of Physiology. Endocrinology and Metabolism*, *287*, E497–E505.

Sullivan, K. A., Hayes, J. M., Wiggin, T. D., Backus, C., Su Oh, S., Lentz, S. I., et al. (2007). Mouse models of diabetic neuropathy. *Neurobiology of Disease, 28*, 276–285.
Szkudelski, T. (2012). Streptozotocin-nicotinamide-induced diabetes in the rat. Characteristics of the experimental model. *Experimental Biology and Medicine, 237*, 481–490.
Taborsky, G. J., Jr., Mei, Q., Hackney, D. J., Figlewicz, D. P., LeBoeuf, R., & Mundinger, T. O. (2009). Loss of islet sympathetic nerves and impairment of glucagon secretion in the NOD mouse: Relationship to invasive insulitis. *Diabetologia, 52*, 2602–2611.
Terata, K., Coppey, L. J., Davidson, E. P., Dunlap, J. A., Gutterman, D., & Yorek, M. A. (1999). Acetylcholine-induced arteriolar dilation is reduced in streptozotocin-induced diabetic rats with motor nerve dysfunction. *British Journal of Pharmacology, 128*, 837–843.
Tesch, G. H., & Allen, T. J. (2007). Rodent models of streptozotocin-induced diabetic nephropathy. *Nephrology, 12*, 261–266.
Tirabassi, R. S., Flanagan, J. F., Wu, T., Kislauskis, E. H., Birckbichler, P. J., & Guberski, D. L. (2004). The BBZDR/Wor rat model for investigating the complications of type 2 diabetes mellitus. *ILAR Journal, 45*, 292–302.
Ueta, K., Ishihara, T., Matsumoto, Y., Oku, A., Nawano, M., Fujita, T., et al. (2005). Long-term treatment with the Na^+-glucose cotransporter inhibitor T-1095 causes sustained improvement in hyperglycemia and prevents diabetic neuropathy in Goto-Kakizaki rats. *Life Sciences, 76*, 2655–2668.
Ullevig, S., Zhao, Q., Zamora, D., & Asmis, R. (2011). Ursolic acid protects diabetic mice against monocyte dysfunction and accelerated atherosclerosis. *Atherosclerosis, 219*, 409–416.
Vareniuk, I., Pavlov, I. A., Drel, V. R., Lyzogubov, V. V., Ilnystska, O., Bell, S. R., et al. (2007). Nitrosative stress and peripheral diabetic neuropathy in leptin-deficient (ob/ob) mice. *Experimental Neurology, 205*, 425–436.
Wada, R., Koyama, M., Mizukami, H., Odaka, H., Ikeda, H., & Yagihashi, S. (1999). Effects of long-term treatment with alpha-glucosidase inhibitor on the peripheral nerve function and structure in Goto-Kakizaki rats: A genetic model for type 2 diabetes. *Diabetes/Metabolism Research and Reviews, 15*, 332–337.
Wang, F., Gao, N., Yin, J., & Yu, F. S. (2012). Reduced innervation and delayed re-innervation after epithelial wounding in type 2 diabetic Goto-Kakizaki rats. *The American Journal of Pathology, 181*, 2058–2066.
Wang, Z., Hsu, C., Lin, H., & Chen, J. (2014). Antidiabetic effects of Carassius auratus complex formula in high fat diet combined streptozotocin-induced diabetic mice. *Evidence Based Complementary and Alternative Medicine, 2014*. 628473.
Wang, B., Li, Y., Liu, X., Liu, S., & Sun, C. (2011). Effect of the duration of high-fat diet and the dosage of streptozotocin on establishing experimental animal model of type 2 diabetes mellitus. *Wei Sheng Yan Jiu, 40*, 99–102.
Wang, C., Li, J., Lv, X., Zhang, M., Song, Y., Chen, L., et al. (2009). Ameliorative effect of berberine on endothelial dysfunction in diabetic rats induced by high-fat diet and streptozotocin. *European Journal of Pharmacology, 620*, 131–137.
Whalen, B. J., Mordes, J. P., & Rossini, A. A. (2001). The BB rat as a model of human insulin-dependent diabetes mellitus. *Current Protocols in Immunology* (Suppl. 19). Chapter 15, 15.3.1–15.3.15.
Wiese, T. J., Matsushita, K., Lowe, W. L., Jr., Stokes, J. B., & Yorek, M. A. (1996). Localization and regulation of renal Na^+/myo-inositol cotransporter in diabetic rats. *Kidney International, 50*, 1202–1211.
Xue, J., Ding, W., & Liu, Y. (2010). Anti-diabetic effects of emodin involved in the activation of PPAR gamma on high-fat diet-fed and low dose of streptozotocin-induced diabetic mice. *Fitoterapia, 81*, 173–177.

Yaguchi, M., Nagashima, K., Izumi, T., & Okamoto, K. (2003). Neuropathological study of C57BL/6Akita mouse, type 2 diabetic model: Enhanced expression of alphaB-crystallin in oligodendrocytes. *Neuropathology, 23*, 44–50.

Yamaguchi, T., Sasase, T., Mera, Y., Tomimoto, D., Tadaki, H., Kemmochi, Y., et al. (2012). Diabetic peripheral neuropathy in spontaneously diabetic Torii-Lepr (fa) (SDT fatty) rats. *The Journal of Veterinary Medical Science, 74*, 1669–1673.

Yang, B., & Chon, K. H. (2011). Assessment of diabetic cardiac autonomic neuropathy in type 1 diabetic mice. In *Annual international conference IEEE engineering medicine biology society* (pp. 6560–6563).

Yorek, M. S., Obrosov, A., Shevalye, H., Holmes, A., Harper, M. M., Kardon, R. H., et al. (2015). Effect of diet-induced obesity or type 1 or type 2 diabetes on corneal nerves and peripheral neuropathy in C57Bl/6J mice. *Journal of the Peripheral Nervous System, 20*, 24–31.

Zhang, M., Lv, X. Y., Li, J., Xu, Z. G., & Chen, L. (2008). The characterization of high-fat diet and multiple low-dose streptozotocin induced type 2 diabetes rat model. *Experimental Diabetes Research, 2008*. 704045.

Zhang, W., Yorek, M., Pierson, C. R., Murakawa, Y., Breidenbach, A., & Sima, A. A. (2001). Human C-peptide dose dependently prevents early neuropathy in the BB/Wor rat. *International Journal of Experimental Diabetes Research, 2*, 187–193.

Zhu, N., Liu, B., Luo, W., Zhang, Y., Li, H., Li, S., et al. (2014). Vasoconstrictor role of cyclooxygenase-1-mediated prostacyclin synthesis in non-insulin-dependent diabetic mice induced by high-fat diet and streptozotocin. *American Journal of Physiology. Heart and Circulatory Physiology, 307*, H319–H327.

Zhu, C. F., Peng, H. B., Liu, G. Q., Zhang, F., & Li, Y. (2010). Beneficial effects of oligopeptides from marine salmon skin in a rat model of type 2 diabetes. *Nutrition, 26*, 1014–1020.

SECTION III

Mechanisms and Therapies

CHAPTER SEVEN

An Introduction to the History and Controversies of the Pathogenesis of Diabetic Neuropathy

P. Fernyhough*,†,1, N.A. Calcutt‡
*University of Manitoba, Winnipeg, MB, Canada
†St. Boniface Hospital Research Centre, Winnipeg, MB, Canada
‡University of California, San Diego, La Jolla, CA, United States
[1]Corresponding author: e-mail address: pfernyhough@sbrc.ca

Early studies into mechanisms of diabetic peripheral neuropathy focused primarily on hyperglycemia-driven pathways. Central to this approach was strong, hypothesis-driven work showing that excessive glucose flux through the aldose reductase pathway (aka the polyol pathway) generated metabolic disturbances leading to impaired nerve conduction velocity and reduced axonal caliber (Tomlinson & Gardiner, 2008). Diverse inhibitors of aldose reductase (ARIs) were developed that blocked accumulation of the pathway intermediates sorbitol and fructose in diabetic rats and prevented or reversed indices of diabetic neuropathy (Oates, 2008). Unfortunately, subsequent clinical trials failed to demonstrate efficacy acceptable to regulatory agencies. A variety of reasons have been cited for this failure, ranging from drug design issues such as toxic side effects and poor efficacy at blocking AR activity in humans to trial design issues such as performing short duration trials in subjects with severe degenerative neuropathy. With hindsight it is also clear that endpoints used to assess efficacy in these clinical trials may have been suboptimal. For example, structural integrity of distal regions of small sensory fibers was not assessed but can now be measured in skin biopsies or by corneal confocal microscopy. Moreover, the main physiological endpoint used in both preclinical and subsequent clinical studies, large myelinated fiber conduction slowing, develops within weeks of onset of diabetes in animal models and may not accurately replicate the slowly developing NCV deficit in humans with diabetic neuropathy where segmental demyelination and fiber loss contribute. An additional area of interest when discussing ARIs

International Review of Neurobiology, Volume 127
ISSN 0074-7742
http://dx.doi.org/10.1016/bs.irn.2016.03.012

is the true site of action. AR has been localized to Schwann cells of myelinated fibers, epineurial endothelial cells, and axons of sympathetic neurons, but a clear explanation as to how excessive polyol pathway activity drives distal axonal damage in small sensory fibers remains to be fully developed (Jiang, Calcutt, Ramos, & Mizisin, 2006). Yagihashi revisits some of these issues in his contribution entitled "Glucotoxic Mechanisms and Related Therapeutic Approaches."

The clinical failure of ARIs has not, in our opinion, denigrated the central idea that excessive glucose flux through the polyol pathway is a key pathogenic process in diabetic neuropathy. As ARI-related research has stalled, various research groups have explored putative downstream consequences of hyperglycemia, primarily in endothelial cell cultures, to study the impact of excessive glucose metabolism and polyol pathway flux in augmenting ischemia and oxidative stress (Figueroa-Romero, Sadidi, & Feldman, 2008; Nishikawa, Edelstein, & Brownlee, 2000). The role of ischemia due to hyperglycemia-induced impaired nerve blood flow remains a major source of controversy. Diabetic neuropathy is classified by many as a microvascular disease. However, the pathology of nerve damage in diabetes does not reflect an ischemic process. For example, in ischemic nerve damage large fibers are preferentially targeted, which does not appear to be the case in diabetic neuropathy (Fujimura, Lacroix, & Said, 1991). In the contribution "Sensory Neurodegeneration in Diabetes: Beyond Glucotoxicity" by Zochodne, the inconsistencies in the vascular hypothesis as a cogent explanation of nerve damage in diabetes are briefly discussed.

Oxidative stress clearly occurs in the nerve of diabetic rodents and diverse antioxidants are protective in various animal models of the disease (Figueroa-Romero et al., 2008). Unfortunately, clinical trials of antioxidants in diabetic neuropathy have been underwhelming, although some support has been gained from use of agents such as alpha lipoic acid that includes antioxidant activity among its properties (Ziegler et al., 2011). The caveats separating successful preclinical studies from demonstrable clinical efficacy, as discussed above for ARI, may equally apply to antioxidants with the additional problem that the financial incentive for drug development in this area is limited. A much more cohesive understanding of the multiple sources of ROS in nerve tissue is also required. For example, endothelial cells may indeed produce excessive ROS via high glucose flux-driven aberrant overactivity of the proximal aspect of the mitochondrial electron transport chain (Nishikawa, Edelstein, Du, et al., 2000). However, it is becoming recognized that neurons and glia have distinct energy demands and produce

ATP by operating flexible regulation of the relative levels of glycolysis and oxidative phosphorylation (Halim et al., 2010). Indeed, in adult sensory neurons maintained under hyperglycemic conditions the flux of electrons via the electron transport system is significantly diminished and this is accompanied by downregulation of the mitochondrial biogenesis pathway (Chowdhury et al., 2010; Zhang, Zhao, Blagg, & Dobrowsky, 2012). This reduced electron transport capacity of sensory neurons in diabetes is also associated with lowered ROS production by the mitochondria—the opposite of what happens in endothelial cells maintained under high glucose concentration (Akude et al., 2011). It appears that hyperglycemia, or nutrient stress, drives neurons toward relying on glycolysis for energy production and diminishes the contribution of mitochondria (Chowdhury, Dobrowsky, & Fernyhough, 2011; Chowdhury, Smith, & Fernyhough, 2013). Thus, excess substrate, as provided by diabetes, appears to impact mitochondrial function and ROS formation in a cell-specific manner.

Interesting headway has been made in identifying mechanisms for the generation of other potentially neurotoxic molecules. For example, diabetes is also associated with dyslipidemia and studies suggest that certain forms of oxidized lipoproteins are also toxic to sensory neurons and peripheral nerve function (Vincent, Hayes, et al., 2009; Vincent, Hinder, Pop-Busui, & Feldman, 2009). The original competing hypothesis to polyol pathway overactivity was hyperglycemia-driven nonenzymatic glycation of proteins leading to their dysfunction and aberrant turnover (Vlassara, Brownlee, & Cerami, 1981). This has subsequently expanded into an appreciation that extracellular advanced glycation end products can also produce adverse intracellular effects, including toxic ROS formation, via receptor-mediated signaling (Thornalley, 2002). Modification of ion channels by intermediates of glucose metabolism such as methylglyoxal or by glycosylation of ion channels has also been explored, primarily in the context of diabetes-induced changes to neuronal excitability that can produce loss of function prior to degeneration or instability and excess activity (Bierhaus et al., 2012; Orestes et al., 2013). The chapter by Todorovic "Painful Diabetic Neuropathy: Prevention or Suppression?" introduces pathways of aberrant modification of ion channels as part of the complex etiology of pain in diabetic neuropathy. Investigation of glucotoxic mechanisms of nerve degeneration and pain in diabetes over the past 10 years has greatly expanded sites for potential therapeutic intervention. While this is encouraging, the plethora of plausible mechanisms and apparently effective interventions identified in animal models does beg the inconvenient question of how so many highly selective interventions can be so successful when the majority of other pathogenic mechanisms remain operational.

An alternative approach to intervening against the apparently complex etiology of diabetic neuropathy has been to side-step the issue of hyperglycemia and its heterogeneous pathogenic mechanisms and to instead focus on repairing the nervous system. This began with the discovery of endogenous neurotrophic factors and their use to enhance in-built regenerative pathways that can repair nerve damage in diabetes (Calcutt, Jolivalt, & Fernyhough, 2008). This approach was initially presented as exogenous replacement of lost endogenous trophic support, and in some cases loss of trophic support was shown to be downstream of aldose reductase activity (Brewster, Fernyhough, Diemel, Mohiuddin, & Tomlinson, 1994; Mizisin, Calcutt, DiStefano, Acheson, & Longo, 1997). Neurotrophic factors such as nerve growth factor (NGF), ciliary neurotrophic factor, and neurotrophin-3 were extensively studied using in vitro and in vivo models (Mizisin, Vu, Shuff, & Calcutt, 2004; Tomlinson, Fernyhough, & Diemel, 1997). Unfortunately, as discussed in an earlier volume (Apfel, 2002), a flawed clinical trial with NGF failed to meet its goals and interest in this approach diminished. In parallel, the role of insulin and insulin-like growth factors as neurotrophic factors for sensory and sympathetic neurons was introduced by Douglas Ishii, with the appealing implication that hyperglycemia was not the sole pathogenic insult in diabetes (Ishii, 1995). The non-glucotoxic components of diabetic neuropathy are being increasingly recognized, and the chapter "Sensory Neurodegeneration in Diabetes: Beyond Glucotoxicity" by Zochodne reviews the latest findings demonstrating that insulin is a very effective treatment for preventing various molecular and structural defects associated with diabetic neuropathy independent of glycemic regulation. This chapter also introduces other novel glucose-independent pathways that drive nerve regeneration in diabetes, including signal transduction routes involving the HSP-27, PTEN, and Rb1 proteins. Dobrowsky in chapter "Promoting Neuronal Tolerance of Diabetic Stress: Modulating Molecular Chaperones" summarizes the state of play with regard to mobilization of the heat shock response to repair protein damage caused by diabetes and subsequently enhance nerve recovery. The novel pathways to encourage nerve repair introduced in these chapters provide promising new avenues for drug discovery that operate outside the need to manipulate glucose and thus may offer a complimentary approach to the fine-tuning of glycemic control and blockade of glucotoxic mechanisms.

REFERENCES

Akude, E., Zherebitskaya, E., Chowdhury, S. K., Smith, D. R., Dobrowsky, R. T., & Fernyhough, P. (2011). Diminished superoxide generation is associated with respiratory chain dysfunction and changes in the mitochondrial proteome of sensory neurons from diabetic rats. *Diabetes*, *60*, 288–297.

Apfel, S. C. (2002). Nerve growth factor for the treatment of diabetic neuropathy: What went wrong, what went right, and what does the future hold? *International Review of Neurobiology*, *50*, 393–413.

Bierhaus, A., Fleming, T., Stoyanov, S., Leffler, A., Babes, A., Neacsu, C., et al. (2012). Methylglyoxal modification of Nav1.8 facilitates nociceptive neuron firing and causes hyperalgesia in diabetic neuropathy. *Nature Medicine*, *18*, 926–933.

Brewster, W. J., Fernyhough, P., Diemel, L. T., Mohiuddin, L., & Tomlinson, D. R. (1994). Diabetic neuropathy, nerve growth factor and other neurotrophic factors. *Trends in Neurosciences*, *17*, 321–325.

Calcutt, N. A., Jolivalt, C. G., & Fernyhough, P. (2008). Growth factors as therapeutics for diabetic neuropathy. *Current Drug Targets*, *9*, 47–59.

Chowdhury, S. K., Dobrowsky, R. T., & Fernyhough, P. (2011). Nutrient excess and altered mitochondrial proteome and function contribute to neurodegeneration in diabetes. *Mitochondrion*, *11*, 845–854.

Chowdhury, S. K., Smith, D. R., & Fernyhough, P. (2013). The role of aberrant mitochondrial bioenergetics in diabetic neuropathy. *Neurobiology of Disease*, *51*, 56–65.

Chowdhury, S. K., Zherebitskaya, E., Smith, D. R., Akude, E., Chattopadhyay, S., Jolivalt, C. G., et al. (2010). Mitochondrial respiratory chain dysfunction in dorsal root ganglia of streptozotocin-induced diabetic rats and its correction by insulin treatment. *Diabetes*, *59*, 1082–1091.

Figueroa-Romero, C., Sadidi, M., & Feldman, E. L. (2008). Mechanisms of disease: The oxidative stress theory of diabetic neuropathy. *Reviews in Endocrine & Metabolic Disorders*, *9*, 301–314.

Fujimura, H., Lacroix, C., & Said, G. (1991). Vulnerability of nerve fibres to ischaemia. A quantitative light and electron microscope study. *Brain*, *114*(Pt. 4), 1929–1942.

Halim, N. D., McFate, T., Mohyeldin, A., Okagaki, P., Korotchkina, L. G., Patel, M. S., et al. (2010). Phosphorylation status of pyruvate dehydrogenase distinguishes metabolic phenotypes of cultured rat brain astrocytes and neurons. *Glia*, *58*, 1168–1176.

Ishii, D. N. (1995). Implication of insulin-like growth factors in the pathogenesis of diabetic neuropathy. *Brain Research. Brain Research Reviews*, *20*, 47–67.

Jiang, Y., Calcutt, N. A., Ramos, K. M., & Mizisin, A. P. (2006). Novel sites of aldose reductase immunolocalization in normal and streptozotocin-diabetic rats. *Journal of the Peripheral Nervous System*, *11*, 274–285.

Mizisin, A. P., Calcutt, N. A., DiStefano, P. S., Acheson, A., & Longo, F. M. (1997). Aldose reductase inhibition increases CNTF-like bioactivity and protein in sciatic nerves from galactose-fed and normal rats. *Diabetes*, *46*, 647–652.

Mizisin, A. P., Vu, Y., Shuff, M., & Calcutt, N. A. (2004). Ciliary neurotrophic factor improves nerve conduction and ameliorates regeneration deficits in diabetic rats. *Diabetes*, *53*, 1807–1812.

Nishikawa, T., Edelstein, D., & Brownlee, M. (2000). The missing link: A single unifying mechanism for diabetic complications. *Kidney International. Supplement*, *77*, S26–S30.

Nishikawa, T., Edelstein, D., Du, X. L., Yamagishi, S., Matsumura, T., Kaneda, Y., et al. (2000b). Normalizing mitochondrial superoxide production blocks three pathways of hyperglycaemic damage. *Nature*, *404*, 787–790.

Oates, P. J. (2008). Aldose reductase, still a compelling target for diabetic neuropathy. *Current Drug Targets*, *9*, 14–36.

Orestes, P., Osuru, H. P., McIntire, W. E., Jacus, M. O., Salajegheh, R., Jagodic, M. M., et al. (2013). Reversal of neuropathic pain in diabetes by targeting glycosylation of Ca(V) 3.2T-type calcium channels. *Diabetes*, *62*, 3828–3838.

Thornalley, P. J. (2002). Glycation in diabetic neuropathy: Characteristics, consequences, causes, and therapeutic options. *International Review of Neurobiology*, *50*, 37–57.

Tomlinson, D. R., Fernyhough, P., & Diemel, L. T. (1997). Role of neurotrophins in diabetic neuropathy and treatment with nerve growth factors. *Diabetes*, *46*(Suppl. 2), S43–S49.

Tomlinson, D. R., & Gardiner, N. J. (2008). Glucose neurotoxicity. *Nature Reviews. Neuroscience*, *9*, 36–45.

Vincent, A. M., Hayes, J. M., McLean, L. L., Vivekanandan-Giri, A., Pennathur, S., & Feldman, E. L. (2009). Dyslipidemia-induced neuropathy in mice: The role of oxLDL/LOX-1. *Diabetes*, *58*, 2376–2385.

Vincent, A. M., Hinder, L. M., Pop-Busui, R., & Feldman, E. L. (2009). Hyperlipidemia: A new therapeutic target for diabetic neuropathy. *Journal of the Peripheral Nervous System*, *14*, 257–267.

Vlassara, H., Brownlee, M., & Cerami, A. (1981). Nonenzymatic glycosylation of peripheral nerve protein in diabetes mellitus. *Proceedings of the National Academy of Sciences of the United States of America*, *78*, 5190–5192.

Zhang, L., Zhao, H., Blagg, B. S., & Dobrowsky, R. T. (2012). C-terminal heat shock protein 90 inhibitor decreases hyperglycemia-induced oxidative stress and improves mitochondrial bioenergetics in sensory neurons. *Journal of Proteome Research*, *11*, 2581–2593.

Ziegler, D., Low, P. A., Litchy, W. J., Boulton, A. J., Vinik, A. I., Freeman, R., et al. (2011). Efficacy and safety of antioxidant treatment with alpha-lipoic acid over 4 years in diabetic polyneuropathy: The NATHAN 1 trial. *Diabetes Care*, *34*, 2054–2060.

CHAPTER EIGHT

Glucotoxic Mechanisms and Related Therapeutic Approaches

S. Yagihashi[1,2]
Hirosaki University Graduate School of Medicine, Hirosaki, Japan
[2]Corresponding author: e-mail address: yagihasi@hirosaki-u.ac.jp

Contents

Abstract

Neuropathy is the earliest and commonest complication of diabetes. With increasing duration of diabetes, frequency and severity of neuropathy are worsened. Long-term

[1] Present address: The Nukada Institute of Medical and Biological Research, 4-16 Inage-machi, Inage-ku, Chiba, Japan. E-mail: soroku@nukada.ac.jp.

International Review of Neurobiology, Volume 127
ISSN 0074-7742
http://dx.doi.org/10.1016/bs.irn.2016.03.006

hyperglycemia is therefore implicated in the development of this disorder. Nerve tissues require glucose energy to function and survive. Upon excessive glucose entry into the peripheral nerve, the glycolytic pathway and collateral glucose-utilizing pathways are overactivated and initiate adverse effects on nerve tissues. During hyperglycemia, flux through the polyol pathway, formation of advanced glycation end-products, production of free radicals, flux into the glucosamine pathway, and protein kinase C activity are all enhanced to negatively influence nerve function and structure. Suppression of these aberrant metabolic pathways has succeeded in prevention and inhibition of the development of neuropathy in animal models with diabetes. Satisfactory results were not attained, however, in patients with diabetes and further clinical trials are required. In this review, the author summarizes the hitherto proposed theories on the pathogenesis of diabetic neuropathy related to glucose metabolism and future prospects for the effective treatment of neuropathy.

1. INTRODUCTION

There is a long history of research to clarify the pathogenetic mechanism of diabetic polyneuropathy (DPN), but it remains unsettled. Since the prevalence of DPN is primarily dependent on the duration of diabetes and the degree of blood glucose control, long-term metabolic aberration is considered to be a major cause of DPN (Pirart, 1978). A large prospective study disclosed that hyperglycemia, duration of diabetes, hypertension, hyperlipidemia, and smoking are significant risk factors for the development of DPN in patients with type 1 diabetes (Tesfaye et al., 2005). Indeed, the 10-year follow-up of patients with type 1 diabetes by the Diabetic Complication and Control Trial (DCCT) demonstrated that meticulous blood glucose control by intensive insulin therapy suppressed the incidence of DPN (DCCT Research Group, 1993). The effects of tight glycemic control persisted for a further 8 years after the termination of the trial as legacy effects termed glucose or metabolic memory (Martin et al., 2006). Although the role of hyperglycemia in driving the progression of DPN in type 2 diabetes is less clear (UKPDS Group, 1998), continuous lowering of glycated hemoglobin below 7% for 6 years was found to suppress the deterioration of DPN in patients with type 2 diabetes (Ohkubo et al., 1995). Despite the ample epidemiological data, there remains no clear-cut explanation why long-term hyperglycemia leads to DPN.

It is known that the most distal portions of somatic nerves are preferentially affected in diabetes. Both anatomical and biochemical characteristics of the peripheral nervous system seem to contribute to the distal and sensory predominant nerve lesions in diabetes (Dyck et al., 1984; Dyck, Lais, Karnes, O'Brien, & Rizza, 1986). Structurally, the sensory neurons extend extremely long axons from their cell bodies in the dorsal root ganglia (DRG). The relative imbalance in this architecture between the soma and axon can result in insufficient supply of nerve nutrients and energy from the cell body to the distal processes. The anatomic organization of the vascular supply to the peripheral nerve may also impose negative impacts on the peripheral nerve because of the relative paucity of endoneurial blood vessels and their lack of autoregulation (Kihara, Schmelzer, et al., 1991; Smith, Kobrine, & Rizzoli, 1977). Such a vascular system likely renders the peripheral nerve ischemic, resulting in severe pathology when ischemia is further exaggerated by vascular occlusion (Nukada, McMorran, Baba, Ogasawara, & Yagihashi, 2011), though ischemic diabetic nerve is temporarily resistant to nerve conduction failure (Low, Schmelzer, & Ward, 1986). Endoneurial microangiopathic changes characteristic of diabetes may further augment nerve damage due to increased permeability and ischemia (Dyck & Giannini, 1996; Malik et al., 2005; Thrainsdottir et al., 2003). Features such as the strong expression of aldose reductase (AR) in peripheral nerve and the abundance of interstitial collagen implicate the polyol pathway and glycation as etiological factors in DPN.

Glucose metabolism in the peripheral nerve is not completely understood. Peripheral nerves express the GLUT-1 glucose transporter at the perineurial and endothelial blood:nerve barriers (Stark, Carlstedt, Cullheim, & Risling, 2000), and glucose uptake into nerve is therefore insulin independent so that endoneurial glucose concentrations will follow plasma glucose concentrations. Under normoglycemic conditions, the majority of glucose entering the cytoplasm is used by mitochondria to produce adenosine triphosphate (ATP) through the tricarboxylic acid (TCA) cycle, downstream from the glycolytic pathway. In neurons, synthesized proteins, cytoplasmic organelles, cargoes of microtubules and neurofilaments, as well as neurotransmitters, are all transported in an energy-requiring process to the distal nerve endings. It has been proposed that if the energy production in the axon is deficient then the distal portion is first affected by the insufficient supply of these materials (Scott, Clark, & Zochodne, 1999; Yagihashi, Kamijo, & Watanabe, 1990). Additionally, hyperglycemia makes glucose available for other uses and perturbed glycolytic pathways may lead to impaired nerve function and disruption of nerve structure. In this review, I will list the collateral glycolytic pathways operating downstream of hyperglycemia and discuss their possible implication in the onset and development of DPN.

2. GLUCOSE-METABOLIZING PATHWAYS RELEVANT TO DPN

There are a number of metabolic pathways that operate collaterally to glycolysis that may contribute to development of DPN as a downstream consequence of hyperglycemia (Fig. 1).

(1) The *polyol pathway* uses AR to produce sorbitol, which is then converted to fructose by sorbitol dehydrogenase (SDH) (Gabbay, 1975; Kinoshita, 1990). AR has a relatively low affinity for D-glucose

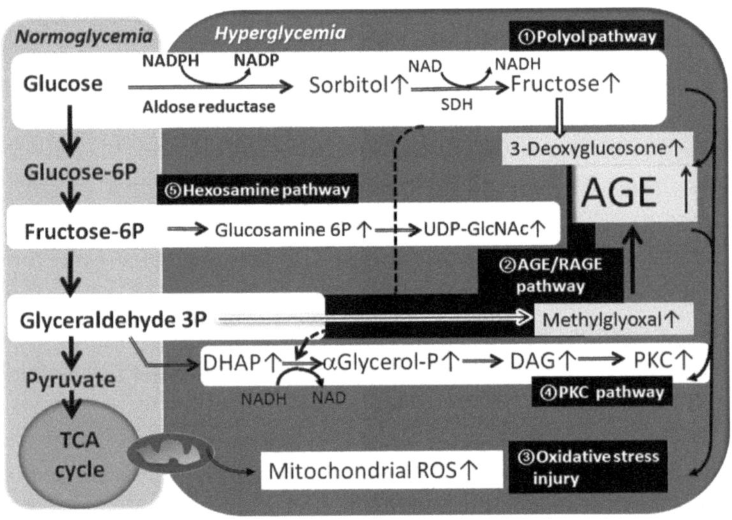

Fig. 1 Glycolytic pathways in hyperglycemia proposed for the cause of diabetic complications. In normoglycemia, intracellular glucose undergoes glycolysis and enters the TCA cycle in mitochondria to produce ATP. Hyperglycemia induces an increased flux through the polyol pathway in endothelial and Schwann cells, nonenzymatic glycation with AGE formation, superoxide production in endothelial cell mitochondria, increased activity of protein kinase C, and enhancement of the hexosamine pathway. These complicated glycolytic pathways are proposed to jointly lead to functional and structural abnormalities encountered in diabetic complications. It should be of note, however, that this theory is mainly built from data on vascular complications in diabetes, and not uniformly applied to neuropathy. For example, reactive oxygen species production in mitochondria is reduced in diabetic neurons (see text).

so this metabolic process is enhanced only when glucose concentrations are high in the cytoplasm.

(2) Hyperglycemia leads to activation of the *glycation or advanced glycation end-products (AGEs) pathway*. Excessive glucose attaches to the amino base of proteins in a nonenzymatic manner, resulting in the production of AGEs (Brownlee, 1992; Yamagishi, Nakamura, & Imaizumi, 2005). AGE then binds with the receptor for AGE (RAGE), giving rise to negative influences on cells and tissues.

(3) Activation of *oxidative stress pathways* that produce reactive oxygen species (ROS), in part through enhanced mitochondrial respiration in endothelial cells, is considered to be a major pathway for vascular and neural tissue damage in diabetes (Baynes & Thorpe, 1999; Brownlee, 1992; Leinninger, Edwards, Lipshaw, & Feldman, 2006).

(4) Activity of the *PKC pathway* is augmented by hyperglycemia. With increased production of diacylglycerol, the protein kinase C (PKC)-β isoform was shown to be activated and to impair vascular function with presumed downstream consequences for neuronal function (Geraldes & King, 2010; King & Brownlee, 1996).

(5) Glycosylation: recent studies have introduced the *glucosamine pathway* as a potentially causative factor in DPN (Hardivillé & Hart, 2014; James et al., 2002). With excessive entry of glucose into the cell, a glucose-attached amine, termed glucosamine (GlcN), accumulates. Since the glucose modification of amine groups is mediated by the enzyme glutamine-fructose-6-phosphate amidotransferase (GFAT) (synonym; glucosamine-6-phosphate synthase), this process is distinct from the process of nonenzymatic glycation. Other glycosylation events and glucose-derived products such as methylglyoxal (MG) have also been implicated in aspects of diabetic neuropathy.

I will now deal in more detail with the processes outlined in each pathway as listed earlier.

3. POLYOL (AR OR SORBITOL–FRUCTOSE) PATHWAY

3.1 Metabolic Sequelae After Increased Flux Through the Polyol Pathway

A complete mechanism of how the polyol pathway is involved in development of diabetic neuropathy remains elusive. Sorbitol is an important osmoregulator and does not readily permeate the cell membrane. Accumulation of sorbitol therefore causes increased osmotic pressure within the cell

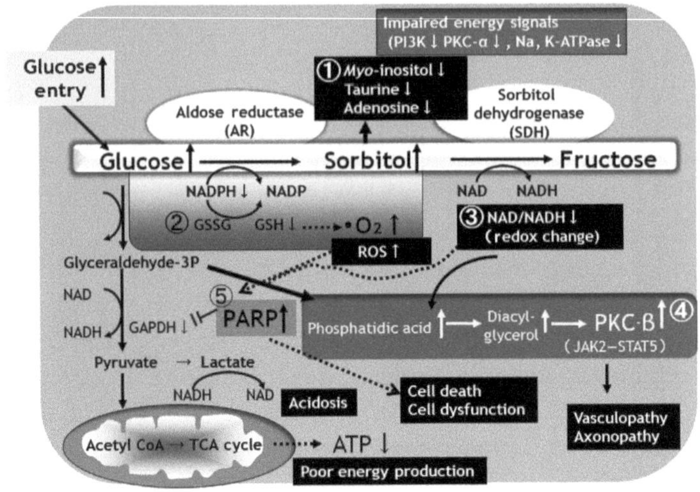

Fig. 2 Implication of polyol pathway in the pathogenesis of diabetic neuropathy. With excessive flux of polyol pathway, accumulation of sorbitol as osmolytes elicits increased osmotic pressure within the cytoplasm, resulting in depletion of other osmolytes such as *myo*-inositol, taurine, and adenosine. Since *myo*-inositol and taurine are important substrates for energy production of phosphatidylinositol, downstream metabolic aberration results in insufficient energy production with reduced activity of protein kinase C-α and Na,K-ATPase activity (1). Conversion of glucose to sorbitol by aldose reductase requires consumption of coenzyme of NADPH which is also utilized for the production of reduced glutathione. Therefore, depletion of reduced glutathione results in excessive free radical formation (2). On the other hand, the conversion from sorbitol to fructose induces redox changes of NADH/NAD (3), which in turn increases phosphatidic acid and diacylglycerol, resulting in activation of protein kinase C-β (4). Oxidative stress also activates poly(ADP)-ribose polymerase (PARP), which causes nuclear DNA damage, caspase activation, and cell death. It also causes inhibition of GAPDH to increase the level of glyceraldehyde, leading to AGE formation and PKC-β activation (5).

(Fig. 2). The original osmotic theory proposed that increased polyol pathway flux during hyperglycemia caused intracellular hyperosmolarity by accumulation of sorbitol, resulting in the expansion of cells and ultimately cell lysis (Gabbay, 1975; Kinoshita, 1990). However, although increased sorbitol concentrations were demonstrated in nerves of persons with diabetes, there was no evidence of nerve edema, swollen cells, or correlation with nerve fiber pathology (Dyck et al., 1980, 1988).

Greene and his associates proposed the poor energy utilization theory as a surrogate of the osmotic theory (Greene, Lattimer, & Sima, 1987; Greene, Sima, Stevens, Feldman, & Lattimer, 1992). With accumulation of sorbitol, other osmolytes including *myo*-inositol, taurine, and adenosine are depleted in the cytoplasm. In turn, *myo*-inositol deficiency results in depletion of phosphatidylinositol, poor production of ATP, and reduced activities of Na,K-ATPase and protein kinase C (PKC)-α. This theory was criticized due to the fact there was no clear *myo*-inositol depletion in nerves of persons with diabetes (Dyck et al., 1980, 1988). However, total nerve *myo*-inositol levels may not necessarily be low in the setting of increased polyol flux with critical decreases occurring in biologically active subpools (Simmons, Kern, Winegrad, & Martin, 1986; Winegrad, 1987). The precise relationship between the polyol flux and energy utilization in diabetes is still open to question.

Subsequent studies have suggested that adverse effects of increased polyol pathway flux are mediated by increased oxygen radicals and poor nitric oxide (NO) production (Greene, Stevens, Obrosova, & Feldman, 1999; Obrosova, 2009; Pop-Busui, Sima, & Stevens, 2006). Conversion of glucose to sorbitol depletes the coenzyme nicotinamide adenine dinucleotide phosphate (NADPH). NADPH is used for the production of reduced glutathione, a major scavenger of free radicals, and is also required for the production of NO during the conversion of arginine to citrulline. Excessive polyol pathway flux and subsequent exhaustion of NADPH can therefore cause increased free radicals and poor production of NO (Stevens et al., 1994) so that tissues suffer from free radical injury and impaired vascular function due to NO depletion.

The second half of polyol pathway, involving conversion of sorbitol to fructose by SDH, also elicits redox changes when driven to excess, with an exaggerated ratio of NADH/NAD leading to lower oxygen levels, similar to the condition of hypoxia. These conditions have been proposed as a pseudohypoxia that accounts for early neural and vascular dysfunction in diabetes (Ido et al., 2010; Williamson et al., 1993). However, studies with inhibitors of SDH that block the second part of the polyol pathway while allowing flux through AR to proceed have not shown consistent efficacy against neuropathy (Schmidt et al., 2001). Conversely, feeding animals galactose, an alternative hexose substrate for AR which converts it to galactitol (dulcitol), produces a significant slowing of nerve conduction and nerve fiber atrophy (Calcutt, Tomlinson, & Biswas, 1990; Nukada, Dyck, Low, Lais, & Sparks, 1986) that parallels aspects of neuropathy reported in streptozotocin (STZ)-induced diabetes. However, some downstream aspects of peripheral nerve metabolism in galactose-fed animals are opposite to those reported in STZ-diabetic rats. For example, nerve from galactose-fed rats shows endoneurial oedema and increased Na,K-ATPase activity (Lambourne, Tomlinson, Brown, & Willars, 1987), whereas Na,K-ATPase activity is widely reported as being reduced in STZ-diabetic rats. Exaggerated flux through the entire polyol pathway may be necessary to precipitate the full repertoire of disorders associated with diabetic neuropathy.

3.2 Studies in Transgenic and Knockout Mice

We have constructed transgenic mice overexpressing human AR (hAR-Tg) in the search for a cause of DPN. hAR-Tg mice develop severe neuropathy when they were fed galactose (Yagihashi et al., 1996). Thus, without hyperglycemia or insulin deficiency, increased flux through the first part of the polyol pathway caused peripheral nerve dysfunction and myelinated fiber changes that are similar to findings in diabetic animals. The study was also extended to STZ-induced diabetes, and there was more severe nerve conduction slowing and reduced Na,K-ATPase activity in diabetic hAR-Tg mice with an enhanced accumulation of sorbitol and fructose, compared with nontransgenic diabetic mice, despite comparable levels of hyperglycemia (Yagihashi et al., 2001). The functional decay in nerve was accompanied by structural changes and alterations of neuropeptide expression in DRG (Uehara, Yamagishi, Otsuki, Chin, & Yagihashi, 2004). Concurrently, diabetic hAR-Tg mice showed a reduction of PKC activity in the endoneurium with decreased membranous expression of PKC-α and a relative increase in the expression of the PKC-β isoform (Yamagishi, Uehara, Otsuki & Yagihashi, 2003). These neuropathic changes in diabetic hAR-Tg mice were markedly improved by the AR inhibitor (ARI), fidarestat.

The pertinence of AR activity to diabetic neuropathy in rodents was further established by studies in AR-deficient mice made diabetic by STZ (Ho et al., 2006). In this model, glutathione levels were preserved and peripheral nerve function and structure remained normal, despite protracted hyperglycemia. These studies convincingly confirm that AR contributes to the development of DPN. In contrast, mice lacking the second enzyme of the polyol pathway, SDH, continued to develop nerve conduction slowing during STZ-induced diabetes that matched that of STZ-diabetic wild-type mice, despite exaggerated sorbitol accumulation (Ng et al., 1998). Together, these data place focus on exaggerated flux through AR as being the major pathogenic component of polyol pathway overactivity that contributes to the nerve conduction deficit during hyperglycemia, rather than simple sorbitol accumulation.

3.3 Effects of AR Inhibition

Early studies showed that inhibition of polyol pathway flux by an ARI succeeded in protection of peripheral nerve function, structure, and axonal transport in rats with experimental diabetes (Tomlinson, Moriarty, & Mayer, 1984; Yagihashi, Kamijo, Ido, & Mirrlees, 1990). Subsequent studies demonstrated that oxidative and nitrosative stress contributed tissue damage in STZ-diabetic rats (Obrosova, Drel, et al., 2005) and animals fed with a high-fat diet (Obrosova et al., 2007) and that an ARI was protective, suggesting that these potentially toxic stressors were downstream of polyol pathway flux. AR inhibition also suppressed oxidative stress-induced poly(ADP-ribose) polymerase (PARP) activation, which was closely associated with a neuropathic phenotype in diabetic animals (Obrosova, Pacher, et al., 2005). Interestingly, a potential pathogenic contribution of the SDH component of the polyol pathway has recently been reemphasized by studies of the relationship between enhanced polyol pathway flux and Sirtuin 2 expression (Schartner, Chowdhury, Smith, & Fernyhough, 2015). Sirtuin is an NAD-dependent protein deacetylase that participates in diverse biological processes such as cell cycle control, genomic integrity, cell differentiation, metabolic networks, and aging. Sirtuin 2 expression is reduced in DRG of diabetic animals. Using DRG neuron cultures derived from adult rodents and grown under a high glucose concentration, it was demonstrated that blockade of the polyol pathway using an SDH inhibitor restored Sirtuin 2 expression, optimized neurite outgrowth, and augmented mitochondrial function.

Hyperglycemia-induced exaggerated flux through the polyol pathway also induces structural changes in autonomic nerves as indicated by axonal dystrophy and a reduction of synapses (Schmidt, 2002). AR inhibition significantly reduced the frequency of dystrophies, whereas treatment with an SDH inhibitor had no protective effect, indicating specific involvement of the first segment of the polyol pathway in structural damage to autonomic nerves (Schmidt et al., 2001; Schmidt, Plurad, Sherman, Williamson, & Tilton, 1989). This finding is contradictory to the reported beneficial effects of an SDH inhibitor on vascular dysfunction in diabetes (Ido et al., 2010; Williamson et al., 1993). The effects of polyol inhibition therefore appear to be different between neural and vascular tissues and could be accounted for by a difference in the tissue distribution of the two enzymes. We, and others, have reported that AR was highly expressed in Schwann cells of somatic nerves (Kasajima, Yamagishi, Sugai, Yagihashi, & Yagihashi, 2001) and in axons of autonomic nerves (Jiang, Calcutt, Ramos, &

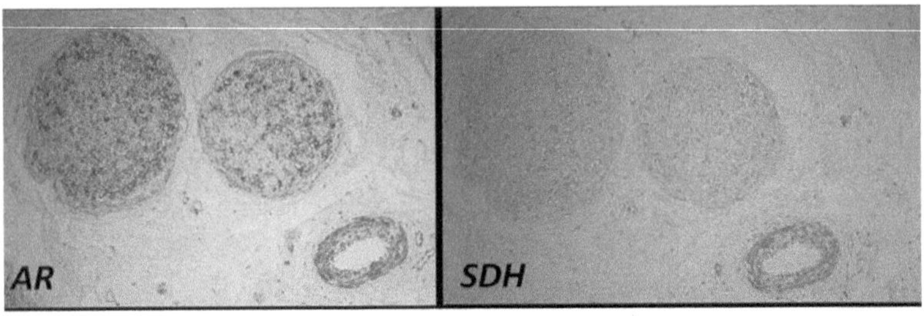

Fig. 3 Topographic difference in the localization of the enzymes that catalyze polyol pathway. Aldose reductase locates in the endoneurium, mainly Schwann cells and the wall of epineurial artery. On the other hand, sorbitol dehydrogenase (SDH) expresses mainly in the wall of epineurial artery, but not much in the endoneurium. Such difference may contribute to the distinct downstream of metabolic signals such as protein kinase C and redox changes. *Cited from Kasajima, H., Yamagishi, S., Sugai, S., Yagihashi, N., & Yagihashi, S. (2001). Enhanced in situ expression of aldose reductase in peripheral nerve and renal glomeruli in diabetic patients.* Virchows Arch, *439, 46–54.*

Mizisin, 2006), whereas SDH was more highly expressed in vascular tissues (Fig. 3). Such anatomical differences in the localization of polyol pathway enzymes may also be involved in the differential effects of diabetes on PKC activity between nerve and vascular tissues, where it is increased in vascular tissues and decreased in neural tissues (Yamagishi et al., 2003).

3.4 Clinical Application of ARIs

Based on preclinical studies, many ARIs have been developed. In early trials, patients treated with ARI (sorbinil or zenarestat) for a year showed increased nerve fiber density in the sural nerve biopsy specimens compared to the untreated group (Greene, Arezzo, & Brown, 1999; Sima et al., 1988). Unfortunately, safety concerns of sorbinil and zenarestat did not allow the approval of the compounds for clinical applications. The majority of other ARI clinical trials have been unsuccessful in meeting efficacy requirements for regulatory approval. Factors contributing to this lack of success include dose efficacy, short trial durations, and inappropriate patient selection. Currently, epalrestat (ONO2235) is the only approved ARI, and it is licensed only in Japan. Epalrestat was approved after a 3-month double-blinded trial, which showed improvement of symptoms and nerve conduction (Goto, Hotta, Shigeta, Sakamoto, & Kikkawa, 1995). Further clinical studies confirmed that epalrestat treatment significantly suppressed progressive nerve conduction slowing over a 3-year period (Hotta et al., 2006, 2008) and appearance of negative symptoms such as sensory loss (Baba, Kimura, Suda, Yagihashi, & Aomori Diabetic Study (ADNS) Group, 2006). In a long-term follow-up, positive symptoms such as pain or paresthesia were not influenced by ARI (Baba et al., 2006). The ARI effects on nerve conduction were most marked in patients with early neuropathy and exhibiting modestly elevated levels of glycated hemoglobin (Hotta et al., 2006). The ARI ranirestat has also been shown to produce modest, but statistically significant, improvements in peroneal motor nerve conduction velocity, again in patients with relatively mild neuropathy (Bril & Buchanan, 2006). In a more recent study, however, comparison of its effects with a placebo group of well-controlled diabetic subjects faced challenges to obtain significant differences (Polydefkis et al., 2015).

Although these studies confirmed the critical role of AR in contributing to the development of DPN, it seems to be clear from the clinical experience of ARI trials that the polyol pathway is unlikely to be the only causative factor in DPN.

3.5 Polyol Pathway in Ischemia/Reperfusion Injury

Emerging evidence suggests that AR is strongly implicated in ischemia/reperfusion injury in nondiabetic conditions (Ramasamy & Goldberg, 2010). In experimental studies, ARI alleviated the pathological lesions observed during infarction of the heart (Iwata, Matsuno, Nishinaka, Persson, & Yabe-Nishimura, 2006; Ramasamy & Goldberg, 2010), the brain (Yeung et al., 2010), and the retina (Cheung, Lo, So, Chung, & Chung, 2007). Our study also found that acute kidney injury caused by renal ischemia (Yagihashi et al., 2010) or experimental septicemia (Takahashi et al., 2012) was rescued by ARI. Because diabetic nerves are susceptible to ischemia/reperfusion injury (Nukada et al., 2011), it is possible that ischemia/reperfusion is also involved in the progression or exacerbation of DPN and that ARI might be effective in treating this disorder.

4. NONENZYMATIC GLYCATION AND AGEs

4.1 Glycation of Proteins in Diabetes

Increased nonenzymatic glycation of proteins is considered to be a major contributor to a variety of diabetic complications including micro- and macroangiopathy (Brownlee, 1992; Yamagishi et al., 2005).

In a hyperglycemic environment, glucose is attached nonenzymatically to amino residues of proteins forming glucose adducts as initial glycation metabolites that progress from Schiff base to Amadori products (Thornalley, 2002). These products are unstable and likely to transform to MG or 3-deoxyglucosone (3DG) as intermediate glycation products, both of which are toxic to neural tissues (Kikuchi et al., 1999; Sekido et al., 2004) (Fig. 1). Increased levels of MG, an intermediate glycation product also arising from intermediates of glycolysis, have also been associated with pain and hyperalgesia in diabetic subjects (Bierhaus et al., 2012). MG has been reported to bind to the Nav1.8 sodium channel of nociceptive neurons. However, others have emphasized that relative activity of glyoxalase I, the degrading enzyme of MG, may be the critical parameter (Jack, Ryals, & Wright, 2011). Intermediate glycation products may also crosslink together, forming large molecules of AGEs. AGE deposition has been found in nerves of diabetic patients and animal models of diabetes. Every component of peripheral nerve tissue is affected including axoplasm, Schwann cells, endoneurial vessels, and stromal collagen (Sugimoto, Nishizawa, Horiuchi, & Yagihashi, 1997). Glycation of tubulin or neurofilament is suggested to trigger disturbed axoplasmic flow of these proteins leading to axonal atrophy and degeneration (McLean, Pekiner, Cullum, & Casson, 1992; Ryle, Leow, & Donaghy, 1997; Williams, Howarth, Devenny, & Bitensky, 1982). Glycated laminin, fibronectin, and collagen are also invoked as contributing to the impairment of nerve regeneration reported in diabetic animals (Duran-Jimenez et al., 2009). Thus, glycation alters the basic structure of tissue components, resulting in pathologic alterations of the peripheral nerve, with the intensity of AGE deposition detected by carboxymethyllysine immunoreactions correlating well with reduced myelinated fiber density (Sugimoto et al., 1997; Sugimoto, Yasujima, & Yagihashi, 2008).

4.2 Glycation and AGE in DPN

In diabetes, serum concentrations of AGE are elevated, and clearance is disturbed in cases with renal insufficiency. In vascular endothelial cells, AGE binds with its RAGE, resulting in activation of a downstream cascade including NADPH oxidase with subsequent release of free radicals and activation of the transcription factor, nuclear factor-κB (NF-κB) (Haslbeck et al., 2005) (Fig. 4). RAGE is expressed by many cells in

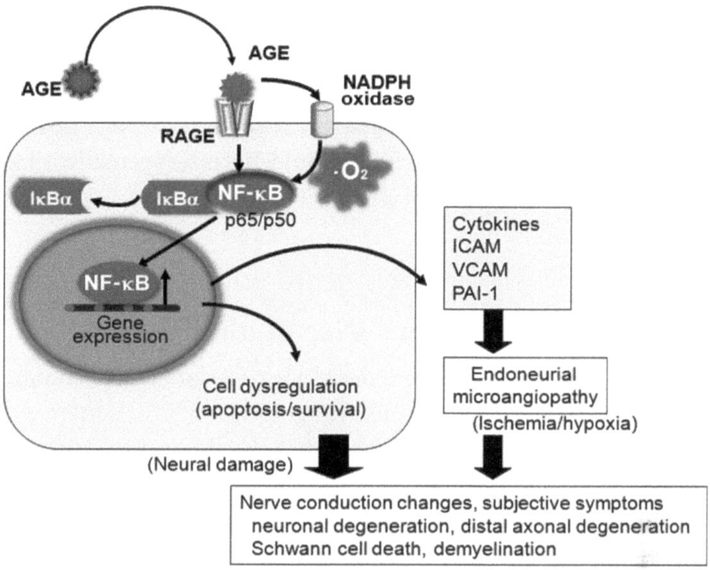

Fig. 4 Biological reactions elicited by the binding of AGE with RAGE in endothelial cells and neural tissues. AGE exerts activation of NADPH oxidase to release oxygen radicals after binding with RAGE. Concurrently, nuclear factor-κB (NF-κB) is transferred into the nuclei to promote gene activation. This results in cellular dysfunction and activation of cell death signals.

peripheral nerve including vascular endothelial cells, Schwann cells, and neurons (Wada & Yagihashi, 2005). Therefore, high AGE in situ may elicit adverse effects on peripheral nerve tissues by signaling through RAGE. Indeed, embryonic sensory neurons underwent functional degeneration and apoptosis when exposed to AGE via production of oxidative stress (Vincent, McLean, Backus, & Feldman, 2005), while endoneurial endothelial cells exposed to exogenous AGE show high expression of NF-κBp65 together with swollen and vacuolar changes at the ultrastructural levels (Nishizawa et al., 2010). AGE-induced toxicity has also been reported in vivo, as animals given exogenous AGE developed vascular dysfunction and a neuropathy resembling that caused by diabetes, with reduced nerve Na,K-ATPase activity, fiber size, and conduction velocity (Nishizawa et al., 2010).

The association of AGE formation and neuropathy has been tested using agents that inhibit AGE formation. Inhibition of AGE accumulation by aminoguanidine or OPB9135 was found to suppress the development of neuropathy, confirming the pathogenic role of AGE in DPN (Wada et al., 2001; Yagihashi, Kamijo, Baba, Yagihashi, & Nagai, 1992). The amelioration of neuropathy by aminoguanidine was also associated with improvement of nerve blood flow (Kihara, Weerasuriya, & Low, 1991). However, effects of aminoguanidine should be interpreted with caution, since it is also a selective inhibitor of inducible nitric oxide synthase (Corbett & McDaniel, 1996).

4.3 Transgenic and Knockout Mice Studies

Animals transgenic for RAGE were developed to clarify the biological consequences of AGE/RAGE reactions (Inagi et al., 2006; Yamamoto et al., 2001). RAGE overexpression by endothelial cells induced characteristic pathology in the kidney in diabetes (Yamamoto et al., 2001). Overexpression of RAGE in vascular endothelial cells also produced a more severe neuropathy phenotype after induction of diabetes by STZ when compared to nontransgenic diabetic mice (Wada & Yagihashi, 2005). In contrast, diabetic mice lacking the RAGE gene were protected against the development of neuropathy and showed enhanced regeneration after nerve injury (Juranek et al., 2013). These findings support the role of AGE/RAGE in the functional and structural development of DPN.

4.4 Clinical Application of Antiglycation Agents

The potential pathogenic role of AGE may extend to subjective symptoms of diabetic neuropathy as sensory loss in diabetic subjects was closely related to the increased concentrations of AGE and its downstream signals of NF-κB (Bierhaus et al., 2004). Aminoguanidine was tested as an antiglycation agent in a clinical trial in which the primary target was nephropathy. It was initially reported to be effective at inhibiting nephropathy, as indicated by a reduction in proteinuria, but no significant improvement was found at the study end (Bolton et al., 2004). Information for neuropathy is not available for aminoguanidine. Benfotiamine, a vitamin B1 derivative with AGE clearing and antioxidant properties, has shown some efficacy on neuropathy symptom scores, most prominently on pain scores, in patients with diabetes following studies of 3–6 weeks duration (Haupt, Ledermann, & Köpcke, 2005; Stracke, Gaus, Achenbach, Federlin, & Bretzel, 2008). The combination of benfotiamine with vitamins B6 and B12 also improved nerve conduction in diabetic subjects following 9 months of treatment (Stracke, Lindemann, &

Federlin, 1996). However, since these studies used a rather small number of subjects, more definitive trials are required.

Other agents that intervene against AGE accumulation or the biological reactions downstream of AGE/RAGE interaction have also been developed, including pyridoxamine/vitamin B6 (Metz, Alderson, Thorpe, & Baynes, 2003), AGE breakers (Vasan, Foiles, & Founds, 2003), and soluble RAGE (Nagai, Shirakawa, Ohno, Moroishi, & Nagai, 2014), although at present there is no information regarding the effects of these agents against diabetic neuropathy in clinical trials.

5. OXIDATIVE STRESS

5.1 Production of Oxidative Stress by Hyperglycemia

The generation of free radicals due to increased flux of glucose through glycolysis has been proposed as a major contributor to diabetic complications, including DPN. This hypothesis was originally raised by the short-term experiments mostly on in vitro endothelial cells and is still not confirmed in neural tissues. During hyperglycemia, increased mitochondrial oxidation of NADH and FADH2 leads to excessive formation of superoxide ions, a form of ROS (Brownlee, 2005) (Fig. 5). Reduced expression of manganese–superoxide dismutase also contributes to production of ROS (Boyle, Newsom, Janssen, Lappas, & Friedman, 2013; Santos, Tewari, Goldberg, & Kowluru, 2011). Thus, in endothelial cells superoxide production, possibly via enhanced mitochondrial electron transport, could induce DNA injury, which in turn drives increased expression of poly(ADP)-ribose polymerase (PARP) to repair DNA damage. Activated PARP inhibits the action of glyceraldehyde-3-phosphate dehydrogenase (GAPDH) (Obrosova, Drel, et al., 2005). Consequently, upstream glycolytic metabolites before GAPDH will be increased in the cytoplasm, which in turn induces enhancement of four glycolytic pathways during hyperglycemia (Baynes & Thorpe, 1999; Brownlee, 1992). However, recent studies do not support the oxidative stress having a mitochondrial origin in sensory neurons in experimental diabetes (Akude et al., 2011; Zherebitskaya, Akude, Smith, & Fernyhough, 2009). These authors found reduced electron transport in mitochondria which was associated with reduced superoxide production, which makes sense since less electron flux equates with less electron leakage. The Dobrowsky group also detected reductions in the rates of electron transport in adult sensory neurons exposed to hyperglycemia or derived from diabetic mice (Urban et al., 2012; Zhang, Zhao, Blagg, & Dobrowsky, 2012). Studies

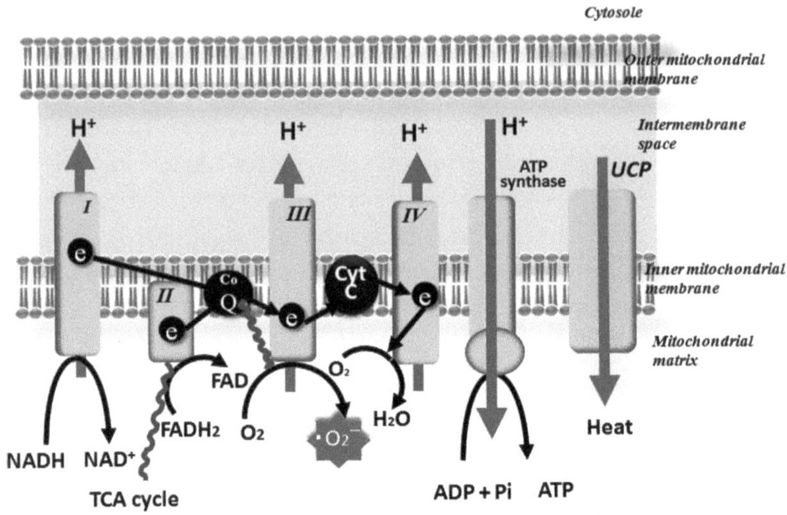

Fig. 5 Mitochondrial electron transport and superoxide production. For the production of ATP NADH is first oxidized to NAD with release of electrons at the site of complex I of inner mitochondrial membrane. At the site of complex II, conversion of succinate to fumarate is mediated with FADH2 conversion to FAD. Then, electrons are directly transferred at the complex III from coenzyme Q (Co-Q) to cytochrome C (Cyt-C). If this step is augmented by hyperglycemia in endothelial cells or blocked by drugs such as actinomycin D, superoxide radical formation is enhanced. Recent data suggest that this theory is only applicable to vascular tissues, but not neurons (see text). *Modified from Brownlee, M. (2005). The pathobiology of diabetic complications: A unifying mechanism.* Diabetes, 54, *1615–1625.*

in cultured Schwann cells showed no augmentation of superoxide under high glucose concentration (Zhang et al., 2010). These studies in neuronal cells suggest a diversity in the role of mitochondria and contribution to oxidative stress in various tissues.

5.2 Oxidative Stress and Diabetic Neuropathy

There are numerous data that have shown oxidative stress-induced tissue injury in the peripheral nerve in experimental diabetes, though its origin is still controversial (Baynes & Thorpe, 1999; Brownlee, 1992; Rabbani & Thornalley, 2015; Thornalley, 2002; Yagihashi, Mizukami, & Sugimoto, 2011). Based on this background, attempts have been made to inhibit DPN with antioxidants. In particular, α-lipoic acid has been used for suppression of oxidative stress in experimental diabetic rats and found to protect nerve conduction, nerve blood flow, and nerve structure (Cameron, Cotter, Archibald, Dines, & Maxfield, 1994; Cameron, Cotter, & Maxfield, 1993; Kishi et al., 1999).

Obrosova et al. (2004) demonstrated that PARP-deficient mice exhibited no nerve conduction delay or impaired nerve blood flow in STZ-induced diabetes. They further demonstrated that inhibition of PARP significantly prevented the development of DPN by suppression of oxidative stress (Lupachyk, Shevalye, Maksimchyk, Drel, & Obrosova, 2011). In diabetic rats, hyperglycemia stimulated NO production and increased peroxynitrate by superoxide reaction, which impaired function in epineurial and endoneurial arterioles (Obrosova, Drel, et al., 2005). Conversely, degradation of peroxynitrates ameliorated the nerve conduction delay or pain sensation in diabetic mice. Since NO or peroxynitrate cannot be measured directly because of their instability, nitrotyrosine as nitrate or peroxynitrate-derived metabolites of NO were used as markers of nitrosylation of proteins, or biochemical marker of nitrosative stress (Ilnytska et al., 2006).

It is interesting that a small amount of insulin, which does not alter systemic blood glucose levels, was shown to improve the impaired mitochondrial membrane potential and delayed nerve conduction in STZ-diabetic rats (Chowdhury et al., 2010; Huang et al., 2003). Impaired signaling from adenosine monophosphate-activated protein kinase is also found to associate with mitochondrial dysfunction and neuropathy in STZ-diabetic rats, in which reduced production of ROS in mitochondria was characteristic in DRG (Akude et al., 2011; Roy Chowdhury et al., 2012). Thus, the derivation and possible role of oxidative stress in diabetic neuropathy seems to be different from vascular tissues and should be interpreted with caution.

5.3 Effects of Antioxidants on Diabetic Neuropathy

α-Lipoic acid is the most extensively studied antioxidant in humans with diabetic neuropathy and is marketed in Germany (Ziegler & Gries, 1997; Ziegler, Reljanovic, Mehnert, & Gries, 1999). In experimental animal models, positive effects of α-lipoic acid were found on nerve blood flow, conduction velocity, and peripheral nerve structure (Cameron, Cotter, Horrobin, & Tritschler, 1998; Kishi et al., 1999; Low, Nickander, & Tritschler, 1997; Stevens, Obrosova, Cao, Van Huysen, & Greene, 2000). A short-term clinical study reported that α-lipoic acid alleviated neuropathic symptoms in patients with diabetes (Ametov et al., 2003), and this finding was replicated in the NATHAN study of 4 years duration (Ziegler et al., 2011). However, the results concerning efficacy against nerve conduction slowing and other objective endpoints were inconclusive due to minimal progression of neuropathy over the study period.

6. PKC ACTIVITY

PKC plays a role in maintaining normal nerve function, and its aberrant regulation may contribute to the pathogenesis of DPN (Way, Katai, & King, 2001). PKC has several isoforms from α, β, γ, δ, and so on. Under ambient hyperglycemia, glucose is converted to glyceraldehydes-3-phosphate and phosphatidic acid, which in turn changes into diacylglycerol that serves as a substrate of PKC-β. Vascular tissues in diabetes thus show increased PKC activity, leading to increased permeability and dysfunction (Geraldes & King, 2010). An inhibitor of PKC-β was found to be effective for correction of vascular abnormalities in diabetic animals (King & Brownlee, 1996). Based on the premise that microangiopathy in nerve may contribute to development of neuropathy, preclinical and clinical trials of PKC-β inhibitor were conducted. Experimental studies demonstrated beneficial effects of a PKC-β inhibitor on peripheral nerve dysfunction and nerve blood flow in diabetic rats (Cameron & Cotter, 2002). However, the alterations of PKC activity are complex in nerves and their supportive vascular systems, in part due to the fact that the major enzymes of collateral glycolytic pathways are different between these two tissues (Fig. 3) (Kasajima et al., 2001). Such nonuniform tissue composition might explain inconsistent findings regarding PKC activity in diabetic nerves. Thus, Nakamura et al. (1999) did not find any significant change of PKC activity in homogenates of whole peripheral nerve from STZ-diabetic rats, although a PKC-β-specific inhibitor improved nerve conduction slowing and nerve blood flow. In contrast, in our studies on STZ-induced diabetic mice, we separated the peripheral nerve tissues into endoneurium and epineurium for the measurement of PKC activity, the latter of which was rich in microvessels (Yamagishi et al., 2003). We found that the former showed decreased PKC activity with significantly decreased expression of the membranous PKC-α isoform, whereas the latter showed increased PKC activity with enhanced expression of the PKC-β isoform. This increase of PKC activity in the epineurial compartment is consistent with reports in other systemic vascular tissues, while the decrease in the endoneurial compartment is consistent with a report that Schwann cells exposed to high glucose in vitro show reduced expression of PKC-α and reduced PKC activity (Kamiya et al., 2003). Given these findings, the application of PKC-β-specific inhibitor would be expected to be a useful for the treatment of diabetic vascular complications, but not any disorders related to reduced endoneurial expression of PKC-α. A 6-month clinical trial of the PKC-β-specific inhibitor Ruboxistaurin in subjects with DPN showed that it increased skin blood flow and mitigated some sensory symptoms but was without effect on objective measures of neuropathy such as large-fiber nerve conduction velocity slowing, reduced small sensory fiber nerve density in the epidermis, or impaired autonomic function (Casellini et al., 2007).

7. GLYCOSYLATION

7.1 The Hexosamine Pathway

Under conditions of elevated intracellular glucose and flux into glycolysis, excess fructose 6-phosphate is converted to glucosamine 6-phosphate (GlcN-6-phosphate) by glutamine-fructose-6-phosphate amidotransferase (GFAT). GlcN-6-phosphate is then modified to UDP-*N*-acetylglucosamine (UDP-GlcNAc) (Du et al., 2000; Issad, Masson, & Pagesy, 2010) (Fig. 6). UDP-GlcNAc has an affinity for cell membrane or nucleic acid or transcription factors and excessive modification of these substrates by posttranslational modification of serine and threonine residues (termed O-GlcNAcylation) can perturb cellular functions and may contribute to both the pathogenesis of type 2 diabetes by promoting insulin resistance and also to the complications of diabetes (Peterson & Hart, 2016). Evidence is strongest for a role of the hexosamine pathway in predominantly vascular complications such as retinopathy and nephropathy. For example, inhibition of GFAT suppresses expression of genes for transforming growth factor (TGF)-α and TGF-β1 gene and also the activation of plasminogen activator inhibitor-I (PAI-I). As increased expression and activity of these factors are known to be induced by hyperglycemia, the hexosamine pathway is therefore implicated in vascular complications of diabetes. UDP-GlcNAc modification of Akt activation sites in vascular endothelial cells also reduced eNOS activity. Further, GFAT in vascular smooth muscle cells is activated during hyperglycemia and muscle proteins are excessively modified by UDP-GlcNAc (Heath et al., 2014). In contrast to the plentiful data in vascular tissues, it is not yet entirely clear what kinds of peripheral nerve proteins may be modified by the hexosamine pathway during diabetes, although there is emerging evidence that imbalances in

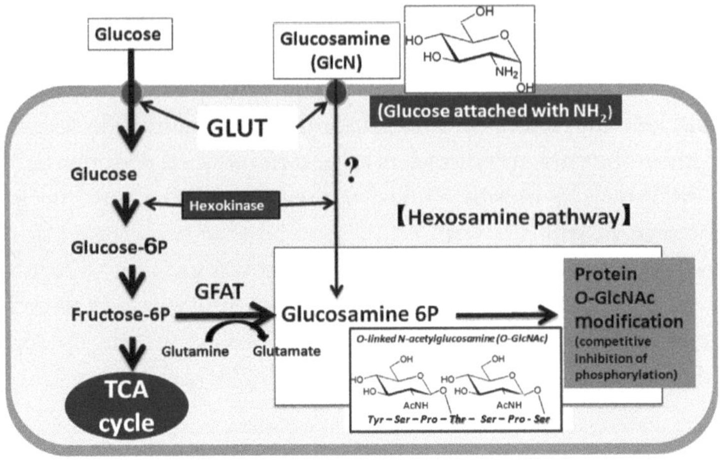

GFAT, glutamine-fructose-6-phosphate amidotransferase

Fig. 6 Hexosamine (synonym of glucosamine) pathway as a route to diabetic complications. This pathway is activated in hyperglycemia. As a consequence, *N*-acetylglucosamine (GlcNAc) is attached with proteins by O-glycosylation (O-GlcNAc modification), resulting in suppression of competitive phosphorylation. It is proposed that O-GlcNAc modification is implicated in the beta-cell dysfunction or insulin action, or endothelial dysfunction. Glucosamine is popular as a dietary supplement and metabolized by hexokinase, but the downstream metabolism is not known.

the O-GlcNAcylation state of nervous system proteins is associated with neurodegenerative disease (Hart, 2014). In experimental studies, it was shown that GlcN was toxic to neuronal cells, eliciting apoptosis by induction of death signals (Lim et al., 2010). This process involved oxidative stress and was inhibited by an antioxidant of *N*-acetyl-cysteine. The contribution of the hexosamine pathway and dysregulation of O-GlcNAcylation to diabetic neuropathy remains to be explored.

7.2 Other Glycosylation Events

Posttranslational glycosylation of the asparagine residues of neuronal proteins such as the T-type calcium channel (Cav3.2) can influence their function and therefore membrane stability and axonal excitability. There is recent evidence that aberrant glycosylation of neuronal ion channels during hyperglycemia may contribute to painful diabetic neuropathy (Todorovic, 2015), and this currently evolving hypothesis is covered in detail elsewhere in this volume.

8. CONCLUSION

It is well established that long-term hyperglycemia contributes to the genesis and development of DPN. However, the process is complicated and the precise mechanisms of how high glucose affects the assorted cell types within a nerve are still mysterious. Epidemiological studies suggest that many factors are responsible for DPN, and the targeting of a single collateral pathway may not be sufficient for the protection and suppression of DPN. Experimental studies have identified multiple pathways that appear to be involved in the functional and structural alterations in the peripheral nerves in animal models of diabetes and there may yet be other undiscovered pathways. As the clinical signs and symptoms of DPN differ from case to case, the dominant pathogenic mechanisms may also vary based on individual genotype and phenotype. If this is the case, in addition to meticulous blood glucose control, it will be important to identify which collateral pathway is most relevant to specific clinical manifestations. Future investigations should perhaps explore in more detail the relationships between each collateral pathway and specific clinical manifestations of DPN.

ACKNOWLEDGMENTS

The author appreciates the previous collaborators who contributed to the published works from the author's laboratory. The author also is grateful for the research funding from the Japanese government of the Ministry of Science, Culture, Education, and Sports, the Ministry of Health and Welfare, and the Juvenile Diabetes Foundation International, New York.

Duality of Interest: There is no conflict of interest for the author regarding the content of this review.

REFERENCES

Akude, E., Zherebitskaya, E., Chowdhury, S. K., Smith, D. R., Dobrowsky, R. T., & Fernyhough, P. (2011). Diminished superoxide generation is associated with respiratory chain dysfunction and changes in the mitochondrial proteome of sensory neurons from diabetic rats. *Diabetes*, *60*, 288–297.

Ametov, A. S., Barinov, A., Dyck, P. J., Hermann, R., Kozlova, N., Litchy, W. J., et al. (2003). The sensory symptoms of diabetic polyneuropathy are improved with alpha-lipoic acid: The SYDNEY trial. *Diabetes Care*, *26*, 770–776.

Baba, M., Kimura, K., Suda, T., Yagihashi, S., & Aomori Diabetic Study (ADNS) Group. (2006). Three-year inhibition of aldose reductase on development of symptomatic neuropathy in diabetic patients. *Journal of the Peripheral Nervous System*, *11*, 176–178.

Baynes, J. W., & Thorpe, S. R. (1999). Role of oxidative stress in diabetic complications: A new perspective on an old paradigm. *Diabetes*, *48*, 1–9.

Bierhaus, A., Fleming, T., Stoyanov, S., Leffler, A., Babes, A., Neacsu, C., et al. (2012). Methylglyoxal modification of Nav1.8 facilitates nociceptive neuron firing and causes hyperalgesia in diabetic neuropathy. *Nature Medicine, 18*, 926–933.

Bierhaus, A., Haslbeck, K. M., Humpert, P. M., Liliensiek, B., Dehmer, T., Morcos, M., et al. (2004). Loss of pain perception in diabetes is dependent on a receptor of the immunoglobulin superfamily. *The Journal of Clinical Investigation, 114*, 1741–1751.

Bolton, W. K., Cattran, D. C., Williams, M. E., Adler, S. G., Appel, G. B., Cartwright, K., et al. (2004). Randomized trial of an inhibitor of formation of advanced glycation end products in diabetic nephropathy. *American Journal of Nephrology, 24*, 32–40.

Boyle, K. E., Newsom, S. A., Janssen, R. C., Lappas, M., & Friedman, J. E. (2013). Skeletal muscle MnSOD, mitochondrial complex II, and SIRT3 enzyme activities are decreased in maternal obesity during human pregnancy and gestational diabetes mellitus. *The Journal of Clinical Endocrinology and Metabolism, 98*, E1601–E1609.

Bril, V., & Buchanan, R. A. (2006). Long-term effects of ranirestat (AS-3201) on peripheral nerve function in patients with diabetic sensorimotor polyneuropathy. *Diabetes Care, 29*, 68–72.

Brownlee, M. (1992). Nonenzymatic glycosylation of macromolecules. Prospects of pharmacologic modulation. *Diabetes, 41*(Suppl. 2), 57–60.

Brownlee, M. (2005). The pathobiology of diabetic complications: A unifying mechanism. *Diabetes, 54*, 1615–1625.

Calcutt, N. A., Tomlinson, D. R., & Biswas, S. (1990). Coexistence of nerve conduction deficit with increased Na(+)-K(+)-ATPase activity in galactose-fed mice. Implications for polyol pathway and diabetic neuropathy. *Diabetes, 39*, 663–666.

Cameron, N. E., & Cotter, M. A. (2002). Effects of protein kinase C-β inhibition on neurovascular dysfunction in diabetic rats: Interaction with oxidative stress and essential fatty acid dysmetabolism. *Diabetes/Metabolism Research and Reviews, 18*, 315–323.

Cameron, N. E., Cotter, M. A., Archibald, V., Dines, K. C., & Maxfield, E. K. (1994). Anti-oxidant and pro-oxidant effects on nerve conduction velocity, endoneurial blood flow and oxygen tension in non-diabetic and streptozotocin-diabetic rats. *Diabetologia, 37*, 449–459.

Cameron, N. E., Cotter, M. A., Horrobin, D. H., & Tritschler, H. J. (1998). Effects of alpha-lipoic acid on neurovascular function in diabetic rats: Interaction with essential fatty acids. *Diabetologia, 41*, 390–399.

Cameron, N. E., Cotter, M. A., & Maxfield, E. K. (1993). Anti-oxidant treatment prevents the development of peripheral nerve dysfunction in streptozotocin-diabetic rats. *Diabetologia, 36*, 299–304.

Casellini, C. M., Barlow, P. M., Rice, A. L., Casey, M., Simmons, K., Pittenger, G., et al. (2007). A 6-month, randomized, double-masked, placebo-controlled study evaluating the effects of the protein kinase C-β inhibitor ruboxistaurin on skin microvascular blood flow and other measures of diabetic peripheral neuropathy. *Diabetes Care, 30*, 896–902.

Cheung, A. K., Lo, A. C., So, K. F., Chung, S. S., & Chung, S. K. (2007). Gene deletion and pharmacological inhibition of aldose reductase protect against retinal ischemic injury. *Experimental Eye Research, 85*, 608–616.

Chowdhury, S. K., Zherebitskaya, E., Smith, D. R., Akude, E., Chattopadhyay, S., Jolivalt, C. G., et al. (2010). Mitochondrial respiratory chain dysfunction in dorsal root ganglia of streptozotocin-induced diabetic rats and its correction by insulin treatment. *Diabetes, 59*, 1082–1091.

Corbett, J. A., & McDaniel, M. L. (1996). Selective inhibition of inducible nitric oxide synthase by aminoguanidine. *Methods in Enzymology, 268*, 398–408.

Diabetes Control and Complications Trial Research Group. (1993). The effect of intensive treatment of diabetes on the development and progression of long-term complications in insulin-dependent diabetes mellitus. *The New England Journal of Medicine, 329*, 977–986.

Du, X. L., Edelstein, D., Rossetti, L., Fantus, I. G., Goldberg, H., Ziyadeh, F., et al. (2000). Hyperglycemia-induced mitochondrial superoxide overproduction activates the hexosamine pathway and induces plasminogen activator inhibitor-1 expression by increasing Sp1 glycosylation. *Proceedings of the National Academy of Sciences of the United States of America*, *97*, 12222–12226.

Duran-Jimenez, B., Dobler, D., Moffatt, S., Rabbani, N., Streuli, C. H., Thornalley, P. J., et al. (2009). Advanced glycation end products in extracellular matrix proteins contribute to the failure of sensory nerve regeneration in diabetes. *Diabetes*, *58*, 2893–2903.

Dyck, P. J., & Giannini, C. (1996). Pathologic alterations in the diabetic neuropathies of humans: A review. *Journal of Neuropathology and Experimental Neurology*, *55*, 1181–1193.

Dyck, P. J., Karnes, J., O'Brien, P., Nukada, H., Lais, A., & Low, P. (1984). Spatial pattern of nerve fiber abnormality indicative of pathologic mechanism. *The American Journal of Pathology*, *117*, 225–238.

Dyck, P. J., Lais, A., Karnes, J. L., O'Brien, P., & Rizza, R. (1986). The spatial distribution of fiber loss in diabetic polyneuropathy suggests ischemia. *Annals of Neurology*, *19*, 440–449.

Dyck, P. J., Sherman, W. R., Hallcher, L. M., Service, F. J., O'Brien, P. C., Grina, L. A., et al. (1980). Human diabetic endoneurial sorbitol, fructose, and myo-inositol related to sural nerve morphometry. *Annals of Neurology*, *8*, 590–596.

Dyck, P. J., Zimmerman, B. R., Vilen, T. H., Minnerath, S. R., Karnes, J. L., Yao, J. K., et al. (1988). Nerve glucose, fructose, sorbitol, myo-inositol, and fiber degeneration and regeneration in diabetic neuropathy. *The New England Journal of Medicine*, *319*, 542–548.

Gabbay, K. H. (1975). Hyperglycemia, polyol metabolism, and complications of diabetes mellitus. *Annual Review of Medicine*, *26*, 521–536.

Geraldes, P., & King, G. L. (2010). Activation of protein kinase C isoforms and its impact on diabetic complications. *Circulation Research*, *106*, 1319–1331.

Goto, Y., Hotta, N., Shigeta, Y., Sakamoto, N., & Kikkawa, R. (1995). Effects of an aldose reductase inhibitor, epalrestat, on diabetic neuropathy. Clinical benefit and indication for the drug assessed from the results of a placebo-controlled double-blind study. *Biomedicine & Pharmacotherapy*, *49*, 269–277.

Greene, D. A., Arezzo, J. C., & Brown, M. B. (1999). Effect of aldose reductase inhibition on nerve conduction and morphometry in diabetic neuropathy. Zenarestat Study Group. *Neurology*, *53*, 580–591.

Greene, D. A., Lattimer, S. A., & Sima, A. A. (1987). Sorbitol, phosphoinositides, and sodium-potassium-ATPase in the pathogenesis of diabetic complications. *The New England Journal of Medicine*, *316*, 599–606.

Greene, D. A., Sima, A. A., Stevens, M. J., Feldman, E. L., & Lattimer, S. A. (1992). Complications: Neuropathy, pathogenetic considerations. *Diabetes Care*, *15*, 1902–1925.

Greene, D. A., Stevens, M. J., Obrosova, I., & Feldman, E. L. (1999). Glucose-induced oxidative stress and programmed cell death in diabetic neuropathy. *European Journal of Pharmacology*, *375*, 217–223.

Hardivillé, S., & Hart, G. W. (2014). Nutrient regulation of signaling, transcription, and cell physiology by O-GlcNAcylation. *Cell Metabolism*, *20*, 208–213.

Hart, G. W. (2014). Minireview series on the thirtieth anniversary of research on O-GlcNAcylation of nuclear and cytoplasmic proteins: Nutrient regulation of cellular metabolism and physiology by O-GlcNAcylation. *The Journal of Biological Chemistry*, *289*, 34422–34433.

Haslbeck, K. M., Schleicher, E., Bierhaus, A., Nawroth, P., Haslbeck, M., Neundörfer, B., et al. (2005). The AGE/RAGE/NF-(kappa)B pathway may contribute to the pathogenesis of polyneuropathy in impaired glucose tolerance (IGT). *Experimental and Clinical Endocrinology & Diabetes*, *113*, 288–291.

Haupt, E., Ledermann, H., & Köpcke, W. (2005). Benfotiamine in the treatment of diabetic polyneuropathy—A three-week randomized, controlled pilot study (BEDIP study). *International Journal of Clinical Pharmacology & Therapeutics*, *43*, 71–77.

Heath, J. M., Sun, Y., Yuan, K., Bradley, W. E., Litovsky, S., Dell'Italia, L. J., et al. (2014). Activation of AKT by O-linked N-acetylglucosamine induces vascular calcification in diabetes mellitus. *Circulation Research, 114*, 1094–1102.
Ho, E. C., Lam, K. S., Chen, Y. S., Yip, J. C., Arvindakshan, M., Yamagishi, S., et al. (2006). Aldose reductase-deficient mice are protected from delayed motor nerve conduction velocity, increased c-Jun NH2-terminal kinase activation, depletion of reduced glutathione, increased superoxide accumulation, and DNA damage. *Diabetes, 55*, 1946–1953.
Hotta, N., Akanuma, Y., Kawamori, R., Matsuoka, K., Oka, Y., Shichiri, M., et al. (2006). Long-term clinical effects of epalrestat, an aldose reductase inhibitor, on diabetic peripheral neuropathy: The 3-year, multicenter, comparative Aldose Reductase Inhibitor-Diabetes Complications Trial. *Diabetes Care, 29*, 1538–1544.
Hotta, N., Kawamori, R., Atsumi, Y., Baba, M., Kishikawa, H., Nakamura, J., et al. (2008). Stratified analyses for selecting appropriate target patients with diabetic peripheral neuropathy for long-term treatment with an aldose reductase inhibitor, epalrestat. *Diabetic Medicine, 25*, 818–825.
Huang, T. J., Price, S. A., Chilton, L., Calcutt, N. A., Tomlinson, D. R., Verkhratsky, A., et al. (2003). Insulin prevents depolarization of the mitochondrial inner membrane in sensory neurons of type 1 diabetic rats in the presence of sustained hyperglycemia. *Diabetes, 52*, 2129–2136.
Ido, Y., Nyengaard, J. R., Chang, K., Tilton, R. G., Kilo, C., Mylari, B. L., et al. (2010). Early neural and vascular dysfunctions in diabetic rats are largely sequelae of increased sorbitol oxidation. *Antioxidants & Redox Signaling, 12*, 39–51.
Ilnytska, O., Lyzogubov, V. V., Stevens, M. J., Drel, V. R., Mashtalir, N., Pacher, P., et al. (2006). Poly(ADP-ribose) polymerase inhibition alleviates experimental diabetic sensory neuropathy. *Diabetes, 55*, 1686–1694.
Inagi, R., Yamamoto, Y., Nangaku, M., Usuda, N., Okamato, H., Kurokawa, K., et al. (2006). A severe diabetic nephropathy model with early development of nodule-like lesions induced by megsin overexpression in RAGE/iNOS transgenic mice. *Diabetes, 55*, 356–366.
Issad, T., Masson, E., & Pagesy, P. (2010). O-GlcNAc modification, insulin signaling and diabetic complications. *Diabetes & Metabolism, 36*, 423–435.
Iwata, K., Matsuno, K., Nishinaka, T., Persson, C., & Yabe-Nishimura, C. (2006). Aldose reductase inhibitors improve myocardial reperfusion injury in mice by a dual mechanism. *Journal of Pharmacological Sciences, 102*, 37–46.
Jack, M. M., Ryals, J. M., & Wright, D. E. (2011). Characterisation of glyoxalase I in a streptozocin-induced mouse model of diabetes with painful and insensate neuropathy. *Diabetologia, 54*, 2174–2182.
James, L. R., Tang, D., Ingram, A., Ly, H., Thai, K., Cai, L., et al. (2002). Flux through the hexosamine pathway is a determinant of nuclear factor κB-dependent promoter activation. *Diabetes, 51*, 1146–1156.
Jiang, Y., Calcutt, N. A., Ramos, K. M., & Mizisin, A. P. (2006). Novel sites of aldose reductase immunolocalization in normal and streptozotocin-diabetic rats. *Journal of the Peripheral Nervous System, 11*, 274–285.
Juranek, J. K., Geddis, M. S., Song, F., Zhang, J., Garcia, J., Rosario, R., et al. (2013). RAGE deficiency improves postinjury sciatic nerve regeneration in type 1 diabetic mice. *Diabetes, 62*, 931–943.
Kamiya, H., Nakamura, J., Hamada, Y., Nakashima, E., Naruse, K., Kato, K., et al. (2003). Polyol pathway and protein kinase C activity of rat Schwannoma cells. *Diabetes/Metabolism Research and Reviews, 19*, 131–139.
Kasajima, H., Yamagishi, S., Sugai, S., Yagihashi, N., & Yagihashi, S. (2001). Enhanced in situ expression of aldose reductase in peripheral nerve and renal glomeruli in diabetic patients. *Virchows Archiv, 439*, 46–54.

Kihara, M., Schmelzer, J. D., Poduslo, J. F., Curran, G. L., Nickander, K. K., & Low, P. A. (1991). Aminoguanidine effects on nerve blood flow, vascular permeability, electrophysiology, and oxygen free radicals. *Proceedings of the National Academy of Sciences of the United States of America*, *88*, 6107–6111.

Kihara, M., Weerasuriya, A., & Low, P. A. (1991). Endoneurial blood flow in rat sciatic nerve during development. *The Journal of Physiology*, *439*, 351–360.

Kikuchi, S., Shinpo, K., Moriwaka, F., Makita, Z., Miyata, T., & Tashiro, K. (1999). Neurotoxicity of methylglyoxal and 3-deoxyglucosone on cultured cortical neurons: Synergism between glycation and oxidative stress, possibly involved in neurodegenerative diseases. *Journal of Neuroscience Research*, *57*, 280–289.

King, G. L., & Brownlee, M. (1996). The cellular and molecular mechanisms of diabetic complications. *Endocrinology and Metabolism Clinics of North America*, *25*, 255–270.

Kinoshita, J. H. (1990). A thirty year journey in the polyol pathway. *Experimental Eye Research*, *50*, 567–573.

Kishi, Y., Schmelzer, J. D., Yao, J. K., Zollman, P. J., Nickander, K. K., Tritschler, H. J., et al. (1999). Alpha-lipoic acid: Effect on glucose uptake, sorbitol pathway, and energy metabolism in experimental diabetic neuropathy. *Diabetes*, *48*, 2045–2051.

Lambourne, J. E., Tomlinson, D. R., Brown, A. M., & Willars, G. B. (1987). Opposite effects of diabetes and galactosaemia on adenosine triphosphatase activity in rat nervous tissue. *Diabetologia*, *30*, 360–362.

Leinninger, G. M., Edwards, J. L., Lipshaw, M. J., & Feldman, E. (2006). Mechanisms of disease: Mitochondria as new therapeutic targets in diabetic neuropathy. *Nature Clinical Practice. Neurology*, *2*, 620–628.

Lim, J. G., Lee, J. J., Park, S. H., Park, J. H., Kim, S. J., Cho, H. C., et al. (2010). Glucagon-like peptide-1 protects NSC-34 motor neurons against glucosamine through Epac-mediated glucose uptake enhancement. *Neuroscience Letters*, *479*, 13–17.

Low, P. A., Nickander, K. K., & Tritschler, H. J. (1997). The roles of oxidative stress and antioxidant treatment in experimental diabetic neuropathy. *Diabetes*, *46*(Suppl. 2), S38–S42.

Low, P. A., Schmelzer, J. D., & Ward, K. K. (1986). The effect of age on energy metabolism and resistance to ischaemic conduction failure in rat peripheral nerve. *The Journal of Physiology*, *374*, 263–271.

Lupachyk, S., Shevalye, H., Maksimchyk, Y., Drel, V. R., & Obrosova, I. G. (2011). PARP inhibition alleviates diabetes-induced systemic oxidative stress and neural tissue 4-hydroxynonenal adduct accumulation: Correlation with peripheral nerve function. *Free Radical Biology & Medicine*, *50*, 1400–1409.

Malik, R. A., Tesfaye, S., Newrick, P. G., Walker, D., Rajbhandari, S. M., Siddique, I., et al. (2005). Sural nerve pathology in diabetic patients with minimal but progressive neuropathy. *Diabetologia*, *48*, 578–585.

Martin, C. L., Albers, J., Herman, W. H., Cleary, P., Waberski, B., Greene, D. A., et al. (2006). Neuropathy among the diabetes control and complications trial cohort 8 years after trial completion. *Diabetes Care*, *29*, 340–344.

McLean, W. G., Pekiner, C., Cullum, N. A., & Casson, I. F. (1992). Posttranslational modifications of nerve cytoskeletal proteins in experimental diabetes. *Molecular Neurobiology*, *6*, 225–237.

Metz, T. O., Alderson, N. L., Thorpe, S. R., & Baynes, J. W. (2003). Pyridoxamine, an inhibitor of advanced glycation and lipoxidation reactions: A novel therapy for treatment of diabetic complications. *Archives of Biochemistry and Biophysics*, *419*, 41–49.

Nagai, R., Shirakawa, J., Ohno, R., Moroishi, N., & Nagai, M. (2014). Inhibition of AGEs formation by natural products. *Amino Acids*, *46*, 261–266.

Nakamura, J., Kato, K., Hamada, Y., Nakayama, M., Chaya, S., Nakashima, E., et al. (1999). A protein kinase C-beta-selective inhibitor ameliorates neural dysfunction in streptozotocin-induced diabetic rats. *Diabetes*, *48*, 2090–2095.

Ng, T. F., Lee, F. K., Song, Z. T., Calcutt, N. A., Lee, A. Y., Chung, S. S., et al. (1998). Effects of sorbitol dehydrogenase deficiency on nerve conduction in experimental diabetic mice. *Diabetes*, *47*, 961–966.

Nishizawa, Y., Wada, R., Baba, M., Takeuchi, M., Hanyu-Itabashi, C., & Yagihashi, S. (2010). Neuropathy induced by exogenously administered advanced glycation end-products. *Journal of Diabetes Investigation*, *1*, 40–49.

Nukada, H., Dyck, P. J., Low, P. A., Lais, A. C., & Sparks, M. F. (1986). Axonal caliber and neurofilaments are proportionately decreased in galactose neuropathy. *Journal of Neuropathology and Experimental Neurology*, *45*, 140–150.

Nukada, H., McMorran, P. D., Baba, M., Ogasawara, S., & Yagihashi, S. (2011). Increased susceptibility to ischemia and macrophage activation in STZ-diabetic rat nerve. *Brain Research*, *1373*, 172–182.

Obrosova, I. G. (2009). Diabetes and the peripheral nerve. *Biochimica et Biophysica Acta*, *1792*, 931–940.

Obrosova, I. G., Drel, V. R., Pacher, P., Ilnytska, O., Wang, Z. Q., Stevens, M. J., et al. (2005). Oxidative-nitrosative stress and poly(ADP-ribose) polymerase (PARP) activation in experimental diabetic neuropathy: The relation is revisited. *Diabetes*, *54*, 3435–3441.

Obrosova, I. G., Ilnytska, O., Lyzogubov, V. V., Pavlov, I. A., Mashtalir, N., Nadler, J. L., et al. (2007). High-fat diet induced neuropathy of pre-diabetes and obesity: Effects of "healthy" diet and aldose reductase inhibition. *Diabetes*, *56*, 2598–2608.

Obrosova, I. G., Li, F., Abatan, O. I., Forsell, M. A., Komjáti, K., Pacher, P., et al. (2004). Role of poly(ADP-ribose) polymerase activation in diabetic neuropathy. *Diabetes*, *53*, 711–720.

Obrosova, I. G., Pacher, P., Szabó, C., Zsengeller, Z., Hirooka, H., Stevens, M. J., et al. (2005). Aldose reductase inhibition counteracts oxidative-nitrosative stress and poly(ADP-ribose) polymerase activation in tissue sites for diabetes complications. *Diabetes*, *54*, 234–242.

Ohkubo, Y., Kishikawa, H., Araki, E., Miyata, T., Isami, S., Motoyoshi, S., et al. (1995). Intensive insulin therapy prevents the progression of diabetic microvascular complications in Japanese patients with non-insulin-dependent diabetes mellitus: A randomized prospective 6-year study. *Diabetes Research and Clinical Practice*, *28*, 103–117.

Peterson, S. B., & Hart, G. W. (2016). New insights: A role for O-GlcNAcylation in diabetic complications. *Critical Reviews in Biochemistry and Molecular Biology*, *51*, in press. http://dx.doi.org/10.3109/10409238.2015.1135102.

Pirart, J. (1978). Diabetes mellitus and its degenerative complications: A prospective study of 4400 patients observed between 1947 and 1973. *Diabetes Care*, *1*, 168–188.

Polydefkis, M., Arezzo, J., Nash, M., Bril, V., Shaibani, A., Gordon, R. J., et al. (2015). Safety and efficacy of ranirestat in patients with mild-to-moderate diabetic sensorimotor polyneuropathy. *Journal of the Peripheral Nervous System*, *20*, 363–371.

Pop-Busui, R., Sima, A., & Stevens, M. (2006). Diabetic neuropathy and oxidative stress. *Diabetes/Metabolism Research and Reviews*, *22*, 257–273.

Rabbani, N., & Thornalley, P. J. (2015). Dicarbonyl stress in cell and tissue dysfunction contributing to ageing and disease. *Biochemical and Biophysical Research Communications*, *458*, 221–226.

Ramasamy, R., & Goldberg, I. J. (2010). Aldose reductase and cardiovascular diseases, creating human-like diabetic complications in an experimental model. *Circulation Research*, *106*, 1449–1458.

Roy Chowdhury, S. K., Smith, D. R., Saleh, A., Schapansky, J., Marquez, A., Gomes, S., et al. (2012). Impaired adenosine monophosphate-activated protein kinase signalling in dorsal root ganglia neurons is linked to mitochondrial dysfunction and peripheral neuropathy in diabetes. *Brain*, *135*, 1751–1766.

Ryle, C., Leow, C. K., & Donaghy, M. (1997). Nonenzymatic glycation of peripheral and central nervous system proteins in experimental diabetes mellitus. *Muscle & Nerve, 20*, 577–584.

Santos, J. M., Tewari, S., Goldberg, A. F., & Kowluru, R. A. (2011). Mitochondrial biogenesis and the development of diabetic retinopathy. *Free Radical Biology & Medicine, 51*, 1849–1860.

Schartner, E., Chowdhury, S. R., Smith, D., & Fernyhough, P. (2015). High glucose concentration down-regulates SIRT2 expression in sensory neurons via a sorbitol dehydrogenase-dependent pathway. In *(Abstract) Neurodiab 2015, 25th annual meeting of the diabetic neuropathy study group of the EASD, September 11–13, Elsinore, Denmark (Abstract O29)* (p. 48).

Schmidt, R. E. (2002). Neuropathology and pathogenesis of diabetic autonomic neuropathy. *International Review of Neurobiology, 50*, 257–292.

Schmidt, R. E., Dorsey, D. A., Beaudet, L. N., Plurad, S. B., Parvin, C. A., Yarasheski, K. E., et al. (2001). Inhibition of sorbitol dehydrogenase exacerbates autonomic neuropathy in rats with streptozotocin-induced diabetes. *Journal of Neuropathology and Experimental Neurology, 60*, 1153–1169.

Schmidt, R. E., Plurad, S. B., Sherman, W. R., Williamson, J. R., & Tilton, R. G. (1989). Effects of aldose reductase inhibitor sorbinil on neuroaxonal dystrophy and levels of myo-inositol and sorbitol in sympathetic autonomic ganglia of streptozocin-induced diabetic rats. *Diabetes, 38*, 569–579.

Scott, J. N., Clark, A. W., & Zochodne, D. W. (1999). Neurofilament and tubulin gene expression in progressive experimental diabetes: Failure of synthesis and export by sensory neurons. *Brain, 122*, 2109–2118.

Sekido, H., Suzuki, T., Jomori, T., Takeuchi, M., Yabe-Nishimura, C., & Yagihashi, S. (2004). Reduced cell replication and induction of apoptosis by advanced glycation end products in rat Schwann cells. *Biochemical and Biophysical Research Communications, 320*, 241–248.

Sima, A. A., Bril, V., Nathaniel, V., McEwen, T. A., Brown, M. B., Lattimer, S. A., et al. (1988). Regeneration and repair of myelinated fibers in sural-nerve biopsy specimens from patients with diabetic neuropathy treated with sorbinil. *The New England Journal of Medicine, 319*, 548–555.

Simmons, D. A., Kern, E. F., Winegrad, A. I., & Martin, D. B. (1986). Basal phosphatidylinositol turnover controls aortic Na+/K+ ATPase activity. *The Journal of Clinical Investigation, 77*, 503–513.

Smith, D. R., Kobrine, A. I., & Rizzoli, H. V. (1977). Absence of autoregulation in peripheral nerve blood flow. *Journal of the Neurological Sciences, 33*, 347–352.

Stark, B., Carlstedt, T., Cullheim, S., & Risling, M. (2000). Developmental and lesion-induced changes in the distribution of the glucose transporter Glut-1 in the central and peripheral nervous system. *Experimental Brain Research, 131*, 74–84.

Stevens, M. J., Dananberg, J., Feldman, E. L., Lattimer, S. A., Kamijo, M., Thomas, T. P., et al. (1994). The linked roles of nitric oxide, aldose reductase, and (Na+, K+)-ATPase in the slowing of nerve conduction in the streptozotocin diabetic rat. *The Journal of Clinical Investigation, 94*, 853–859.

Stevens, M. J., Obrosova, I., Cao, X., Van Huysen, C., & Greene, D. A. (2000). Effects of DL-alpha-lipoic acid on peripheral nerve conduction, blood flow, energy metabolism, and oxidative stress in experimental diabetic neuropathy. *Diabetes, 49*, 1006–1015.

Stracke, H., Gaus, W., Achenbach, U., Federlin, K., & Bretzel, R. G. (2008). Benfotiamine in diabetic polyneuropathy (BENDIP): Results of a randomised, double blind, placebo-controlled clinical study. *Experimental and Clinical Endocrinology & Diabetes, 116*, 600–605.

Stracke, H., Lindemann, A., & Federlin, K. (1996). A benfotiamine-vitamin B combination in treatment of diabetic polyneuropathy. *Experimental and Clinical Endocrinology & Diabetes, 104*, 311–316.

Sugimoto, K., Nishizawa, Y., Horiuchi, S., & Yagihashi, S. (1997). Localization in human diabetic peripheral nerve of N(epsilon)-carboxymethyllysine-protein adducts, an advanced glycation endproduct. *Diabetologia, 40*, 1380–1387.

Sugimoto, K., Yasujima, M., & Yagihashi, S. (2008). Role of advanced glycation end products in diabetic neuropathy. *Current Pharmaceutical Design, 14*, 953–961.

Takahashi, K., Mizukami, H., Kamata, K., Inaba, W., Kato, N., Hibi, C., et al. (2012). Amelioration of acute kidney injury in lipopolysaccharide-induced systemic inflammatory response syndrome by an aldose reductase inhibitor, fidarestat. *PLoS One, 7*, e30134. http://dx.doi.org/10.1371/journal.pone.0030134.

Tesfaye, S., Chaturvedi, N., Eaton, S. E., Ward, J. D., Manes, C., Ionescu-Tirgoviste, C., et al. (2005). Vascular risk factors and diabetic neuropathy. *The New England Journal of Medicine, 352*, 341–350.

Thornalley, P. J. (2002). Glycation in diabetic neuropathy: Characteristics, consequences, causes, and therapeutic options. *International Review of Neurobiology, 50*, 37–57.

Thrainsdottir, S., Malik, R. A., Dahlin, L. B., Wiksell, P., Eriksson, K. F., Rosén, I., et al. (2003). Endoneurial capillary abnormalities presage deterioration of glucose tolerance and accompany peripheral neuropathy in man. *Diabetes, 52*, 2615–2622.

Todorovic, S. M. (2015). Is diabetic nerve pain caused by dysregulated ion channels in sensory neurons? *Diabetes, 64*, 3987–3989.

Tomlinson, D. R., Moriarty, R. J., & Mayer, J. H. (1984). Prevention and reversal of defective axonal transport and motor nerve conduction velocity in rats with experimental diabetes by treatment with the aldose reductase inhibitor Sorbinil. *Diabetes, 33*, 470–476.

Uehara, K., Yamagishi, S., Otsuki, S., Chin, S., & Yagihashi, S. (2004). Effects of polyol pathway hyperactivity on protein kinase C activity, nociceptive peptide expression, and neuronal structure in dorsal root ganglia in diabetic mice. *Diabetes, 53*, 3239–3247.

UK Prospective Diabetes Study Group. (1998). Intensive blood-glucose control with sulphonylureas or insulin compared with conventional treatment and risk of complications in patients with type 2 diabetes (UKPDS33). *Lancet, 352*, 837–853.

Urban, M. J., Pan, P., Farmer, K. L., Zhao, H., Blagg, B. S., & Dobrowsky, R. T. (2012). Modulating molecular chaperones improves sensory fiber recovery and mitochondrial function in diabetic peripheral neuropathy. *Experimental Pathology, 235*, 388–396.

Vasan, S., Foiles, P., & Founds, H. (2003). Therapeutic potential of breakers of advanced glycation end product-protein crosslinks. *Archives of Biochemistry and Biophysics, 419*, 89–96.

Vincent, A. M., McLean, L. L., Backus, C., & Feldman, E. L. (2005). Short-term hyperglycemia produces oxidative damage and apoptosis in neurons. *The FASEB Journal, 19*, 638–640.

Wada, R., Nishizawa, Y., Yagihashi, N., Takeuchi, M., Ishikawa, Y., Yasumura, K., et al. (2001). Effects of OPB-9195, anti-glycation agent, on experimental diabetic neuropathy. *European Journal of Clinical Investigation, 31*, 513–520.

Wada, R., & Yagihashi, S. (2005). Role of advanced glycation end products and their receptors in development of diabetic neuropathy. *The Annals of the New York Academy of Sciences, 1043*, 598–604.

Way, K. J., Katai, N., & King, G. L. (2001). Protein kinase C and the development of diabetic vascular complications. *Diabetic Medicine, 18*, 945–959.

Williams, S. K., Howarth, N. L., Devenny, J. J., & Bitensky, M. W. (1982). Structural and functional consequences of increased tubulin glycosylation in diabetes mellitus. *Proceedings of the National Academy of Sciences of the United States of America, 79*, 6546–6550.

Williamson, J. R., Chang, K., Frangos, M., Hasan, K. S., Ido, Y., Kawamura, T., et al. (1993). Hyperglycemic pseudohypoxia and diabetic complications. *Diabetes, 42*, 801–813.

Winegrad, A. I. (1987). Banting lecture 1986. Does a common mechanism induce the diverse complications of diabetes? *Diabetes, 36*, 396–406.

Yagihashi, S., Kamijo, M., Baba, M., Yagihashi, N., & Nagai, K. (1992). Effect of aminoguanidine on functional and structural abnormalities in peripheral nerve of STZ-induced diabetic rats. *Diabetes, 41*, 47–52.

Yagihashi, S., Kamijo, M., Ido, Y., & Mirrlees, D. J. (1990a). Effects of long-term aldose reductase inhibition on development of experimental diabetic neuropathy. Ultrastructural and morphometric studies of sural nerve in streptozocin-induced diabetic rats. *Diabetes, 39*, 690–696.

Yagihashi, S., Kamijo, M., & Watanabe, K. (1990b). Reduced myelinated fiber size correlates with loss of axonal neurofilaments in peripheral nerve of chronically streptozotocin diabetic rats. *The American Journal of Pathology, 136*, 1365–1373.

Yagihashi, S., Mizukami, H., Ogasawara, S., Yamagishi, S., Nukada, H., Kato, N., et al. (2010). The role of the polyol pathway in acute kidney injury caused by hindlimb ischaemia in mice. *The Journal of Pathology, 220*, 530–541.

Yagihashi, S., Mizukami, H., & Sugimoto, K. (2011). Mechanism of diabetic neuropathy: Where are we now and where to go? *Journal of Diabetes Investigation, 2*, 18–32.

Yagihashi, S., Yamagishi, S. I., Wada, R., Baba, M., Hohman, T. C., Yabe-Nishimura, C., et al. (2001). Neuropathy in diabetic mice overexpressing human aldose reductase and effects of aldose reductase inhibitor. *Brain, 124*, 2448–2458.

Yagihashi, S., Yamagishi, S., Wada, R., Sugimoto, K., Baba, M., Wong, H. G., et al. (1996). Galactosemic neuropathy in transgenic mice for human aldose reductase. *Diabetes, 45*, 56–59.

Yamagishi, S., Nakamura, K., & Imaizumi, T. (2005). Advanced glycation end products (AGEs) and diabetic vascular complications. *Current Diabetes Reviews, 1*, 93–106.

Yamagishi, S., Uehara, K., Otsuki, S., & Yagihashi, S. (2003). Differential influence of increased polyol pathway on protein kinase C expressions between endoneurial and epineurial tissues in diabetic mice. *Journal of Neurochemistry, 87*, 497–507.

Yamamoto, Y., Kato, I., Doi, T., Yonekura, H., Ohashi, S., Takeuchi, M., et al. (2001). Development and prevention of advanced diabetic nephropathy in RAGE-overexpressing mice. *The Journal of Clinical Investigation, 108*, 261–268.

Yeung, C. M., Lo, A. C., Cheung, A. K., Chung, S. S., Wong, D., & Chung, S. K. (2010). More severe type 2 diabetes-associated ischemic stroke injury is alleviated in aldose reductase-deficient mice. *Journal of Neuroscience Research, 88*, 2026–2034.

Zhang, L., Yu, C., Vasquez, F. E., Galeva, N., Onyango, I., Swerdlow, R. H., et al. (2010). Hyperglycemia alters the Schwann cell mitochondrial proteome and decreased coupled respiration in the absence of superoxide production. *Journal of Proteome Research, 9*, 458–471.

Zhang, L., Zhao, H., Blagg, B. S., & Dobrowsky, R. T. (2012). C-terminal heat shock protein 90 inhibitor decreases hyperglycemia-induced oxidative stress and improves mitochondrial bioenergetics in sensory neurons. *Journal of Proteome Research, 11*, 2581–2593.

Zherebitskaya, E., Akude, E., Smith, D. R., & Fernyhough, P. (2009). Development of selective axonopathy in adult sensory neurons isolated from diabetic rats: Role of glucose-induced oxidative stress. *Diabetes, 58*, 1356–1364.

Ziegler, D., & Gries, F. A. (1997). Alpha-lipoic acid in the treatment of diabetic peripheral and cardiac autonomic neuropathy. *Diabetes, 46*(Suppl. 2), S62–S66.

Ziegler, D., Low, P. A., Litchy, W. J., Boulton, A. J., Vinik, A. I., Freeman, R., et al. (2011). Efficacy and safety of antioxidant treatment with α-lipoic acid over 4 years in diabetic polyneuropathy: The NATHAN 1 trial. *Diabetes Care, 34*, 2054–2060.

Ziegler, D., Reljanovic, M., Mehnert, H., & Gries, F. A. (1999). Alpha-lipoic acid in the treatment of diabetic polyneuropathy in Germany: Current evidence from clinical trials. *Experimental and Clinical Endocrinology & Diabetes, 107*, 421–430.

CHAPTER NINE

Sensory Neurodegeneration in Diabetes: Beyond Glucotoxicity

D.W. Zochodne[1]

Neuroscience and Mental Health Institute and Alberta Diabetes Institute, University of Alberta, Edmonton, AB, Canada

[1]Corresponding author: e-mail address: zochodne@ualberta.ca

Contents

Abstract

Diabetic polyneuropathy in humans is of gradual, sometimes insidious onset, and is more likely to occur if glucose control is poor. Arguments that the disorder arises chiefly from glucose toxicity however ignore the greater complexity of a unique neurodegenerative disorder. For example, sensory neurons regularly thrive in media with levels of glucose at or exceeding those of poorly controlled diabetic persons. Also, all of the linkages between hyperglycemia and neuropathy develop in the setting of altered insulin availability or sensitivity. Insulin itself is recognized as a potent growth, or trophic factor for adult sensory neurons. Low doses of insulin, insufficient to alter blood glucose levels, reverse features of diabetic neurodegeneration in animal models. Insulin resistance, as occurs in diabetic adipose tissue, liver, and muscle, also develops in sensory neurons, offering a mechanism for neurodegeneration in the setting of normal or elevated insulin levels. Other interventions that "shore up" sensory neurons prevent features of diabetic polyneuropathy from developing despite persistent hyperglycemia. More recently evidence has emerged that a series of subtle molecular changes in sensory neurons can be

International Review of Neurobiology, Volume 127
ISSN 0074-7742
http://dx.doi.org/10.1016/bs.irn.2016.03.007

linked to neurodegeneration including epigenetic changes in the control of gene expression. Understanding the new complexity of sensory neuron degeneration may give rise to therapeutic strategies that have a higher chance of success in the clinical trial arena.

1. TERMINALS AT RISK: SENSORY NEURODEGENERATION IN DIABETES

Persons that develop diabetic polyneuropathy (DPN) may have underlying type 1 or type 2 diabetes mellitus (DM). With rare exceptions, DPN develops gradually and insidiously, targeting the distal terminals of the sensory neurons first (Zochodne, Ramji, & Toth, 2008). This usually translates into damage to the longest sensory neurons, innervating the toes. Newer approaches, however, have also identified early nerve terminal retraction in shorter neurons, such as trigeminal axons that innervate the cornea, assessed by corneal confocal microscopy (Malik, Kallinikos, Abbott, et al., 2003). Most persons experiencing symptoms of DPN first report tingling, pain, or loss of sensation in their toes. With time, and particularly if glucose control is poor, neuropathy advances with sensory loss involving more of the extremities and even in the central chest, the terminal zones of the intercostal sensory nerves. Epidermal biopsies have confirmed that distal extremity loss of sensory axon terminals is greater than that observed at more proximal sites, such as the thigh (Polydefkis, Hauer, Griffin, & McArthur, 2001). This pattern of onset, shared with other chronic polyneuropathies, bespeaks a unique pattern of gradual neurodegeneration with initial manifestations in distal sensory territories.

The gradual loss of distal terminals during chronic diabetes is not accompanied, to our knowledge, by early or substantial loss of parent sensory neurons in dorsal root ganglia (DRGs). While recent investigations into DRG pathology in human diabetes are lacking, original work suggested a lack of significant parent neuron damage (Greenbaum, Richardson, Salmon, & Urich, 1964; Schmidt, Dorsey, et al., 1997). In several animal models, despite loss of distal footpad epidermal axons, neuron numbers in DRGs are preserved (Kamiya, Zhangm, & Sima, 2005; Zochodne, Verge, Cheng, Sun, & Johnston, 2001). While not discussed in this chapter, similar preservation of autonomic primary neurons also appears to be the case (Schmidt et al., 1998; Schmidt, Dorsey, et al., 1997). Autonomic dysfunction, a separate form of diabetic targeting, does offer many parallels to sensory loss and often accompanies it. Collectively, these findings have indicated that neurons do not experience classical apoptosis early in diabetes and that substantial dropout is not a feature of the disease. Thus, acute hyperglycemia or "glucotoxicity" in humans does not appear to lead to early and massive neuronopathy. Overall, this information is important news for DPN since it indicates that once DM is reversed by therapy or islet transplantation, preserved parent neurons are available to reestablish connections with terminals. Improvement in the sensory deficits of DPN may be possible.

Why chronic DM exhibits this topographical distribution of change has generated considerable thought. That it is a generalized feature of neurodegeneration is supported by similar, albeit more rapid changes in ALS, where motor terminals are retracted before the loss of anterior horn cells. Several ideas are relevant to understanding this form of neuronal alteration. Loss of distal terminals in models accompanies early key, but nonlethal alteration of the perikaryal (cell body) gene output and protein synthesis (Cheng, Kobayashi, Martinez, et al., 2015). For example, declines in the gene expression of neurofilament polymer subunits in the cell body accompany parallel declines of similar magnitude in neurofilament investment of distal axons (Scott, Clark, & Zochodne, 1999). A decline in neurofilaments, part of the key internal structural lattice of the axon, within distal nerves accompanies distal axon atrophy. However, declines in neurofilament investment probably do not completely account for a number of other features of DPN, such as conduction velocity slowing. In mice lacking axonal neurofilaments, diabetes caused accelerated conduction abnormalities, indicating a separate alteration in membrane excitability (Zochodne, Sun, Cheng, & Eyer, 2004). Since axons themselves are now recognized to translate proteins, declines in perikaryal mRNAs targeted toward distal axonal ribosomes may similarly connect subtle changes in cell body function with the behavior and later retraction of terminals (Willis & Twiss, 2006). Three additional mechanisms might account for distal axon targeting by diabetes. First, the intimate relationship between Schwann cells (SCs) and axons, including the exchange of ribosomes, illustrates an important codependence (Court, Hendriks, MacGillavry, Alvarez, & van Minnen, 2008). Axons support SCs with mitogenic molecules such as neuregulin and CGRP, whereas SCs supply extracellular basement membrane molecules that support axon growth and trophic factors that are taken up and signal axons and their parent neurons. Second, considerable attention has been devoted toward explaining DPN as a mitochondriopathy with neuronal energy failure arising from oxidative stress and other mechanisms (Bennett, Doyle, & Salvemini, 2014; Fernyhough & McGavock, 2014; Ma, Farmer, Pan, et al., 2014; Ma, Pan, Anyika, Blagg, & Dobrowsky, 2015; Sivitz & Yorek, 2010). Mitochondrial dysfunction may be important in the development of chemotherapy-related polyneuropathies and there are specific alterations documented in heredity motor and sensory polyneuropathies, such as mutations in mitofusin-2, a mitochondrial fusion protein. Why distal mitochondria, transported from the cell body, would be selectively targeted is uncertain and why patients with diabetes do not exhibit myopathy, a more common target of mitochondriopathies, is also unexplained. Despite these questions, important connections between mitochondria and axon viability may be critical to understanding axonopathy.

Third, distal targeting of nerve terminals in diabetes may occur because of their demands to undergo ongoing growth and plasticity. In the skin, normal shedding and egress of keratinocytes are part of a microenvironment that is constantly changing and remodeling. To maintain their role within the skin, and perhaps other structures, persistent growth and branching are required. For example, epidermal nerve terminals express higher levels of GAP43, a growth protein, than do "stable" more proximal axons (Cheng, Guo, Martinez, Singh, & Zochodne, 2010). Local trophic interactions between the skin and these terminals may be quite active and include roles for NGF, HGF, and others.

Overall, the neurodegenerative phenotype of sensory neurons in diabetes is unique and is probably paralleled by autonomic neurons and later by motor neurons. Attribution of this phenotype to glucotoxicity alone is problematic, given that at least in the short term, adult sensory neurons grown in hyperglycemic (20–30 mmol/L or higher) conditions fare remarkably well with no evidence of neuronal death or terminal retraction (Huang, Price, Chilton, et al., 2003; Singh, Singh, Krishnan, et al., 2014). While toxicity has been described at higher concentrations of glucose (Russell, Sullivan, Windebank, Herrmann, & Feldman, 1999), these studies reflect conditions not experienced chronically by patients, if at all.

2. NOT NECESSARILY A MICROVASCULAR DISEASE

DPN continues to be listed as a "microvascular" complication of DM, despite equivocal evidence for this description. Part of the confusion arises because DPN prevalence correlates in human epidemiological studies with that of cardiovascular risk factors including atherosclerosis (Tesfaye, Chaturvedi, Eaton, et al., 2005). These correlations however do not prove cause and effect. Against the "microvascular" hypothesis is evidence that DPN can develop in children without long-term disease of their blood vessels, in as short a period as 3 months after the development of DM (Rundles, 1945). There has also been no evidence that specific cardiovascular interventions blunt DPN prevalence in humans. The selective involvement of peripheral nerves, especially sensory axons, is unexplained by this hypothesis since vasculopathy would be more likely to target more ischemia-prone tissues such as brain and spinal cord. Peripheral nerves have a redundant vascular supply and only relatively severe ischemia leads to axonal degeneration or alterations in nerve function (Schmelzer, Zochodne, & Low, 1989). Nerve blood vessels are likely to exhibit enough structural change to make human nerves ischemic only later in the disease; the role of microangiopathy in the development of DPN remains controversial. Finally the more recent identification of loss of corneal nerves seems incompatible with an ischemic etiology in this anatomical distribution (Tavakoli, Quattrini, Abbott, et al., 2010).

Some older experimental models, largely in rat, have identified reductions in nerve blood flow (NBF) in DPN (Cameron, Cotter, & Low, 1991). This literature was extensively reviewed in a previous edition of this text, concluding that there is significant uncertainty of their importance (Zochodne, 2002). Reductions in NBF do not appear in all models of DM and some long-term models in rats had normal NBF (Zochodne & Ho, 1992). A large number of papers linking improvement in electrophysiological indices of DPN with better NBF have offered an unrealistic palette of potential vascular treatments for the disorder from simple vasodilators to cannabinoids. Independent actions on axons and blood vessels could not be excluded in most of this work, and key physiological variations, including potential changes in nerve temperature, could account for improved nerve conduction in some. Thus, technical issues such as lack of temperature control of preparations, reliance on single nonquantitative measures of NBF, inappropriate intraneural recording electrodes, and other factors have been problematic in the literature. Morphological studies of microvessels in diabetic models with DPN have not identified loss of vessels or attenuation of vessel calibers (Zochodne & Nguyen, 1999). Instead, our own group has identified rises in luminal caliber or evidence of angiogenesis (Zochodne & Nguyen, 1999). Finally, declines in whole nerve trunk blood flow do not adequately explain distal terminal retraction or selective involvement of sensory or autonomic axons.

In humans with advanced DPN, reductions in nerve vascular transit times and nerve hypoxia have been reported (Newrick, Wilson, Jakubowski, Boulton, & Ward, 1986; Tesfaye, Harris, Jakubowski, et al., 1993). However, these patients had evidence of advanced disease, reflecting parallel microvascular and nerve disease. In patients undergoing nerve biopsy for research purposes in a DPN trial, direct measures of blood flow did not identify declines in NBF (Theriault, Dort, Sutherland, & Zochodne, 1997). Indeed, NBF measures tended toward higher levels in patients who proved to have more severe DPN, as evidenced by fiber loss and recordings of sensory nerve action potentials.

Despite the difficulties in linking the development, or cause of DPN with initial changes in NBF or microvessels, there is clear evidence that microangiopathy does eventually develop in DM. A variety of epineurial and endoneurial alterations have been carefully described in human nerve biopsy samples, most often in patients with well-established DM and DPN (Korthals, Gieron, & Dyck, 1988; Malik, Newrick, Sharma, et al., 1989; Malik, Veves, Masson, et al., 1992; Yasuda & Dyck, 1987). Taken together, it is likely that microangiopathy develops in parallel with early perikaryal and axon changes in DM and both become prominent in later disease. Functional microvascular changes in the properties of vasa nervorum have been identified in models and may emerge, for example, after nerve injury when rises in blood flow are expected (Kennedy & Zochodne, 2002). The balance of these changes suggests impaired mechanisms of vasodilatation (Thomsen, Rubin, & Lauritzen, 2002).

3. ALTERED INSULIN SIGNALING

Disentangling intimate connections between glycemic control and insulin administration in trials, such as DCCT, is difficult (The Diabetes Control and Complications Trial Research Group, 1993). While patients with better control of hyperglycemia fared better after 5 years, they were also exposed to more physiological and often larger doses of insulin. The robust trophic actions of insulin on neurons, independent of its glycemic actions, have been recognized for several decades. For example, Frazier and colleagues (Frazier, Angeletti, & Bradshaw, 1972) described structural and possible evolutionary parallels between insulin and nerve growth factor (NGF) in 1972. Fernyhough, Tomlinson, and colleagues subsequently described dose-dependent rises in neurite outgrowth of adult sensory neurons exposed to insulin in vitro (Fernyhough, Willars, Lindsay, & Tomlinson, 1993). Insulin receptors are expressed in most sensory neurons and at nodes of Ranvier (Brussee, Cunningham, & Zochodne, 2004; Sugimoto, Murakawa, & Sima, 2002; Sugimoto, Murakawa, Zhang, Xu, & Sima, 2000; Xu et al., 2004) (Fig. 1). Insulin also acts as a growth factor in vivo following nerve injury. Small systemic doses of insulin improved the reinnervation of foot interosseous endplates by motor axons after sciatic nerve transection, increased numbers of regenerating myelinated axons, and enhanced recovery of mouse hindpaw function (Xu et al., 2004). In separate work, low-dose intrathecal administration of insulin also improved structural and functional indices of distal axon regeneration (Toth et al., 2006). Collectively these findings establish a major role for insulin trophic support of neurons and axons during regeneration.

Given all of these observations, it is plausible that sensory neurodegeneration is related to altered trophic support from deficits of insulin signaling. In type 1 DM, there is absence of the insulin ligand. While these patients receive insulin supplementation, its administration may have been inadequate in dose and nonphysiological, given the intermittent regimens many diabetic persons use. In type 2 DM, circulating insulin levels may be normal, elevated, or later lowered. However, these patients usually have insulin resistance involving muscle, liver, or adipose tissue. In type 2 DM models, high doses of insulin fail to reverse the DPN phenotype. We now recognize however that neurons, like muscle, liver, and fat, may be resistant to the trophic properties of insulin, a concept first proposed by Singh et al. (Grote, Morris, Ryals, Geiger, & Wright, 2011; Kim, McLean, Philip, & Feldman, 2011; Singh et al., 2009; Singh, Xu, McLaughlin, et al., 2012). This is important because it indicates that DPN might develop as a result of the downstream failure of insulin to activate growth pathways that support neurons, including their requirement for constant terminal remodeling. Grote et al. noted that ob/ob mice with leptin deficiency and type 2 DM had blunted Akt activation with insulin and IGF-1 and had elevated levels of JNK, a mediator of insulin resistance in other tissues (Grote et al., 2013). Singh et al. (2012) noted that neurons exposed to high doses of insulin, even briefly, failed to respond to subsequent exogenous insulin. Moreover, chronic low-dose insulin exposure similarly blunted subsequent challenges of insulin to support growth signaling. Potential mechanisms for neuronal insulin resistance, with some evidence supporting them, have included upregulated levels of GSK-3β, a known "brake" on axon growth, with downregulated pGSK-3β and

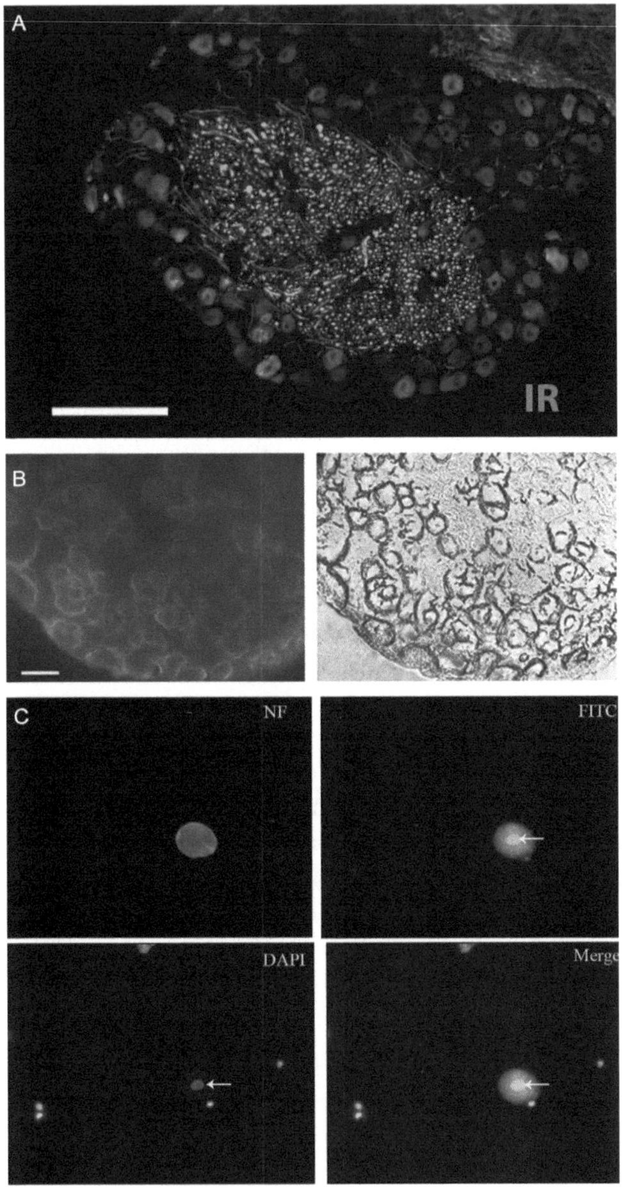

Fig. 1 Insulin directly ligates receptors on sensory neurons. In (A) sensory neurons and axons in the lumbar dorsal root ganglia (DRG) of rats are labeled with an antibody directed against IRβ, a subunit of the insulin receptor (IR) (bar = 200 μm). In (B) DRG neurons are labeled by FITC-labeled insulin that attached to the cell surface. The panel to the *right* is the identical section viewed under light microscopy (bar = 20 μm). Labeled insulin was delivered by intrathecal injection and improved phenotypic features of

lowered levels of pAkt. Exogenous insulin also downregulated expression of its own receptor, IRβ. Additional work has noted that insulin resistance in neurons may be linked to IRS2 serine phosphorylation (Grote et al., 2011).

In addition to problems of inadequate or nonphysiological dosing and resistance, insulin may have difficulty accessing neurons because of the blood–brain barrier so that CSF insulin levels may be different from the blood (Folli, Bonfanti, Renard, Kahn, & Merighi, 1994). To address this possibility, we administered low doses of insulin intrathecally, using chronic catheters, in diabetic rats (Brussee et al., 2004). This route and dosing of insulin did not influence systemic glucose levels. Intrathecal insulin offered dose-dependent improvements in motor and sensory conduction velocities. It corrected axonal atrophy and improved epidermal innervation of the footpads. A control insulin infusion, otherwise identical, into the subcutaneous skin of the back of the diabetic rat did not improve neuropathy. The findings are particularly remarkable given this evidence that "central" administration, at the level of the spinal fluid that accesses DRG, was able to correct electrophysiological and structural changes in distal axons. This is evidence of a robust connection between perikaryal function and axons. In additional work, we showed that intrathecal administration of antiinsulin antibody, an intervention designed to sequester CSF insulin, generated conduction abnormalities and axonal atrophy in nondiabetic rats, resembling the changes of DM (Brussee et al., 2004). Dosing with antialbumin antibody in identical doses had no impact. This work indicated that ongoing, or ambient insulin signaling through its elaboration in spinal fluid that bathes DRG within root sleeves, is important for the ongoing maintenance of sensory neurons. Interruption of this support elicits neuropathic abnormalities.

As an alternative route to direct intrathecal infusion, intranasal administration may be an option for administering low doses to access DRG neurons (Reger, Watson, Green, et al., 2008). This route, tested in human Alzheimer's disease, allows CSF access of the peptide and there is evidence it may improve DPN (C. de la Hoz & D. Zochodne, unpublished data; Francis, Martinez, Liu, et al., 2011; Toth et al., 2007).

experimental diabetic neuropathy. In (C) dissociated adult rat sensory neurons are triple labeled with neurofilament antibody (*red; gray* in the print version), FITC-labeled insulin (*green; gray* in the print version) added to the media, and DAPI (*blue; gray* in the print version) illustrating that insulin is taken up by neurons and labels both the cytoplasm and nucleus of sensory neurons. *Reproduced with permission from the American Diabetes Association and Brussee, V., Cunningham, F. A., & Zochodne, D.W., 2004. Direct insulin signaling of neurons reverses diabetic neuropathy.* Diabetes, 53, *1824–1830.*

The earlier work has established that insulin ligation of receptors at the level of the perikaryon, or cell body influences the behavior of the entire neuronal tree, repairing damage in distal axon segments. There is additional evidence that distal axons also express the insulin receptor. During injury and regeneration, receptors were identified on regrowing axons just beyond the injury site (Xu et al., 2004). Singhal, Cheng, Sun, and Zochodne (1997) injected low subhypoglycemic doses of insulin around sciatic nerves of rats with experimental diabetes and compared their function with contralateral limb sciatic nerves exposed to carrier. Despite ongoing evidence of hyperglycemia, and progressive DPN in the opposite limb, nerves exposed to local insulin had improvements in electrophysiology and had a rise in the number of small myelinated axons. Dermal and epidermal axons also express the insulin receptor and mRNAs for downstream insulin transduction pathways, IRS-1 and IRS-2, are found in skin. As discussed earlier, this supports the idea that this "normal" population of axons is in a growth state, constantly remodeling to keep apace of keratinocyte turnover. Guo, Kan, Martinez, and Zochodne (2011) injected small dermal doses of insulin, insufficient to alter systemic glucose levels into the hind paw of mice with chronic DM of 5 months duration. The contralateral paws in the same mice were injected with carrier. Local insulin, but not carrier, unilaterally improved epidermal innervation of the hind paw over a surprisingly short period of within 1 week. Improvements occurred in both type 1 and 2 diabetic mice. These findings support the concept that ongoing plasticity of skin innervation offers opportunities for short-term manipulation of regrowth to improve neurological function. Chen, Calcutt, and colleagues (Chen et al., 2013) have also demonstrated similar short-term plasticity, over 4 weeks, of nerve fibers in the subbasal plexus of cornea in rats. DM in this model was associated with a progressive decline in corneal innervation, whereas local administration of insulin over the cornea prevented loss of corneal axons.

4. ONGOING GROWTH: OTHER FORMS OF SUPPORT

4.1 C-Peptide

Novel routes of insulin administration may offer feasible and potent therapy for patients in clinical trials. Indeed a pilot Phase I trial of intranasal insulin in type 1 diabetic subjects has been completed to evaluate safety (Korngut, Mawani, Francis, et al., 2013). Work is ongoing. The concept of insulin resistance does add a cautionary note to relying on the insulin peptide alone for chronic therapy in diabetic patients. These patients may have already been exposed to daily systemic injection or they may exhibit systemic insulin resistance. In patients with established DPN, often first presenting to clinicians, reversing DPN has usually not been possible and features of DPN persist despite improvements in glucose control. Experimentally, it also appears difficult to reverse features of DPN in models with established disease (Kan, Guo, Singh, Singh, & Zochodne, 2012). For all of these reasons, moving downstream of insulin, or identifying approaches to supplement its actions may be important. For example, extensive work on C-peptide, the normally cleaved bridging peptide of proinsulin, suggests that it benefits DPN through an insulin-sensitizing action. C-peptide has evidence of benefit in models and early human trials (Ekberg et al., 2003; Kamiya, Zhang, Ekberg, Wahren, & Sima, 2006; Zhang et al., 2001).

4.2 Glucagon-Like Peptide-1

GLP-1 is an incretin peptide, released by the intestine following meal ingestion (Baggio & Drucker, 2007). Its actions include enhancing insulin secretion and sensitivity, of importance in type 2 DM. However, GLP-1 also has direct actions in the nervous system that include trophic actions, impacts on CNS function, and protection from pyridoxine neuropathy (Perry, Holloway, Weerasuriya, et al., 2007). Along with work of two independent groups, we identified impacts of GLP-1 in experimental DPN (Himeno, Kamiya, Naruse, et al., 2011; Jolivalt, Fineman, Deacon, Carr, & Calcutt, 2011; Kan et al., 2012). Exendin-4, isolated from *Heloderma suspectum*, is a GLP-1 agonist that is resistant to dipeptidyl peptidase-4 cleavage allowing a longer half-life. GLP-1 receptors are widely expressed in DRG sensory neurons and exendin-4 had direct trophic actions on adult sensory neurons in vitro, increasing their outgrowth. In models of both experimental type 1 and type 2 DM, exendin-4 offered benefits among established indices of DPN including motor and sensory conduction slowing and mechanical and thermal sensory loss. While the impact was incomplete and did not reverse all facets of the disorder (for example, sensory conduction and mechanical sensation in type 2 DM were not reversed), the improvements differed from those offered by systemic insulin. For some of these indices of DPN, normalizing glucose levels with chronic insulin infusions was unhelpful, whereas exendin-4 provided benefits. Overall, these findings indicated that GLP-1 agonism, like insulin, may improve DPN through direct signaling actions on neurons, but that its actions may be complementary.

4.3 Heat-Shock Proteins

Heat-shock protein (HSP) 27 is a molecular chaperone that helps to refold denatured proteins and inhibits apoptosis. Its levels rise in sensory neurons of experimental DM (Zochodne et al., 2001). Korngut, Ma, Martinez, et al. (2012) studied experimental type 1 DM in mice with overexpression of a human HSP27 transgene in their peripheral neurons. In comparison to wild-type littermates, these mice had attenuation of some of the features of DPN: loss of footpad thermal sensation, mechanical allodynia, loss of epidermal innervation, and slowing of sensory conduction velocity. Protection was more pronounced in female diabetic mice compared to males. RAGE, NFκB, and activated caspase-3 nuclear expression, features of neuronal dysfunction in DM within sensory neurons, were also attenuated. Collectively, these data suggested that upregulation of the HSP27 pathway in sensory neurons protects them from DPN. Since the impact of this approach was incomplete, its actions may need to supplement that of other strategies. The differential impact on female DM mice is unexplained and raises the possibility that gender may have important influences on the development of DPN. Work in the Dobrowsky laboratory has demonstrated that a small molecule modulating another HSP, HSP70, improved the behavioral, electrophysiological, structural, and bioenergetic abnormalities in experimental DPN (Ma et al., 2014).

5. NEURONS "ON EDGE"

In chronic DM sensory neurons might be characterized as being "under stress" without frank cell loss. Unfortunately this is a vague term, usually reserved for a cell at acute risk of apoptotic cell death. In the case of chronic diabetes, several indicators suggest abnormal physiology, although not necessarily imminent demise. DRG sensory neurons express cleaved caspase-3, including nuclear localization (Cheng & Zochodne, 2003). A large literature has linked activated caspase-3 expression as equivalent to apoptosis, a virtual cellular "executioner." However, other markers indicate that the neurons in this long-term model of diabetes were not under imminent threat. Nuclear fragmentation was not a feature of their architecture, TUNEL labeling was negative in neurons, and careful three-dimensional unbiased counts of neurons using the "physical dissector" methodology did not identify neuron dropout, all analyzed in long-duration DM models (12 months in rats). Caspase-3 localization in neurons included a presence in initial axon segments. Schmidt and colleagues, in humans and other models of DPN, have characterized these segments as also harboring prominent neurofilament inclusions (Schmidt, Beaudet, Plurad, & Dorsey, 1997; Schmidt, Dorsey, et al., 1997; Schmidt & Scharp, 1982). Fernyhough and colleagues have identified specific sites of oxidative stress that target axons unevenly (Zherebitskaya, Akude, Smith, & Fernyhough, 2009). While caspase-3 is not usually considered as among the mediators of localized axonal degeneration, mislocalized caspase-3 could potentially offer this kind of role. It is not known whether caspase-3 inhibitors have a potential role in modifying the course of DPN.

PARP (poly (ADP-ribose) polymerase) is a DNA repair molecule downstream of caspase-3 activation. Excessive activation of PARP leads to neuronal death from excessive consumption of ATP. In chronic experimental DPN of rats, we identified rises in the intensity of both nuclear and cytoplasmic PARP. Cytoplasmic expression was not prominent in nondiabetic controls (Cheng & Zochodne, 2003). Similar to caspase-3, unusual localization to initial axon segments was also present. Upregulation of PARP in neurons, however, has not been widely accepted. For example, a separate group did not identify its upregulation in chronic type 1 BB/Wor rats (Kamiya, Zhang, & Sima, 2006; Kamiya et al., 2005). Despite these discrepancies, PARP inhibitors have been noted to protect diabetic mice from DPN. PARP may also be expressed in nerve microvessels, and its inhibition has been thought to provide benefit through inhibition of microangiopathy (Obrosova, Drel, Pacher, et al., 2005; Obrosova, Li, Abatan, et al., 2004).

Chronic hyperglycemia leads to the production and deposition of advanced glycation endproducts (AGEs), irreversible products of nonenzymatic glycation of proteins (Brownlee, Cerami, & Vlassara, 1988). AGEs can be detected in the circulation, the nervous system, and other tissues. Receptors for AGEs (RAGE) are also expressed in the nervous system and specifically in sensory neurons. Moreover, RAGE expression rises in experimental DM and RAGE-null mice appear to be protected from DPN (Cameron, Gibson, Nangle, & Cotter, 2005; C. de la Hoz & D. Zochodne, unpublished data; Toth et al., 2004). AGE–RAGE interactions may therefore have a role in generating neuronal stress and dysfunction. However, like the difficulties described around PARP, the exact role of AGE–RAGE signaling is unclear in the nervous system. For example, inhibiting this pathway in adult sensory neurons in vitro paradoxically impaired adult neuron outgrowth (Saleh, Smith, Tessler, et al., 2013). Inhibitors of AGE formation such as aminoguanidine, metformin, and *n*-phenacylthiazolium bromide (Reddy & Beyaz, 2006) and scavenging of circulating AGEs by soluble RAGE infusion have all been proposed as therapy for DPN and diabetes complications generally. In the case of aminoguanidine, its actions are nonspecific and widespread beyond AGE–RAGE disruption and its use has been associated with significant side effects in humans (Thornalley, 2003).

AGE–RAGE signaling alters neuronal function through the NFκB pathway. NFκB is a transcription factor that is sensitive to varied forms of cellular "stress" (Ahn & Aggarwal, 2005) and signals through its subunits p50 and RelA (p65) that are normally inhibited by cytoplasmic IκB. Degradative phosphorylation of IκB allows divergent p50 and RelA signaling in the nucleus. Downstream, this pathway participates in both neuroprotection and neurodegeneration, and its activity may be attenuated by DM (Fernyhough, Smith, Schapansky, et al., 2005; Massa, Xie, Yang, et al., 2006; Purves & Tomlinson, 2002). Expression studies have also identified rises in NFκB expression in experimental DM (Kan et al., 2012; Korngut et al., 2012; Toth et al., 2004). Therefore, like PARP and AGE–RAGE, NFκB may have varied roles in neuronal survival and DPN.

Nitrergic stress has been listed together with oxidative stress as a mechanism of diabetic complications. This is usually considered to arise from peroxynitrite toxicity, a by-product of the interaction of nitric oxide (NO) and superoxide. In the peripheral nervous system, all three nitric oxide synthase (NOS) isoenzymes are identified: nNOS ("neuronal" NOS) in subsets of sensory neurons, eNOS ("endothelial" NOS) classically described in vascular endothelial cells but also localized to neurons, and iNOS ("inducible" or inflammatory NOS) found in macrophages and other inflammatory cells. NO may contribute to diverse actions in the peripheral nervous system including pain, local vasodilatation, and regeneration (Levy, Hoke, & Zochodne, 1999; Levy, Tal, Hoke, & Zochodne, 2000; Zochodne, Levy, Zwiers, et al., 1999). Mice lacking iNOS, for example, have impaired regeneration linked to abnormal clearance of the products of axonal degeneration, a requirement for subsequent axonal growth (Levy, Kubes, & Zochodne, 2001). In experimental DM, "fingerprints" of NO toxicity were identified in neurons evidenced by nitrotyrosine deposition. There were chronic rises in overall NOS enzymatic activity not accompanied by specific changes in mRNA or immunohistochemical localization of any of the three isoforms. To address the role of NOS isoforms in experimental DM, it was shown that mice lacking iNOS had protection from experimental DPN but those lacking nNOS did not (Vareniuk, Pacher, Pavlov, Drel, & Obrosova,

2009; Vareniuk, Pavlov, & Obrosova, 2008). A peroxynitrite decomposition catalyst was also protective (Drel, Pacher, Vareniuk, et al., 2007).

6. DIABETES, NEURONS, AND EPIGENETICS

A puzzling aspect of sensory neurodegeneration in DM is the range of apparently disconnected phenotypic features that characterize DPN. These include loss of sensory terminal innervation in the epidermis that does not always correlate with uniform and "across the board" behavioral loss of sensation. There are seemingly unrelated electrophysiological changes that include slowing of motor and sensory conduction velocity, the development of resistance to ischemic conduction failure and axonal atrophy. Moreover, these phenotypic features may vary among models and patients. In some models, the behavior abnormalities involve heightened sensitivity with hyperalgesia and allodynia, respectively, rather than loss of sensation. Similarly, on the molecular side, a range of mRNA changes have been observed in differing models, depending on the duration of DM. These include loss of structural protein mRNAs such as neurofilament subunits, and tubulin, loss of growth molecules such as GAP43, rises in injury and stress-related molecules discussed earlier that include NFκB, caspase-3, RAGE, and others, alterations in ion channels (reviewed in Zochodne, 2014), and additional rises in insulin receptor mRNAs. All of these alterations suggest an overriding shift in gene expression that may link the varied manifestations of the disorder. How this might arise however is uncertain. One possibility is that epigenetic alterations of neurons might be layered onto this diverse array of abnormalities. This additional layer of control could be a secondary phenomenon or could represent a primary initial alteration fundamental to the disease.

miRNAs are a key element of epigenetic control. Precursor nucleotides that eventually form miRNAs are exported from the nucleus and processed to form small single-chain endogenous noncoding nucleotides of 19–23 base pair length. These nucleotides then associate with the RISC (RNA-induced silencing complex), a protein complex that cleaves mRNA. miRNAs thus interfere with normal translation through mRNA cleavage but also by binding to mRNAs to impair their translation. A wide family of miRNAs target mRNAs with diverse impacts both in the targets of each miRNA and in overlap of the number of miRNAs that may target a single relevant mRNA. To explore the role of posttranscriptional RNA silencing in the sensory neurons of mice with longstanding and chronic DM, we characterized type 1,

STZ mice untreated for 5 months. These mice exhibit key phenotypic features of DPN including motor and sensory conduction slowing, loss of sensation to thermal and mechanical stimuli, and loss of epidermal innervation. GW/P bodies are ultrastructural cytoplasmic complexes that contain key elements of RISC. Their expression may identify neurons under "stress." We noted that GW/P expression was heightened in the sensory neurons of mice with DM, indicating overt structural evidence of altered mRNA processing as a feature of this chronic disease (Cheng et al., 2015).

In the same cohort of diabetic mice and littermates, we examined 28,869 DRG mRNAs for differential expression of at least a 1.5-fold change in DM samples and within these identified 261 mRNAs that included 91 upregulated and 170 downregulated. Of these 24 achieved a statistical difference between diabetics and nondiabetics of $p<0.05$ (5 down and 19 up). Almost all coded for proteins of unknown function in sensory neurons or diabetes. For example, one upregulated molecule, CWC22, provided a valuable lead to analyze spliceosome function in diabetic sensory neurons, work in progress that offers new ideas about specific molecular deficits in diabetic sensory neurons (Kobayashi, Cheng, de la Hoz, & Zochodne, 2015). We further explored concurrent changes in miRNAs, their supraregulatory companions. As in the case of mRNAs, we identified a number of differentially expressed miRNAs in DRGs from mice with chronic DM and DPN. Of 1042 examined, there were 19 altered that included 12 downregulated and 7 upregulated high-abundance miRNAs. Additionally, there were 123 low-abundance miRNAs altered including 56 downregulated and 67 upregulated. Focusing on miRNAs in the high-abundance group, we studied expression of mmu-let-7i, which was downregulated by 39% in diabetic DRGs, a change also confirmed by qRT-PCR. mmu-let-7i was an interesting miRNA to consider first, given widely divergent impacts on over 900 targets. To begin with, we assessed whether this miRNA was expressed in neurons, rather than DRG satellite cells or vessels. By in situ hybridization, we noted striking expression in most sensory neurons of the DRG (Fig. 2). To determine whether alterations in mmu-let-7i might be associated with an overall phenotype, given its divergent actions, we studied the impact of sensory neuron transfection with an exogenous mmu-let-7i mimic in vitro. The approach was associated with robust trophic actions including an increase in neurite outgrowth and branching. Collectively these findings suggested that the downregulation of mmu-let-7i in chronic diabetic DRG neurons might impair overall supportive or trophic mechanisms.

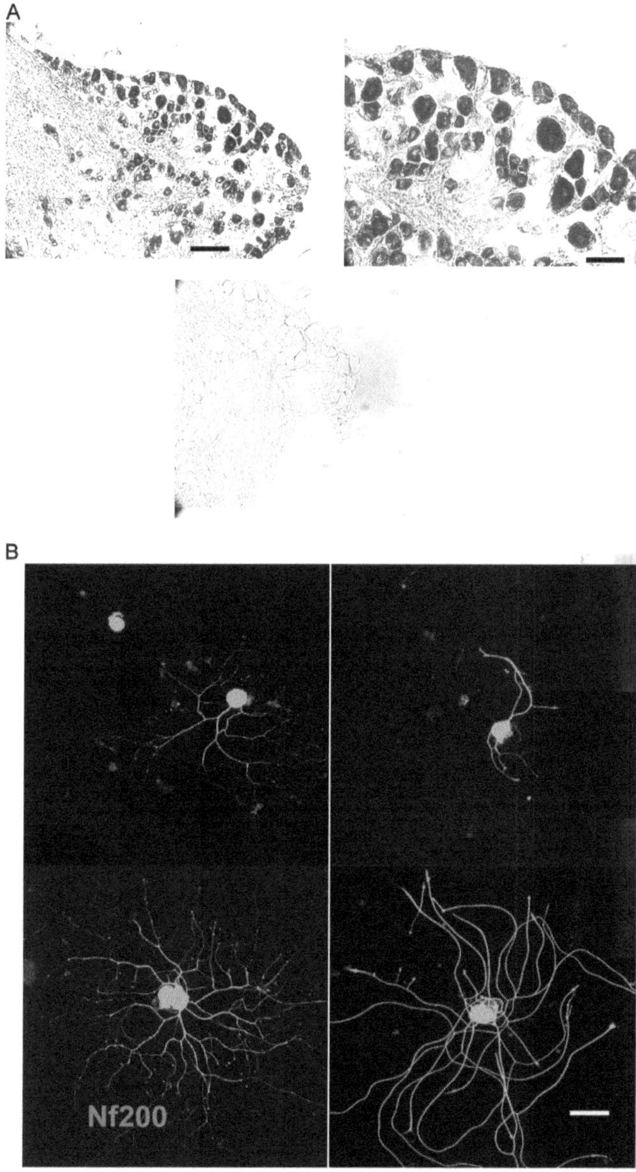

Fig. 2 Epigenetic miRNAs are expressed in sensory neurons and influence their growth properties. In (A), in situ hybridization images of normal mouse DRGs identify mmu-let-7i expressed in sensory neurons at lower (*left*) and higher power (*right*) (bar = 100 and 50 μm, respectively). A control image without label is below. In (B) are illustrated confocal images of preinjured (axotomized) adult rat sensory neurons harvested from lumbar DRG and grown in the presence of control carrier solution (*top images*) or a mimic molecule of mmu-let-7i for 20 h and stained with antineurofilament antibody (bar = 100 μm). *Reproduced with permission from Cheng, C., Kobayashi, M., Martinez, J. A., et al. (2015). Evidence for epigenetic regulation of gene expression and function in chronic experimental diabetic neuropathy. Journal of Neuropathology and Experimental Neurology, 74, 804–817.*

Identifying the specific and relevant target genes responsible for a growth phenotype in sensory neurons is challenging. A comparison of parallel microarrays, in the same diabetic and nondiabetic mouse cohorts, examining statistically significant changes in mRNAs due to diabetes with four of our altered miRNAs identified some predicted changes. For example, these included upregulation of connective tissue growth factor, upregulation of dual-specifity phosphatase 1, upregulation of F3 and TXNIP (a thioredoxin-interacting protein that may mediate oxidative stress), a small rise in insulin receptor, and downregulation of CACNG4, a subunit of the calcium L-type channel. Most have unclarified relationships to sensory neuron function to date. Two previous reports have linked abnormalities in calcium channels to experimental diabetes (Hall, Sima, & Wiley, 1995; Voitenko, Kruglikov, Kostyuk, & Kostyuk, 2000). These predicted changes were only based on four high-abundance miRNA changes, suggesting that much more extensive interactions occur than we explored in this work. However, none of the apparent alterations mediated by these mRNAs offered clear or established candidates for growth enhancement.

The next step in addressing the relevance of miRNA changes in chronic DM is understanding whether these changes represent an epiphenomenon of the disease or whether they effect critical changes in neuron function. Given confirmation that mmu-let-7i was downregulated in DRGs with clear impacts on sensory neuron biology, we tested whether supplementing neurons with exogenous mmu-let-7i, in the form of a mimic molecule, might alter the DPN phenotype. The approach to this was not trivial given that access of the nucleotide to sensory perikarya would be key to its potential actions. A viral vector approach to transfection may be a potent option for stable replenishment of mmu-let-7i. However, with a thought toward downstream human translation, we tested a nonviral approach that involved intranasal administration of an mmu-let-7i mimic molecule. To test whether miRNAs might access the CSF and DRGs through their root sleeves by this approach, we first administered an unrelated miRNA by intranasal delivery—a plant species miRNA–*Arabidopsis thaliana* miR171, not present in mammals. After administration for 6 days, miR171 was detected in both the olfactory bulb and the lumbar DRG of mice indicating access to the CSF through this route. Next we administered a mimic mmu-let-7i by the intranasal route and confirmed that in nondiabetic mice, daily dosing for 1 week could raise mmu-let-7i levels as tested in DRG by qRT-PCR.

Given the evidence that a nonconventional route of miRNA delivery had the potential to access the CNS and spinal spaces including DRG root sleeves, we tested its impact on diabetic mice. Using mice with 5-month duration DM and well-established indices of DPN, we administered intranasal mmu-let-7i mimic over 6 days and retested after a further 3 weeks. The rationale was that alterations in gene expression effected by mmu-let-7i might require at least 2–3 weeks to alter epidermal innervation and other features of DPN. Intranasal mmu-let-7i, but not control treatment, improved impaired diabetic mechanical sensitivity to the levels of mice without DM and improved loss of thermal sensation, both accompanied by an improvement in epidermal innervation. Motor and sensory conduction velocities were also improved by the mmu-let-7i mimic. Additional work examined whether knockdown of a second separate, but in this case, elevated miRNA mmu-341, in diabetic mice might also impact DPN. In contrast to mmu-let7i, mmu-341 was the most robustly upregulated miRNA detected in our array survey. In this case contrary knockdown with an miRNA anti-miR also improved features of DPN, although the results were less robust, influencing only sensory conduction and thermal sensation. Taken together, however, these findings identified a striking impact of nucleotide delivery on structure, electrophysiology, and behavior in diabetic mice.

Epigenetic manipulation of a range of mRNA targets altered in DM is an attractive "single bullet" that might be available to repair neuropathic damage. Albeit promising, particularly in the use of nonviral delivery, the approach needs independent confirmation. The overall possibilities linked to manipulating gene alterations in neurological disease using intranasal access to the CSF are also enormous, but require verification. Despite these caveats, the observations that relatively short-term interventions, presumably targeting altered but not lost parent neurons, might allow recovery from deficits that have developed chronically over several months are exciting. There are other facets of epigenetic regulation in DM that may be equally important, including targeting of SCs, also key targets of DM, that are not discussed in this review.

7. A REGENERATION STRATEGY

The axon terminals of adult sensory neurons might be considered as being in a permanent and ongoing state of growth. Within the epidermis, the constant movement, replenishment, and shedding of keratinocytes require that axons adapt and remodel to retain their roles. Indeed, the trajectories of epidermal axons, while well spaced, are highly irregular, with twists and turns that suggest ongoing growth. More direct evidence arises from their prominent expression of growth proteins such as GAP43 and the finding that interventions, including simple and noninvasive steps such as hair clipping, can dramatically influence the density of innervation (Cheng et al., 2010).

In chronic DM with DPN, epidermal terminals are retracted, requiring regrowth to restore their innervation. Collateral sprouting, distinct from regenerative sprouting, allows reinnervation of denervated territories by invasion from intact neighbor axons. Work by Diamond and colleagues (Diamond, Coughlin, Macintyre, Holmes, & Visheau, 1987; Diamond, Holmes, & Coughlin, 1992) has shown that skin collateral sprouting, but not regenerative sprouting, is NGF dependent. Collateral sprouting does occur in DM (Theriault, Dort, Sutherland, & Zochodne, 1998), but this form of recovery generally may not be sufficiently robust to completely restore sensory fidelity. With this caveat in mind, retracted axons in DM may be called upon to regrow into their previous territories, forestalling ingrowth by neighbors. Taken together, strategies that support both forms of neurological recovery may be required.

In DM, there are other considerations that make regenerative strategies important. Patients with DM are disposed to develop single nerve lesions, or focal neuropathies such as carpal tunnel syndrome or ulnar neuropathy at the elbow (Zochodne, 2007). A more exacting clinical terminology thus refers to diabetic neuropathies, recognizing a range of additional complications involving the peripheral nervous system. Focal neuropathies are common and disabling. Considering the range of neurological complications identified, DM imposes a "double hit," a degenerative polyneuropathy, or DPN, but also a deficit in regenerative capacity. For example, in carpal tunnel syndrome, surgical decompression of the entrapped nerve at the wrist and subsequent recovery has a less satisfactory outcome in DM.

Mechanisms for impaired diabetic nerve regeneration have been considered extensively and have included impaired neurotrophic support, microangiopathy that renders the regenerative microenvironment unsupportive, local oxidative stress, impaired macrophage clearance, accelerated retrograde loss of neurons, polyol flux, mitochondrial dysfunction, nonenzymatic glycosylation of basement membrane regeneration scaffolds, SC dysfunction, and others (Kennedy & Zochodne, 2005). New molecular approaches to coax greater plasticity and regrowth out of reluctant adult neurons have highlighted the potential roles of targeting their intrinsic growth mechanisms. These operate downstream of growth factor signals and may be final arbiters of growth cone behavior. Several "brakes" to regrowth fall within the class of "tumor suppressors," molecules that help block uncontrolled oncogenic growth. The first example in this category is PTEN (phosphatase and tensin homolog deleted on chromosome 10), a tumor suppressor molecule that inhibits PI3K–pAkt signaling. The latter is a critical growth pathway for neurons and its inhibition or "braking" attenuates growth. Unlike oncogenesis however, neurons are characterized by poor growth. Persistent expression of PTEN in neurons, despite injury and regenerative demands, is an example of a counterproductive barrier to growth. In the intact adult nervous system, limitations of growth behavior may be important in protecting fine and detailed connections, especially in the CNS. After injury however, PTEN may attenuate growth and recovery. In adult sensory neurons, inhibition or knockdown of PTEN was associated with robust neurite outgrowth (Christie, Webber, Martinez, Singh, & Zochodne, 2010). Outgrowth was most prominent in preinjured neurons, a remarkable finding. Given that preexisting injury acts as a "conditioning lesion" that ramps up regenerative behavior, further acceleration from PTEN knockdown indicates a synergistic action beyond preconditioning. Regrowth at this level has not previously been described. Following nerve transection in adult rats, PTEN inhibition or knockdown increased the outgrowth of adult axons from the proximal stump. However, PTEN expression in sensory neurons is uneven, with very prominent expression in small caliber nonpeptidergic IB4 neurons. These neurons have been linked with significantly slower growth properties, additional evidence of a close link between PTEN expression and regenerative potential (Guo, Singh, & Zochodne, 2014; Tucker, Rahimtula, & Mearow, 2006). Since this class of neurons also innervates the skin, PTEN's role in diabetic epidermal regrowth may be important.

PTEN is upregulated in diabetic sensory neurons (Singh et al., 2014; Fig. 3). Not only do individual neurons appear to have more intense expression, but also higher expression neurons seem to be more widely distributed in the DRG. Emphasizing the potential importance of PTEN action, in vitro adult sensory neurons from mice with chronic DM that were harvested and analyzed in culture retained the capacity for increased growth following PTEN knockdown. In mice with chronic DM and established indices of DPN, we studied the impact of PTEN knockdown on recovery from superimposed focal nerve injuries (Fig. 3). Nonviral siRNA delivery at the site of crush injury was associated with ipsilateral DRG PTEN mRNA knockdown at the DRG and spinal cord with evidence that the nucleotide

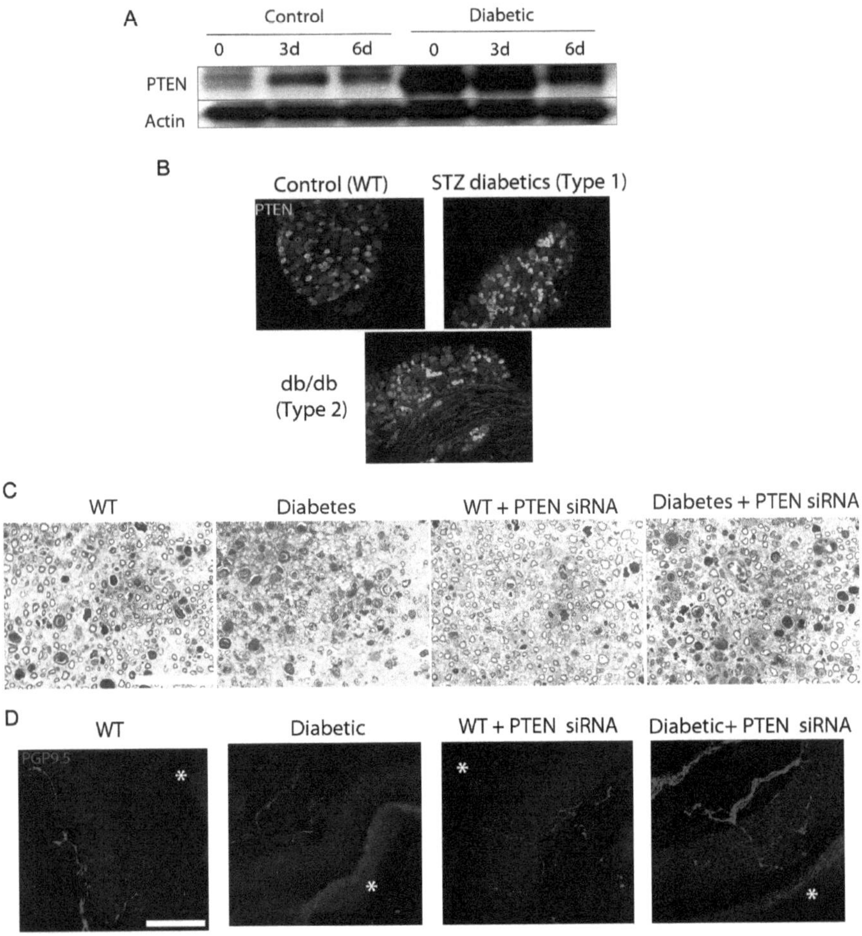

Fig. 3 PTEN, a phosphatase that acts as a "brake" on regeneration, is expressed in adult sensory neurons, upregulated in diabetes. PTEN knockdown improves regeneration. In (A) Western immunoblots are illustrated that are labeled with PTEN (beta-actin control band) in wild type and diabetic dorsal root ganglia before (basal, Day 0) 3 and 6 days after sciatic nerve injury. Expression was analyzed in 3-month type 1 diabetic mice with or without nerve injury and 12-week-old type 2 diabetic (db/db) mice. In (B) are illustrated examples of adult sensory neurons from wild-type, streptozotocin (STZ)-induced type 1 diabetic, and db/db, type 2 diabetic dorsal root ganglia labeled with PTEN (*green*; *gray* in the print version). Note that PTEN is expressed in a larger number of neurons with brighter luminosity in diabetics (bar = 100 μm). In (C) and (D) are illustrated the impact of PTEN siRNA on in vivo regeneration of axons following injury. Analysis is carried out in nondiabetic mice and mice with type 1 diabetes (3 months) treated for an additional 1 month (28 days). In (C) are representative images of semithin sections of regenerating distal tibial nerves (10 mm from the injury site) from wild-type and diabetics 21 days after sciatic nerve injury with and without PTEN siRNA. Diabetic

was retrogradely transported along axons to inhibit gene expression. PTEN knockdown improved the recovery of motor compound muscle action potentials, conduction velocities of regenerating motor and sensory axons, repopulation of myelinated axons distal to the injury site, and reinnervation of the hind paw skin. Taken together the findings demonstrated not only an active inhibitor role of PTEN in diabetic axon regrowth but also a novel approach to unilaterally effect alterations in gene expression of injured nerves.

The RhoA–ROK (RhoA kinase) pathway is an example of an intrinsic regenerative "brake" that causes growth cone retraction. It operates through one of several mechanisms including phosphorylation of cofilin that interferes with actin turnover in growth cones (Ng & Luo, 2004), inhibition of myosin phosphatase that increases myosin ATPase activity (Luo, Jan, & Jan, 1997), and direct phosphorylation of myosin (Giniger, 2002) converging on increased central actin bundle contractility and stability to collapse growth cones (Jin, Guan, Jiang, et al., 2005). RhoA and ROK are expressed in peripheral sensory neurons and ROK inhibition increased the outgrowth of neurites in primary neuronal cultures. ROK inhibition also increased regrowth of axons in vivo following a nerve transection (Cheng, Webber, Wang, et al., 2008). This pathway has not been explored in DPN.

Additional pathways, also yet to be explored in DM, may offer further targets for influencing nerve regeneration. In our laboratory, more recent work has marshaled evidence that a second tumor suppressor molecule, retinoblastoma 1 (Rb1), may operate as a regenerative brake, similar to PTEN and RhoA, to peripheral neuron regrowth (Christie, Krishnan, Martinez, et al., 2014). Rb1 inhibits divergent transcriptional signals by binding to E2F. It is widely expressed in sensory neurons, and its knockdown in vitro and in vivo increases axon outgrowth. Rb1 may mediate its actions through the PPARγ pathway.

animals displayed regeneration deficits with significantly fewer and smaller caliber axons regenerating 3 weeks after injury. PTEN inhibition partially rescued the deficit (bar = 50 μm). In (D) are representative immunohistochemically labeled (PGP 9.5) images of footpads (*external surface of the skin) from wild type and diabetics with and without PTEN siRNA indicating newly regenerating sensory afferents in the epidermis. Reinnervation was reduced in diabetics with few reinnervating axons crossing into the epidermis and improved with PTEN siRNA (bar = 100 μm). *Reproduced with permission from Singh, B., Singh, V., Krishnan, A., et al. Regeneration of diabetic axons is enhanced by selective knockdown of the PTEN gene.* Brain, 137, *1051–1067.*

8. CONCLUSIONS

Exploiting novel molecular approaches to support sensory neurons in chronic DM may offer a palette of therapeutic targets. These are urgently required in a field that has suffered from a series of failed clinical trials. Moving beyond "microvascular" ideas might include approaches to offer direct neuronal insulin signaling through new delivery routes, adding GLP-1 agonism to the mix and supporting neurons with HSP27 or inhibitors of caspase-3, PARP, or AGE–RAGE signaling. New epigenetic understanding of the pathogenesis of this complex disorder is warranted. Finally, new strategies that support the regenerative potential of peripheral neurons damaged by DM may be essential to restore neurological function in patients.

ACKNOWLEDGMENTS

The work highlighted in this review was supported by operating grants from the Canadian Institutes of Health Research (FRN184584), the Canadian Diabetes Association (OG-3-12-3669), the Juvenile Diabetes Research Foundation, and the National Institutes of Health (NIDDK), USA. The author has been supported by the Alberta Heritage Foundation for Medical Research and the University Hospital Foundation, Edmonton. The Zochodne laboratory is supported by the Neuroscience and Mental Health Institute, Alberta Diabetes Institute, Division of Neurology and Department of Medicine, University of Alberta.

REFERENCES

Ahn, K. S., & Aggarwal, B. B. (2005). Transcription factor NF-kappaB: A sensor for smoke and stress signals. *The Annals of the New York Academy of Sciences*, *1056*, 218–233.

Baggio, L. L., & Drucker, D. J. (2007). Biology of incretins: GLP-1 and GIP. *Gastroenterology*, *132*, 2131–2157.

Bennett, G. J., Doyle, T., & Salvemini, D. (2014). Mitotoxicity in distal symmetrical sensory peripheral neuropathies. *Nature Reviews. Neurology*, *10*, 326–336.

Brownlee, M., Cerami, A., & Vlassara, H. (1988). Advanced glycosylation end products in tissue and the biochemical basis of diabetic complications. *The New England Journal of Medicine*, *318*, 1315–1321.

Brussee, V., Cunningham, F. A., & Zochodne, D. W. (2004). Direct insulin signaling of neurons reverses diabetic neuropathy. *Diabetes*, *53*, 1824–1830.

Cameron, N. E., Cotter, M. A., & Low, P. A. (1991). Nerve blood flow in early experimental diabetes in rats: Relation to conduction deficits. *The American Journal of Physiology*, *261*, E1–E8.

Cameron, N. E., Gibson, T. M., Nangle, M. R., & Cotter, M. A. (2005). Inhibitors of advanced glycation end product formation and neurovascular dysfunction in experimental diabetes. *The Annals of the New York Academy of Sciences*, *1043*, 784–792.

Chen, D. K., Frizzi, K. E., Guernsey, L. S., Ladt, K., Mizisin, A. P., & Calcutt, N. A. (2013). Repeated monitoring of corneal nerves by confocal microscopy as an index of peripheral neuropathy in type-1 diabetic rodents and the effects of topical insulin. *Journal of the Peripheral Nervous System*, *18*, 306–315.

Cheng, C., Guo, G. F., Martinez, J. A., Singh, V., & Zochodne, D. W. (2010). Dynamic plasticity of axons within a cutaneous milieu. *The Journal of Neuroscience*, *30*, 14735–14744.

Cheng, C., Kobayashi, M., Martinez, J. A., et al. (2015). Evidence for epigenetic regulation of gene expression and function in chronic experimental diabetic neuropathy. *Journal of Neuropathology and Experimental Neurology*, *74*, 804–817.

Cheng, C., Webber, C. A., Wang, J., et al. (2008). Activated RHOA and peripheral axon regeneration. *Experimental Neurology*, *212*, 358–369.

Cheng, C., & Zochodne, D. W. (2003). Sensory neurons with activated caspase-3 survive long-term experimental diabetes. *Diabetes*, *52*, 2363–2371.

Christie, K. J., Krishnan, A., Martinez, J. A., et al. (2014). Enhancing adult nerve regeneration through the knockdown of retinoblastoma protein. *Nature Communications*, *5*, 3670.

Christie, K. J., Webber, C. A., Martinez, J. A., Singh, B., & Zochodne, D. W. (2010). PTEN inhibition to facilitate intrinsic regenerative outgrowth of adult peripheral axons. *The Journal of Neuroscience*, *30*, 9306–9315.

Court, F. A., Hendriks, W. T., MacGillavry, H. D., Alvarez, J., & van Minnen, J. (2008). Schwann cell to axon transfer of ribosomes: Toward a novel understanding of the role of glia in the nervous system. *The Journal of Neuroscience*, *28*(43), 11024–11029.

Diamond, J., Coughlin, M., Macintyre, L., Holmes, M., & Visheau, B. (1987). Evidence that endogenous beta nerve growth factor is responsible for the collateral sprouting, but not the regeneration, of nociceptive axons in adult rats. *Proceedings of the National Academy of Sciences of the United States of America*, *84*, 6596–6600.

Diamond, J., Holmes, M., & Coughlin, M. (1992). Endogenous NGF and nerve impulses regulate the collateral sprouting of sensory axons in the skin of the adult rat. *The Journal of Neuroscience*, *12*, 1454–1466.

Drel, V. R., Pacher, P., Vareniuk, I., et al. (2007). A peroxynitrite decomposition catalyst counteracts sensory neuropathy in streptozotocin-diabetic mice. *European Journal of Pharmacology*, *569*, 48–58.

Ekberg, K., Brismar, T., Johansson, B. L., Jonsson, B., Lindstrom, P., & Wahren, J. (2003). Amelioration of sensory nerve dysfunction by C-peptide in patients with type 1 diabetes. *Diabetes*, *52*, 536–541.

Fernyhough, P., & McGavock, J. (2014). Mechanisms of disease: Mitochondrial dysfunction in sensory neuropathy and other complications in diabetes. *Handbook of Clinical Neurology*, *126*, 353–377.

Fernyhough, P., Smith, D. R., Schapansky, J., et al. (2005). Activation of nuclear factor-kappaB via endogenous tumor necrosis factor alpha regulates survival of axotomized adult sensory neurons. *The Journal of Neuroscience*, *25*, 1682–1690.

Fernyhough, P., Willars, G. B., Lindsay, R. M., & Tomlinson, D. R. (1993). Insulin and insulin-like growth factor I enhance regeneration in cultured adult rat sensory neurones. *Brain Research*, *607*, 117–124.

Folli, F., Bonfanti, L., Renard, E., Kahn, C. R., & Merighi, A. (1994). Insulin receptor substrate-1 (IRS-1) distribution in the rat central nervous system. *The Journal of Neuroscience*, *14*, 6412–6422.

Francis, G. J., Martinez, J. A., Liu, W. Q., et al. (2011). Motor end plate innervation loss in diabetes and the role of insulin. *Journal of Neuropathology and Experimental Neurology*, *70*, 323–339.

Frazier, W. A., Angeletti, R. H., & Bradshaw, R. A. (1972). Nerve growth factor and insulin. *Science*, *176*, 482–488.

Giniger, E. (2002). How do Rho family GTPases direct axon growth and guidance? A proposal relating signaling pathways to growth cone mechanics. *Differentiation, 70*, 385–396.
Greenbaum, D., Richardson, P. C., Salmon, M. V., & Urich, H. (1964). Pathological observations on six cases of diabetic neuropathy. *Brain, 87*, 201–214.
Grote, C. W., Groover, A. L., Ryals, J. M., Geiger, P. C., Feldman, E. L., & Wright, D. E. (2013). Peripheral nervous system insulin resistance in ob/ob mice. *Acta Neuropathologica Communications, 1*, 15.
Grote, C. W., Morris, J. K., Ryals, J. M., Geiger, P. C., & Wright, D. E. (2011). Insulin receptor substrate 2 expression and involvement in neuronal insulin resistance in diabetic neuropathy. *Experimental Diabetes Research, 2011*, 212571.
Guo, G., Kan, M., Martinez, J. A., & Zochodne, D. W. (2011). Local insulin and the rapid regrowth of diabetic epidermal axons. *Neurobiology of Disease, 43*, 414–421.
Guo, G., Singh, V., & Zochodne, D. W. (2014). Growth and turning properties of adult glial cell-derived neurotrophic factor coreceptor alpha 1 nonpeptidergic sensory neurons. *Journal of Neuropathology and Experimental Neurology, 73*(9), 820–836.
Hall, K. E., Sima, A. A., & Wiley, J. W. (1995). Voltage-dependent calcium currents are enhanced in dorsal root ganglion neurones from the Bio Bred/Worchester diabetic rat. *Journal of Physiology, 486*, 313–322.
Himeno, T., Kamiya, H., Naruse, K., et al. (2011). Beneficial effects of exendin-4 on experimental polyneuropathy in diabetic mice. *Diabetes, 60*, 2397–2406.
Huang, T. J., Price, S. A., Chilton, L., et al. (2003). Insulin prevents depolarization of the mitochondrial inner membrane in sensory neurons of type 1 diabetic rats in the presence of sustained hyperglycemia. *Diabetes, 52*, 2129–2136.
Jin, M., Guan, C. B., Jiang, Y. A., et al. (2005). Ca^{2+}-dependent regulation of rho GTPases triggers turning of nerve growth cones. *The Journal of Neuroscience, 25*, 2338–2347.
Jolivalt, C. G., Fineman, M., Deacon, C. F., Carr, R. D., & Calcutt, N. A. (2011). GLP-1 signals via ERK in peripheral nerve and prevents nerve dysfunction in diabetic mice. *Diabetes, Obesity & Metabolism, 13*, 990–1000.
Kamiya, H., Zhang, W., Ekberg, K., Wahren, J., & Sima, A. A. (2006a). C-peptide reverses nociceptive neuropathy in type 1 diabetes. *Diabetes, 55*, 3581–3587.
Kamiya, H., Zhang, W., & Sima, A. A. (2006b). Degeneration of the Golgi and neuronal loss in dorsal root ganglia in diabetic BioBreeding/Worcester rats. *Diabetologia, 49*, 2763–2774.
Kamiya, H., Zhangm, W., & Sima, A. A. (2005). Apoptotic stress is counterbalanced by survival elements preventing programmed cell death of dorsal root ganglions in subacute type 1 diabetic BB/Wor rats. *Diabetes, 54*, 3288–3295.
Kan, M., Guo, G., Singh, B., Singh, V., & Zochodne, D. W. (2012). Glucagon-like peptide 1, insulin, sensory neurons, and diabetic neuropathy. *Journal of Neuropathology and Experimental Neurology, 71*, 494–510.
Kennedy, J. M., & Zochodne, D. W. (2002). Influence of experimental diabetes on the microcirculation of injured peripheral nerve. Functional and morphological aspects. *Diabetes, 51*, 2233–2240.
Kennedy, J. M., & Zochodne, D. W. (2005). Impaired peripheral nerve regeneration in diabetes mellitus. *Journal of the Peripheral Nervous System, 10*, 144–157.
Kim, B., McLean, L. L., Philip, S. S., & Feldman, E. L. (2011). Hyperinsulinemia induces insulin resistance in dorsal root ganglion neurons. *Endocrinology, 152*, 3638–3647.
Kobayashi, M., Cheng, C., de la Hoz, C., & Zochodne, D. (2015). Diabetic neuropathy and the sensory neuron: New molecular targets. *American Academy of Neurology*, P2.010.
Korngut, L., Ma, C. H., Martinez, J. A., et al. (2012). Overexpression of human HSP27 protects sensory neurons from diabetes. *Neurobiology of Disease, 47*, 436–443.

Korngut, L., Mawani, S., Francis, G., et al. (2013). A pilot study of intranasal insulin for the treatment of diabetic peripheral neuropathy. *Journal of the Peripheral Nervous System, 18*, S59.

Korthals, J. K., Gieron, M. A., & Dyck, P. J. (1988). Intima of epineurial arterioles is increased in diabetic polyneuropathy. *Neurology, 38*, 1582–1586.

Levy, D., Hoke, A., & Zochodne, D. W. (1999). Local expression of inducible nitric oxide synthase in an animal model of neuropathic pain. *Neuroscience Letters, 260*, 207–209.

Levy, D., Kubes, P., & Zochodne, D. W. (2001). Delayed peripheral nerve degeneration, regeneration, and pain in mice lacking inducible nitric oxide synthase. *Journal of Neuropathology and Experimental Neurology, 60*, 411–421.

Levy, D., Tal, M., Hoke, A., & Zochodne, D. W. (2000). Transient action of the endothelial constitutive nitric oxide synthase (ecNOS) mediates the development of thermal hypersensitivity following peripheral nerve injury. *The European Journal of Neuroscience, 12*, 2323–2332.

Luo, L., Jan, L. Y., & Jan, Y. N. (1997). Rho family GTP-binding proteins in growth cone signalling. *Current Opinion in Neurobiology, 7*, 81–86.

Ma, J., Farmer, K. L., Pan, P., et al. (2014). Heat shock protein 70 is necessary to improve mitochondrial bioenergetics and reverse diabetic sensory neuropathy following KU-32 therapy. *The Journal of Pharmacology and Experimental Therapeutics, 348*, 281–292.

Ma, J., Pan, P., Anyika, M., Blagg, B. S., & Dobrowsky, R. T. (2015). Modulating molecular chaperones improves mitochondrial bioenergetics and decreases the inflammatory transcriptome in diabetic sensory neurons. *ACS Chemical Neuroscience, 6*, 1637–1648.

Malik, R. A., Kallinikos, P., Abbott, C. A., et al. (2003). Corneal confocal microscopy: A non-invasive surrogate of nerve fibre damage and repair in diabetic patients. *Diabetologia, 46*, 683–688.

Malik, R. A., Newrick, P. G., Sharma, A. K., et al. (1989). Microangiopathy in human diabetic neuropathy: Relationship between capillary abnormalities and the severity of neuropathy. *Diabetologia, 32*, 92–102.

Malik, R. A., Veves, A., Masson, E. A., et al. (1992). Endoneurial capillary abnormalities in mild human diabetic neuropathy. *Journal of Neurology, Neurosurgery, and Psychiatry, 55*, 557–561.

Massa, S. M., Xie, Y., Yang, T., et al. (2006). Small, nonpeptide p75NTR ligands induce survival signaling and inhibit proNGF-induced death. *The Journal of Neuroscience, 26*, 5288–5300.

Newrick, P. G., Wilson, A. J., Jakubowski, J., Boulton, A. J., & Ward, J. D. (1986). Sural nerve oxygen tension in diabetes. *British Medical Journal, 293*, 1053–1054.

Ng, J., & Luo, L. (2004). Rho GTPases regulate axon growth through convergent and divergent signaling pathways. *Neuron, 44*, 779–793.

Obrosova, I. G., Drel, V. R., Pacher, P., et al. (2005). Oxidative-nitrosative stress and poly(ADP-ribose) polymerase (PARP) activation in experimental diabetic neuropathy: The relation is revisited. *Diabetes, 54*, 3435–3441.

Obrosova, I. G., Li, F., Abatan, O. I., et al. (2004). Role of poly(ADP-ribose) polymerase activation in diabetic neuropathy. *Diabetes, 53*, 711–720.

Perry, T., Holloway, H. W., Weerasuriya, A., et al. (2007). Evidence of GLP-1-mediated neuroprotection in an animal model of pyridoxine-induced peripheral sensory neuropathy. *Experimental Neurology, 203*, 293–301.

Polydefkis, M., Hauer, P., Griffin, J. W., & McArthur, J. C. (2001). Skin biopsy as a tool to assess distal small fiber innervation in diabetic neuropathy. *Diabetes Technology & Therapeutics, 3*, 23–28.

Purves, T. D., & Tomlinson, D. R. (2002). Diminished transcription factor survival signals in dorsal root ganglia in rats with streptozotocin-induced diabetes. *Annals of the New York Academy of Sciences, 973*, 472–476.

Reddy, V. P., & Beyaz, A. (2006). Inhibitors of the Maillard reaction and AGE breakers as therapeutics for multiple diseases. *Drug Discovery Today, 11*, 646–654.

Reger, M. A., Watson, G. S., Green, P. S., et al. (2008). Intranasal insulin improves cognition and modulates B-amyloid in early AD. *Neurology, 70*, 440–448.

Rundles, R. W. (1945). Diabetic neuropathy—General review with report of 125 cases. *Medicine, 24*, 111–160.

Russell, J. W., Sullivan, K. A., Windebank, A. J., Herrmann, D. N., & Feldman, E. L. (1999). Neurons undergo apoptosis in animal and cell culture models of diabetes. *Neurobiology of Disease, 6*, 347–363.

Saleh, A., Smith, D. R., Tessler, L., et al. (2013). Receptor for advanced glycation end-products (RAGE) activates divergent signaling pathways to augment neurite outgrowth of adult sensory neurons. *Experimental Neurology, 249*, 149–159.

Schmelzer, J. D., Zochodne, D. W., & Low, P. A. (1989). Ischemic and reperfusion injury of rat peripheral nerve. *Proceedings of the National Academy of Sciences of the United States of America, 86*, 1639–1642.

Schmidt, R. E., Beaudet, L. N., Plurad, S. B., & Dorsey, D. A. (1997). Axonal cytoskeletal pathology in aged and diabetic human sympathetic autonomic ganglia. *Brain Research, 769*, 375–383.

Schmidt, R. E., Dorsey, D. A., Beaudet, L. N., Plurad, S. B., Parvin, C. A., & Bruch, L. A. (1998). Vacuolar neuritic dystrophy in aged mouse superior cervical sympathetic ganglia is strain-specific. *Brain Research, 806*, 141–151.

Schmidt, R. E., Dorsey, D., Parvin, C. A., Beaudet, L. N., Plurad, S. B., & Roth, K. A. (1997). Dystrophic axonal swellings develop as a function of age and diabetes in human dorsal root ganglia. *Journal of Neuropathology and Experimental Neurology, 56*, 1028–1043.

Schmidt, R. E., & Scharp, D. W. (1982). Axonal dystrophy in experimental diabetic autonomic neuropathy. *Diabetes, 31*, 761–770.

Scott, J. N., Clark, A. W., & Zochodne, D. W. (1999). Neurofilament and tubulin gene expression in progressive experimental diabetes: Failure of synthesis and export by sensory neurons. *Brain, 122*, 2109–2118.

Singh, B., Singh, V., Krishnan, A., et al. (2014). Regeneration of diabetic axons is enhanced by selective knockdown of the PTEN gene. *Brain, 137*, 1051–1067.

Singh, B., Xu, Y., Guo, G. F., Cunningham, S., Martinez, J. A., & Zochodne, D. W. (2009). Insulin signaling and insulin resistance in sensory neurons. *Journal of the Peripheral Nervous System, 13*, 254.

Singh, B., Xu, Y., McLaughlin, T., et al. (2012). Resistance to trophic neurite outgrowth of sensory neurons exposed to insulin. *Journal of Neurochemistry, 121*, 263–276.

Singhal, A., Cheng, C., Sun, H., & Zochodne, D. W. (1997). Near nerve local insulin prevents conduction slowing in experimental diabetes. *Brain Research, 763*, 209–214.

Sivitz, W. I., & Yorek, M. A. (2010). Mitochondrial dysfunction in diabetes: From molecular mechanisms to functional significance and therapeutic opportunities. *Antioxidants & Redox Signaling, 12*, 537–577.

Sugimoto, K., Murakawa, Y., & Sima, A. A. (2002). Expression and localization of insulin receptor in rat dorsal root ganglion and spinal cord. *Journal of the Peripheral Nervous System, 7*, 44–53.

Sugimoto, K., Murakawa, Y., Zhang, W., Xu, G., & Sima, A. A. (2000). Insulin receptor in rat peripheral nerve: Its localization and alternatively spliced isoforms. *Diabetes/Metabolism Research and Reviews, 16*, 354–363.

Tavakoli, M., Quattrini, C., Abbott, C., et al. (2010). Corneal confocal microscopy: A novel noninvasive test to diagnose and stratify the severity of human diabetic neuropathy. *Diabetes Care, 33*, 1792–1797.

Tesfaye, S., Chaturvedi, N., Eaton, S. E., et al. (2005). Vascular risk factors and diabetic neuropathy. *The New England Journal of Medicine, 352*, 341–350.
Tesfaye, S., Harris, N., Jakubowski, J. J., et al. (1993). Impaired blood flow and arteriovenous shunting in human diabetic neuropathy: A novel technique of nerve photography and fluorescein angiography. *Diabetologia, 36*, 1266–1274.
The Diabetes Control and Complications Trial Research Group. (1993). The effect of intensive treatment of diabetes on the development and progression of long-term complications in insulin-dependent diabetes mellitus. The Diabetes Control and Complications Trial Research Group. *The New England Journal of Medicine, 329*, 977–986.
Theriault, M., Dort, J., Sutherland, G., & Zochodne, D. W. (1997). Local human sural nerve blood flow in diabetic and other polyneuropathies. *Brain, 120*, 1131–1138.
Theriault, M., Dort, J., Sutherland, G., & Zochodne, D. W. (1998). A prospective quantitative study of sensory deficits after whole sural nerve biopsies in diabetic and nondiabetic patients. Surgical approach and the role of collateral sprouting. *Neurology, 50*, 480–484.
Thomsen, K., Rubin, I., & Lauritzen, M. (2002). NO- and non-NO-, non-prostanoid-dependent vasodilatation in rat sciatic nerve during maturation and developing experimental diabetic neuropathy. *The Journal of Physiology, 543*, 977–993.
Thornalley, P. J. (2003). Use of aminoguanidine (Pimagedine) to prevent the formation of advanced glycation endproducts. *Archives of Biochemistry and Biophysics, 419*, 31–40.
Toth, C., Brussee, V., Martinez, J. A., McDonald, D., Cunningham, F. A., & Zochodne, D. W. (2006). Rescue and regeneration of injured peripheral nerve axons by intrathecal insulin. *Neuroscience, 139*, 429–449.
Toth, C., Francis, G., Martinez, J. A., Hanson, L., Frey, W., III, & Zochodne, D. W. (2007). Intranasal insulin delivery reduces morphological, electrophysiological, behavioral, and molecular changes of diabetic neuropathy without change in glycemia. *Journal of the Peripheral Nervous System, 12*(Suppl.), 86.
Toth, C., Schmidt, A. M., Tuor, U., Brussee, V., Kaur, J., & Zochodne, D. W. (2004). RAGE in the nervous system: Insights into a new pathophysiological relationship to diabetic complications within the central and peripheral nervous system. *Canadian Journal of Neurological Sciences, 31*(Suppl. 1), S13.
Tucker, B. A., Rahimtula, M., & Mearow, K. M. (2006). Laminin and growth factor receptor activation stimulates differential growth responses in subpopulations of adult DRG neurons. *The European Journal of Neuroscience, 24*, 676–690.
Vareniuk, I., Pacher, P., Pavlov, I. A., Drel, V. R., & Obrosova, I. G. (2009). Peripheral neuropathy in mice with neuronal nitric oxide synthase gene deficiency. *International Journal of Molecular Medicine, 23*, 571–580.
Vareniuk, I., Pavlov, I. A., & Obrosova, I. G. (2008). Inducible nitric oxide synthase gene deficiency counteracts multiple manifestations of peripheral neuropathy in a streptozotocin-induced mouse model of diabetes. *Diabetologia, 51*, 2126–2133.
Voitenko, N. V., Kruglikov, I. A., Kostyuk, E. P., & Kostyuk, P. G. (2000). Effect of streptozotocin-induced diabetes on the activity of calcium channels in rat dorsal horn neurons. *Neuroscience, 95*, 519–524.
Willis, D. E., & Twiss, J. L. (2006). The evolving roles of axonally synthesized proteins in regeneration. *Current Opinion in Neurobiology, 16*, 111–118.
Xu, Q. G., Li, X.-Q., Kotecha, S. A., Cheng, C., Sun, H. S., & Zochodne, D. W. (2004). Insulin as an in vivo growth factor. *Experimental Neurology, 188*, 43–51.
Yasuda, H., & Dyck, P. J. (1987). Abnormalities of endoneurial microvessels and sural nerve pathology in diabetic neuropathy. *Neurology, 37*, 20–28.
Zhang, W., Yorek, M., Pierson, C. R., Murakawa, Y., Breidenbach, A., & Sima, A. A. (2001). Human C-peptide dose dependently prevents early neuropathy in the BB/Wor-rat. *International Journal of Experimental Diabetes Research, 2*, 187–193.

Zherebitskaya, E., Akude, E., Smith, D. R., & Fernyhough, P. (2009). Development of selective axonopathy in adult sensory neurons isolated from diabetic rats: Role of glucose-induced oxidative stress. *Diabetes*, *58*, 1356–1364.

Zochodne, D. W. (2002). Nerve and ganglion blood flow in diabetes: An appraisal. In D. Tomlinson (Ed.), *Neurobiology of diabetic neuropathy* (pp. 161–202). San Diego: Academic Press.

Zochodne, D. W. (2007). Diabetes mellitus and the peripheral nervous system: Manifestations and mechanisms. *Muscle & Nerve*, *36*, 144–166.

Zochodne, D. W. (2014). Mechanisms of diabetic neuron damage: Molecular pathways. *Handbook of Clinical Neurology*, *126*, 379–399.

Zochodne, D. W., & Ho, L. T. (1992). Normal blood flow but lower oxygen tension in diabetes of young rats: Microenvironment and the influence of sympathectomy. *Canadian Journal of Physiology and Pharmacology*, *70*, 651–659.

Zochodne, D. W., Levy, D., Zwiers, H., et al. (1999). Evidence for nitric oxide and nitric oxide synthase activity in proximal stumps of transected peripheral nerves. *Neuroscience*, *91*, 1515–1527.

Zochodne, D. W., & Nguyen, C. (1999). Increased peripheral nerve microvessels in early experimental diabetic neuropathy: Quantitative studies of nerve and dorsal root ganglia. *Journal of the Neurological Sciences*, *166*, 40–46.

Zochodne, D. W., Ramji, N., & Toth, C. (2008). Neuronal targeting in diabetes mellitus: A story of sensory neurons and motor neurons. *The Neuroscientist*, *14*, 311–318.

Zochodne, D. W., Sun, H.-S., Cheng, C., & Eyer, J. (2004). Accelerated diabetic neuropathy in axons without neurofilaments. *Brain*, *127*, 2193–2200.

Zochodne, D. W., Verge, V. M., Cheng, C., Sun, H., & Johnston, J. (2001). Does diabetes target ganglion neurons? Progressive sensory neuron involvement in long term experimental diabetes. *Brain*, *124*, 2319–2334.

CHAPTER TEN

Promoting Neuronal Tolerance of Diabetic Stress: Modulating Molecular Chaperones

S.M. Emery, R.T. Dobrowsky[1]
The University of Kansas, Lawrence, KS, United States
[1]Corresponding author: e-mail address: dobrowsky@ku.edu

Contents

Abstract

The etiology of diabetic peripheral neuropathy (DPN) involves an interrelated series of metabolic and vascular insults that ultimately contribute to sensory neuron degeneration. In the quest to pharmacologically manage DPN, small-molecule inhibitors have targeted proteins and pathways regarded as "diabetes specific" as well as others whose activity are altered in numerous disease states. These efforts have not yielded any significant therapies, due in part to the complicating issue that the biochemical contribution of these targets/pathways to the progression of DPN does not occur with temporal and/or biochemical uniformity between individuals. In a complex, chronic neurodegenerative disease such as DPN, it is increasingly appreciated that effective disease management may not necessarily require targeting a pathway or protein considered to contribute to disease progression. Alternatively, it may prove sufficiently beneficial to pharmacologically enhance the activity of endogenous cytoprotective pathways to aid neuronal tolerance to and recovery from glucotoxic stress. In pursuing

International Review of Neurobiology, Volume 127
ISSN 0074-7742
http://dx.doi.org/10.1016/bs.irn.2016.03.001

this paradigm shift, we have shown that modulating the activity and expression of molecular chaperones such as heat shock protein 70 (Hsp70) may provide translational potential for the effective medical management of insensate DPN. Considerable evidence supports that modulating Hsp70 has beneficial effects in improving inflammation, oxidative stress, and glucose sensitivity. Given the emerging potential of modulating Hsp70 to manage DPN, the current review discusses efforts to characterize the cytoprotective effects of this protein and the benefits and limitations that may arise in drug development efforts that exploit its cytoprotective activity.

1. INTRODUCTION

Over 300 million individuals worldwide are estimated to be affected by diabetes, with projections that exceed half a billion people by 2030 (Whiting, Guariguata, Weil, & Shaw, 2011). About half of these individuals will develop diabetic peripheral neuropathy (DPN). DPN symptoms can be quite varied and manifest in the extremities as weakness/numbness (insensate DPN) or painful tingling, burning, or lancinating sensations (painful DPN). Though the painful aspects of DPN can be quite debilitating, the loss of sensation can lead to life-altering amputations by exacerbating the seriousness of ulcerations that occur in 15% of patients. In fact, diabetes is the leading cause of nonaccidental lower extremity amputations. Current therapies manage painful symptoms with anticonvulsants, antidepressants, or opiates. However, pharmacologic approaches to manage insensate DPN have not realized any significant clinical success. To this point, management of glycemic control is our most effective option in ameliorating the onset and severity of DPN symptoms, but this approach is more effective for Type 1 than Type 2 diabetics (Callaghan, Cheng, Stables, Smith, & Feldman, 2012).

The recurrent hyperglycemia central to the development of DPN results in a host of biochemical and metabolic changes. Elevated blood glucose levels can increase its flux through the polyol and hexosamine pathways, activate protein kinase C (PKC), increase poly(ADP-ribose) polymerase activity, and enhance formation of advanced glycation endproducts (AGE), which activate AGE receptors (RAGE) to heighten oxidative stress (Brownlee, 2001; Edwards, Vincent, Cheng, & Feldman, 2008; Farmer, Li, & Dobrowsky, 2012). Though each pathway is likely to independently contribute to the metabolic alterations that underlie DPN, current theories suggest that the progression of DPN is strongly related to a convergence of these metabolic disturbances on promoting oxidative stress (Obrosova et al.,

2005; Stavniichuk et al., 2011; Vareniuk et al., 2007) and mitochondrial dysfunction (Chowdhury, Smith, & Fernyhough, 2013; Farmer et al., 2012). Additionally, inflammatory signaling, a well-accepted component of Type 2 diabetes, is also fed by oxidative stress and can play a role in exacerbating DPN through cytokine production, immune cell infiltration, and further elevation of oxidative stress (Wei et al., 2009; Wellen, 2005; Wright, Scism-Bacon, & Glass, 2006).

Classic approaches to pharmacologically manage DPN have assumed that targeting a single biochemical insult is sufficient to alter the interwoven molecular cascade that culminates in the onset of symptomatic sensory neuropathy and the eventual dying back of distal axons. If DPN arose from damage to a single cell population from a single biochemical insult, this approach may be highly fruitful. However, the natural history of the disorder develops over decades due to damage to vascular, glial, and neuronal cell populations that are not necessarily affected synchronously by hyperglycemic stress. Indeed, despite the current focus on minimizing oxidative stress, antioxidant therapies have also met with limited success in treating DPN, though the use of α-lipoic acid may have some benefit (Garcia-Alcala et al., 2015; Ziegler et al., 2006, 2011). Thus, realigning our thinking toward pharmacologic approaches to address this problem may prove beneficial.

A rather unexplored and clinically unappreciated alternative approach for managing DPN is to deemphasize that it is necessary to topple a biochemical kingpin to destroy the downstream network of molecular minions whose slavings contribute to disease progression. On the other hand, cells have evolved endogenous mechanisms to tolerate stress. Pharmacologic upregulation of these pathways may facilitate recovery from glucotoxicity and/or enhance the efficacy of agents that target direct pathogenetic mechanisms of DPN. Growing evidence suggests that modulating endogenous cytoprotective responses by targeting the activity and expression of molecular chaperones such as heat shock protein 90 (Hsp90) and Hsp70 may achieve this purpose.

2. MOLECULAR CHAPERONES

Heat shock proteins are divided into families defined by their molecular size: Hsp110, Hsp100, Hsp90, Hsp70, Hsp60, Hsp40, and small Hsps (Kim, Hipp, Bracher, Hayer-Hartl, & Hartl, 2013; Saibil, 2013). This review will focus primarily on the Hsp90 and Hsp70 proteins that can be found in the endoplasmic reticulum (ER), mitochondria, cytosol, and extracellularly (Becker & Craig, 1994; Haas & Wabl, 1983; Hightower & Guidon, 1989; Leustek, Dalie, Amir-Shapira, Brot, & Weissbach, 1989).

Both Hsp90 and Hsp70 have isoforms that are constitutively expressed or induced by thermal, ischemic, oxidative, and metabolic stressors, as well as by exercise (Beckmann, Lovett, & Welch, 1992; Krause et al., 2007; Richard, Kaeffer, & Thuillez, 1996; Yang et al., 1996). Under physiologic conditions Hsp90 binds and inhibits the transcription factor heat shock factor 1 (HSF1). In response to the accumulation of damaged and/or unfolded proteins, HSF1 is released from this complex, undergoes trimerization, and is phosphorylated before being translocated to the nucleus (Neef, Jaeger, & Thiele, 2011; Vihervaara & Sistonen, 2014). Upon binding to heat shock elements, HSF1 upregulates an array of inducible chaperones including forms of Hsp70 (Hsp70.1 and Hsp70.3) and Hsp90 (Hsp90α), as well as antioxidant proteins (Benarroch, 2011). Hsp70 can help refold unfolded proteins, prevent inappropriate protein interactions, manage protein aggregate formation, and direct the degradation of damaged or dysfunctional proteins. Importantly, the efficacy of modulating chaperones is not limited to neurodegenerative diseases driven by the formation of protein aggregates. Hsp70 has also been shown to downregulate inflammatory signaling, decrease oxidative stress, and improve mitochondrial function in DPN and other conditions whose etiology is not linked to the formation of protein aggregates (Ianaro et al., 2003; Jones, Voegeli, Li, Chen, & William Currie, 2011; Li, Ma, Zhao, Blagg, & Dobrowsky, 2012; Ma, Pan, Anyika, Blagg, & Dobrowsky, 2015; Madden, Sandstrom, Lovell, & McNaughton, 2008; Saibil, 2013; Zhang, Zhao, Blagg, & Dobrowsky, 2012). However, an important aspect of Hsp70 biology that must be considered in moving forward in therapy development is how modulating the intra- vs extracellular pools of Hsp70 may affect disease progression.

Opposing effects have been demonstrated for intracellular Hsp70 (iHsp70) vs extracellular Hsp70 (eHsp70) in multiple studies (Krause et al., 2015; Rodrigues-Krause et al., 2012). Extracellular Hsps were first described as glia–axon transfer proteins when giant squid axons were subjected to heat stress (Tytell, Greenberg, & Lasek, 1986). Later studies specifically identified Hsp70 as one of the glia–axon transfer proteins (Hightower & Guidon, 1989). More recently, eHsp70 has been shown in blood where it acts as a paracrine factor released from dendritic, neuronal, and other cell types (De Maio, 2011). eHsp70 is associated with immune system activation and inflammatory/oxidative stress conditions (Whitham & Fortes, 2008) and has been hypothesized to act as a warning to the immune system for combating infections (Campisi, Leem, & Fleshner, 2003). Conversely, iHsp70 acts to suppress immune activation, in line with it also promoting antiinflammatory and antioxidant actions (Ianaro et al., 2003; Johnson & Fleshner, 2006). Elevated eHsp70 expression and decreased iHsp70 expression are linked to diabetes, obesity, and insulin resistance (Chung et al., 2008; Kurucz et al., 2002; Rodrigues-Krause et al., 2012). Considering these conflicting actions, this review will focus on differences in the mechanistic actions of extracellular vs intracellular Hsp70 proteins and what roles their pharmacological manipulation might have in therapies for DPN.

3. EXTRACELLULAR Hsp70

3.1 Import, Export, and Neuronal Support

Hsp70 export was unknowingly first characterized in heat-treated giant squid axon when protein levels of a heat shock-like transferrin protein increased in the axoplasm (Tytell et al., 1986). This transferrin protein was specifically identified as Hsp70 when heat-treated rat embryonic cell cultures (presumably fibroblasts) were shown to release Hsp70 into the media, independent of an increase in cell death (Hightower & Guidon, 1989). Glial eHsp70 secretion has also been confirmed in models of heat-treated glioblastoma cells (Guzhova et al., 2001).

The export mechanisms behind eHsp70 secretion are still not well understood since Hsp70 lacks a secretory signal sequence and inhibiting the secretory pathway has no effect on the release of eHsp70 (Hightower & Guidon, 1989). Some evidence has pointed toward an innate ability of the protein to traverse the cell membrane (Multhoff, 2007), lipid raft-mediated lysosomal release (Hunter-Lavin et al., 2004), and secretory-like granule excretion (Evdonin et al., 2006). However, the largest collection of evidence points to exosome-dependent trafficking; heat-shocked peripheral blood mononuclear cells increased eHsp70 levels and produced Hsp70-loaded exosomes (Lancaster & Febbraio, 2005). Though further study is needed to fully understand this Hsp70 export mechanism, exosomal eHsp70 release has been demonstrated in other models including colon and pancreatic cancer cell lines (Gastpar et al., 2005) as well as breast and leukemic cancer cells (Bausero, Gastpar, Multhoff, & Asea, 2005).

Once excreted eHsp70 can be absorbed and utilized by neighboring cells, typically for the purpose of cell support (Guzhova et al., 2001). The neuroblastoma cell line LAN-5 showed lower levels of cytotoxicity from both heat shock and staurosporine treatment after internalizing exogenously administered Hsp70 (Guzhova et al., 2001). These cytoprotective effects can likewise translate to nonneoplastic neuronal models. Injection of recombinant human Hsp70 into G93A mutant Cu/Zn superoxide dismutase (SOD1) mice helped preserve motor neurons resulting in a delayed onset of amyotrophic lateral sclerosis (ALS) symptoms (Gifondorwa et al., 2007). In vitro and ex vivo experiments using a trophic factor-deprived environment and sciatic nerve axotomy, respectively, confirm eHsp70 decreased motor neuron death (Robinson et al., 2005; Tidwell, Houenou, & Tytell, 2004). Exogenously administered Hsp70 also preserved viability of dorsal root ganglia (DRG) sensory neurons after sciatic nerve axotomy (Tidwell et al., 2004). Though the exact mechanism for cellular protection is not known, eHsp70 may act as iHsp70 upon internalization (Turturici, Sconzo, & Geraci, 2011).

3.2 Immunomodulation, Inflammation, and Oxidative Stress

In contrast to the effects of iHsp70, the actions of receptor-mediated eHsp70 are typically immune stimulating, proinflammatory, and promote oxidative stress (Krause et al., 2015). For example, multiple studies have identified eHsp70 as mediating an initial inflammatory response after tissue injury (Kovalchin et al., 2006; Senf, Howard, Ahn, Ferreira, & Judge, 2013). Senf et al. (2013) reported that cardiotoxin injections into the tibialis anterior (TA) muscles of Hsp70 knockout mice induced severe muscle injury and delayed recovery when compared to control animals. Hsp70 expression vectors electroporated into TA muscles restored recovery time after injury by reinvigorating the initial postinjury inflammatory response. Extracellular Hsp70 release after TA injury was then shown to be integral to restoring the early inflammatory response when cardiotoxin was injected into TA muscle with and without recombinant Hsp70. The exogenously administered Hsp70 increased macrophage and neutrophil infiltration and resultant cytokine production to control levels in almost all measures (Senf et al., 2013). This study summarizes eHsp70's attributes as a chemoattractant, immune cell activator, and proinflammatory mediator, which essentially allows it to act as an immunomodulatory agent.

Focusing first on chemoattraction, macrophages, neutrophils, and natural killer (NK) cells directly respond to eHsp70 chemotaxic signaling (Horn et al., 2007; Ortega, Hinchado, Martin-Cordero, & Asea, 2009; Senf et al., 2013). However, eHsp70 can also indirectly increase chemotaxis of immune cells by stimulating T-cell-mediated beta-chemokine release of Rantes and macrophage inflammatory protein-1β, which drive infiltration of numerous innate and adaptive immune components (Lehner et al., 2000). Once B- and T-lymphocytes have infiltrated the inflammatory site, they may use eHsp70 itself as a chemoattractant. In vitro experiments with and without heat shock have demonstrated lymphocyte export of eHsp70 (Hunter-Lavin et al., 2004). Similarly, increases in macrophage phagocytosis and neutrophil microbicide activity have been reported after exogenous Hsp70 administration (Kovalchin et al., 2006; Ortega et al., 2006). eHsp70 also prompted proinflammatory cytokine release from epithelial, innate immune, and adaptive immune cells, which may explain why eHsp70 is associated with inflammatory cytokine production in multiple mouse and human studies (Asea et al., 2000, 2002; Basu, Binder, Ramalingam, & Srivastava, 2001; Chase et al., 2007; Dvoriantchikova, Santos, Saeed, Dvoriantchikova, & Ivanov, 2014; Dybdahl et al., 2005; Huang, Wang, Chen, Wang, & Zhang, 2013; Qiao, Liu, & Li, 2008). Collectively, these data support a strong capacity of eHsp70 to initiate and escalate an immune-mediated inflammatory response.

A growing list of receptors seem to mediate the immunomodulatory actions of eHsp70 by evoking a host of proinflammatory cytokines (IL-1β, IL-6, IL-8, IL-12, TNF-α) through a CD14-dependent, Toll-like receptor (TLR) 2/4 mechanism (Asea et al., 2000, 2002; Chase et al., 2007; Dvoriantchikova et al., 2014; Dybdahl, 2001). TLR2 has also been linked to eHsp70 stimulation of neutrophil chemotaxis (Ortega et al., 2009). Similarly, CD94, a C-type lectin receptor, has been demonstrated to mediate eHsp70 stimulation of IFN-γ release from NK cells (Gross, Hansch, Gastpar, & Multhoff, 2003). However, a separate report indicated that this response required coculturing the NK cells with dendritic cells that produce NK G2D ligand after eHsp70 treatment (Qiao et al., 2008). The differences in these studies might be clarified by their use of patient-derived NK cells vs the YT leukemic NK cell line. CD40 was determined to be necessary for mycobacterial-Hsp70-driven beta-chemokine release and human eHsp70 was demonstrated to interact with CD40 (Lehner et al., 2000). The CD40-eHsp70-dependent effects were expanded to include internalization of eHsp70–peptide complexes for antigen presentation in macrophages (Becker, 2002; Lehner et al., 2000; Wang et al., 2001). Surprisingly, an analogous study identified CD91, not CD40, as managing antigen presentation in macrophages and dendritic cells through the MHC class I molecule by an eHsp70–peptide complex (Basu et al., 2001). However, Theriault, Mambula, Sawamura, Stevenson, and Calderwood (2005) demonstrated that Hsp70 has little capacity to directly bind any of the receptors mentioned earlier. Instead, they showed a robust ability for eHsp70 to bind LOX-1, a class E scavenger receptor that collects modified lipoproteins such as oxidized low-density lipoprotein. The authors hypothesize that eHsp70 interacts directly with LOX-1 and then indirectly utilizes CD14, CD40, CD91, CD94, and TLR2/4 to coordinate its effects (Theriault et al., 2005). This hypothesis is supported by eHsp70's LOX-1-dependent ability to modulate antigen cross-presentation in dendritic cells (Delneste et al., 2002). With the exception of Theriault et al. (2005), one limitation of these studies is that they utilized antagonism and correlation to confirm the involvement of these receptors in eHsp70 signaling. Becker (2002) and Wang et al. (2001) are the exceptions, as both studies showed specific binding of fluorescently labeled eHsp70 after CD40 transfection in HEK293 and COS-7 cells, respectively. However, these data are in direct contradiction to Theriault et al. (2005) who showed a lack of fluorescent Hsp70 binding in HEK293 cells after CD40 transfection. Unfortunately, the authors were unable to reconcile the differences between these earlier results indicating more work is still needed to clarify how eHsp70 may interact with putative receptor proteins.

Downstream of eHsp70 binding, NF-κB has been implicated in neutrophil chemotaxis, bronchial epithelial cell-mediated inflammation, LPS-induced liver injury, monocyte cytokine production, and ischemia–reperfusion neurotoxic inflammation (Krause et al., 2015). Multiple lines of evidence support NF-κB involvement in these eHsp70-dependent processes: NF-κB nuclear translocation, Iκ-Bα phosphorylation, Bay 11-7082 and parthenolide antagonism, and NF-κB activity measured by various reporter assays (Asea et al., 2000, 2002; Chase et al., 2007; Dvoriantchikova et al., 2014; Huang et al., 2013; Ortega et al., 2009). PKC has also been implicated in the immunomodulatory action of eHsp70 based on antagonism by polymixin B (Dvoriantchikova et al., 2014). However, polymixin B has not been shown to inhibit eHsp70-dependent NF-κB activation, which suggests PKC and NF-κB may function as separate signaling pathways (Asea et al., 2000). One obvious gap in this research is that all evidence for eHsp70 activation of NF-κB signaling has evaluated CD14 and TLR 2/4 as the intermediary receptors. Work is still needed to investigate downstream eHsp70 signaling after LOX-1 binding and what role CD40, CD91, and CD94 might play in that process.

4. INTRACELLULAR Hsp70

4.1 Chaperone Functions

The molecular chaperone activity of intracellular heat shock proteins has been extensively characterized (De Maio, 2011). iHsp70 is an integral component of this chaperone system and the complex cochaperone interactions and molecular mechanisms that facilitate this process have been described in numerous reviews (Becker & Craig, 1994; Kim et al., 2013; Saibil, 2013; Stetler et al., 2010). A major function of this chaperone activity is to maintain intracellular proteostasis by assisting nascent peptide folding, refolding of unfolded proteins, preventing inappropriate protein interactions, managing protein aggregate formation, and degrading dysfunctional proteins (Saibil, 2013). These activities have particular relevance in neurodegenerative diseases, which are often associated with aggregates of misfolded proteins, including Alzheimer's disease, Parkinson's disease, ALS, Huntington's disease, and others (Turturici et al., 2011). It is believed that postmitotic neurons affected in these pathologies are particularly vulnerable due to an inability to diminish misfolded protein concentrations through cell division (Muchowski & Wacker, 2005). Consequently, enhancing iHsp70 action might prove a viable therapeutic approach to treating these pathologies. In vitro, iHsp70 increases the solubility and microtubule interaction of the tau protein, as well as early-stage aggregation of Aβ that leads to amyloid plaque formation in Alzheimer's disease (Dou et al., 2003; Evans, Wisen, & Gestwicki, 2006). However, managing inappropriate protein interactions is only part of the solution to minimizing misfolded protein stress. If a protein cannot be refolded to a functional state, then it must be degraded to prevent accumulation.

SOD1 is a cytosolic free radical scavenger protein that is integral to managing oxidative stress and mutant SOD1 promotes mitochondrial dysfunction that contributes to the onset of familial ALS (Tan, Pasinelli, & Trotti, 2014). CHIP is a C-terminal Hsp70-interacting protein that facilitates ubiquitination and degradation of misfolded iHsp70 client proteins like mutant SOD1. Mutant SOD1 is also degraded by chaperone-mediated autophagy where regulator proteins such as Bcl-2-associated athanogene 3 (BAG-3) and Hsp22 direct iHsp70's chaperone functions to identify, traffic, and present misfolded proteins for autophagic degradation (Kalmar, Lu, & Greensmith, 2014). This orchestrated removal of dysfunctional protein is crucial for neurological homeostasis, as well as metabolic disorders such as diabetes mellitus. In particular, Type 2 diabetes has been specifically linked to ER stress, which results from misfolded proteins accumulating in the ER. Glucose-regulated protein 78, the ER-specific Hsp70 family member, is a well-known mediator of the unfolded protein response (UPR) to manage ER stress (Kalmar et al., 2014; Lee & Ozcan, 2014). However, there is also evidence that iHsp70-mediated cytosolic protein degradation facilitates stabilization of ER stress in addition to the UPR (Goeckeler & Brodsky, 2010).

4.2 Oxidative Stress

A significant portion of misfolded or dysfunctional proteins are caused by oxidative stress. Although the chaperone functions of iHsp70 are essential to managing intracellular damage directly, iHsp70 can also indirectly mitigate disturbances in proteostasis due to a potent ability to manage oxidative stress. In the human premonocytic U937 cell line, heat shock protects from H_2O_2-driven mitochondrial membrane depolarization and structural changes (Polla et al., 1996). Similarly, Hsp70-overexpressing transgenic animals have curtailed age-associated oxidative stress (Broome et al., 2006) and multiple cell lines are protected from ROS-mediated cytotoxicity via heat treatment and/or iHsp70 overexpression (Bellmann, Jaattela, Wissing, Burkart, & Kolb, 1996; Bellmann et al., 1995; Jaattela & Wissing, 1993). Multiple mechanisms have been linked to mediating these protective effects. For example, the superoxide-generating NADPH oxidases (NOX) are negatively regulated by iHsp70. In diabetic sensory neurons, NOX2 mRNA overexpression was reversed in an Hsp70-dependent manner by treatment with the C-terminal Hsp90 inhibitor, KU-596 and this correlated with an improvement in DPN (see Section 5) (Ma et al., 2015). Further, in vascular smooth muscle cells, coimmunoprecipitation and Hsp70 knockdown studies demonstrate iHsp70 trafficking of NOX4 regulates degradation of the protein (Bocanegra, Manucha, Pena, Cacciamani, & Valles, 2010; Gil Lorenzo, Bocanegra, Benardon, Cacciamani, & Valles, 2014). NOX isoforms are a major source of nonmitochondrial ROS and modulating Hsp70 in a mouse model of Type 2 diabetes decreased nonmitochondrial oxygen respiration, indirectly suggesting an effect on decreasing the activity of NOX isoforms (Ma et al., 2014).

Much of total cellular ROS production does not require NOX activity and Hsp70 management of these non-NOX sources requires intervention with antioxidant enzymes such as SODs. In Hsp70 knockout mice, a connection between Hsp70 and SOD1 expression was surmised since cerebrum homogenates from these animals showed a marked decrease in SOD1 protein levels and activity (Choi et al., 2005). Interestingly, this regulation may not be unidirectional, since SOD1-overexpressing transgenic mice have elevated levels of Hsp70 mRNA following cerebral ischemia (Kondo et al., 1996). Nevertheless, iHsp70 seems to have a capacity to modulate SOD1 expression. Evidence for Hsp70 regulation of SOD2 is not so clear-cut, however. Hsp70 knockout mice also have decreased SOD2 activity, but these mice do not show changes in SOD2 protein expression levels (Choi et al., 2005). This difference between SOD2 activity and expression may be explained by two observations. First, iHsp70 has been found to traffic SOD2 to the mitochondria (Afolayan et al., 2014) and second, SOD2 requires additional processing in the mitochondria before it becomes fully functional (Candas & Li, 2014). As a result, genetic deletion of iHsp70 would prevent SOD2 from being efficiently localized to the mitochondria for activation. Consistent with this premise, pharmacological induction of Hsp70 in hyperglycemically stressed rat embryonic sensory neurons increased mitochondrial SOD2 expression, as determined by quantitative mass spectrometry (Zhang et al., 2012).

Another major role iHsp70 plays in managing oxidative stress involves supporting proper mitochondrial function. Mitochondria are a major source of ROS production within the cell, having a concentration 5- to 10-fold higher than in the cytosol (Cadenas & Davies, 2000). Left unchecked, intracellular stress can overload or impair the mitochondrial machinery, decrease protein import, and ultimately result in mitochondrial dysfunction and significant increases in oxidative stress (Baseler et al., 2011; Tomlinson & Gardiner, 2008). For instance, hyperglycemic stress of embryonic sensory neurons decreased the translation of numerous mitochondrial proteins and mitochondrial oxygen consumption while increasing superoxide levels. However, increasing iHsp70 levels with a C-terminal Hsp90 inhibitor decreased superoxide production and improved mitochondrial bioenergetics (Zhang et al., 2012). Further studies from our lab have demonstrated that pharmacologically modulating Hsp70 has a beneficial effect on mitochondrial bioenergetics (Ma et al., 2014; Urban et al., 2012; Zhang et al., 2012). Interestingly, RNA-Seq analysis indicated that the mitochondrial transcriptome was unaltered by pharmacological modulation of Hsp70 in diabetic mice, despite an improvement in mitochondrial bioenergetics. Therefore iHsp70's protection of mitochondrial bioenergetics may not be significantly mediated through gene transcription (Ma et al., 2015).

iHsp70-dependent influx of mitochondrial proteins may offer an explanation for at least part of this mitochondrial bioenergetics protection (Afolayan et al., 2014).

Only 1% of mitochondrial proteins are encoded by mitochondrial DNA, the remaining 99% must be imported after nuclear transcription (Schmidt, Pfanner, & Meisinger, 2010; Wiedemann, Frazier, & Pfanner, 2004). Hsp70 was first shown to bind and facilitate translocation of proteins to mitochondria in yeast (Deshaies, Koch, Werner-Washburne, Craig, & Schekman, 1988; Murakami, Pain, & Blobel, 1988). Translocases of the outer mitochondrial membrane (Tom) form a complex that is composed of various receptors and Tom40, an import pore formed which most nuclear-transcribed proteins must traverse (Schmidt et al., 2010; Wiedemann et al., 2004). Tom70 acts as the mitochondrial receptor for iHsp70 to facilitate mitochondrial protein import (Young, Hoogenraad, & Hartl, 2003). After transit through the outer membrane, the preprotein that iHsp70 was transporting is pulled through the inner mitochondrial membrane using translocases of the inner mitochondrial membrane (Tim). Tim50 helps transition from the outer mitochondrial membrane to the inner, sending the preprotein to a channel formed by Tim23 and Tim17 (Schmidt et al., 2010; Wiedemann et al., 2004). Mitochondrial Hsp70 is anchored to Tim44 and together they act as the major components of an ATP-dependent motor that pulls the preprotein into the inner mitochondrial space (Krimmer, Rassow, Kunau, Voos, & Pfanner, 2000). Once inside, the preprotein is processed to facilitate proper mitochondrial function (Schmidt et al., 2010; Wiedemann et al., 2004).

Mitochondrial dysfunction is only one mechanism by which intracellular stress can drive oxidative stress. Another can be seen in the interaction between hyperglycemia- and thioredoxin (Trx)-interacting protein (TXNIP). It has been speculated that TXNIP plays a role in regulating insulin-independent glucose transport into tissues (Singh, 2013). This is largely based on an observation that TXNIP can regulate glucose transporter expression and localization (Wu et al., 2013). However, prolonged hyperglycemic stress can drive persistent activation of TXNIP, as was shown in the retina of Type 1 diabetic rats (Devi et al., 2012). TXNIP's most well-characterized function is as an inhibitor of the antioxidant- and thiol-reducing protein, Trx, and extended inhibition of Trx by TXNIP drives oxidative stress (Schulze et al., 2004). Further exacerbating the situation, prolonged TXNIP activation can also increase inflammatory signaling, although it is unclear if these oxidative and inflammatory events occur sequentially or in parallel (Devi et al., 2012; Lane, Flam, Lockey, & Kolliputi, 2013). Recent work from our lab has shown that Hsp70 may play a role in affecting the expression of TXNIP. RNA-seq analysis of mRNA from DRG of Type 1 diabetic mice showed a twofold elevated expression of TXNIP expression (Ma et al., 2015) and treating diabetic mice with the chaperone modulator, KU-596 (see Section 5), suppressed TXNIP expression to control levels (Fig. 1A). Moreover, this suppression was Hsp70 dependent as TXNIP mRNA levels remained elevated (~1.4-fold) in DRG from diabetic Hsp70 KO mice that were treated with KU-596 (Fig. 1B). Though it remains unclear whether Hsp70 inhibited TXNIP overexpression directly or by alleviating the initial hyperglycemic/oxidative

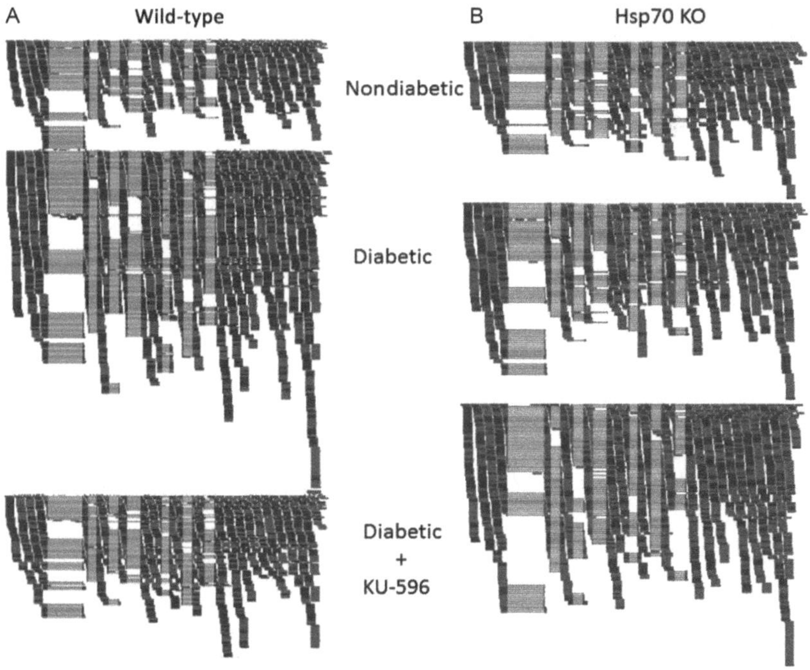

Fig. 1 Modulating Hsp70 with KU-596 decreased the expression of *txnip* in diabetic DRG in an Hsp70-dependent manner. Wild-type (A) and Hsp70 KO (B) mice were rendered diabetic with streptozotocin and after 12 weeks of diabetes, treated with vehicle or weekly doses of KU-596 at 20 mg/kg for 4 weeks. mRNA was harvested from the lumbar DRG and used for RNA-Seq analysis. Shown are representative alignments to the mouse *txnip* gene from nondiabetic, diabetic, and diabetic animals that received KU-596 therapy. *Dark blue* (*black* in the print version) bars indicate sequences that mapped to the forward strand, while *dark red* (*dark gray* in the print version) bars are sequences that mapped to the reverse strand of the eight exons of the *txnip* gene. See Ma et al., 2015 for additional details.

stress insult, these results suggest that Hsp70 attenuates intracellular oxidative stress induced by diabetes, which correlates with an improvement in clinically relevant measures of DPN.

4.3 Inflammation

Another major aspect of iHsp70 biology is its ability to regulate inflammatory signaling. Heat shock prevents TNF-α production after ischemic stress in the renal tubular epithelial cell line LLC-PK1 and in rat liver after induction of sepsis by cecal ligation and puncture (CLP; Chen, Kuo, Wang, Lu, & Yang, 2005; Meldrum, 2003). In vivo, Hsp70 overexpression suppresses increased expression of IL-6 and TNF-α induced by LPS, as well as elevation of IL-6 and cytokine-induced neutrophil chemoattractant-1 in rat blood after CLP (Dokladny, Lobb, Wharton, Ma, & Moseley, 2010; Weiss et al., 2007). Each of these studies reported that an Hsp70-dependent suppression of NF-κB activation mediated the antiinflammatory action. Using DNA binding and nuclear localization, numerous studies have supported this claim with heat shock (Feinstein et al., 1996), liposomal Hsp70 delivery (Meldrum, 2003), and Hsp70 overexpression (Feinstein, Galea, & Reis, 1997). Mechanistically, this effect may be due to Hsp70-dependent decrease in degradation of the NF-κB inhibitor (IκBα), which binds NF-κB to inhibit its nuclear localization (Dokladny et al., 2010; Feinstein et al., 1997; Wong, Ryan, & Wispe, 1997). Precipitating this action is Hsp70's ability to inhibit phosphorylation of the IκB kinase (IKK). IKK phosphorylates IκBα, causing it to dissociate from NF-κB and be degraded. In the absence of IKK phosphorylation, this process is blocked (Chan, Ou, Wang, & Chan, 2004; Chung et al., 2008; Weiss et al., 2007). Immunoprecipitation studies demonstrate that Hsp70 associates with the IKK complex, which hinders the IKKβ subunit of the complex from being integrated (Weiss et al., 2007). Additional studies support Hsp70 binding to IKKγ (NEMO), which prevents its trimerization and function as a structural component of the IKK complex (Agou et al., 2002; Chen et al., 2005; Ran et al., 2004).

A second Hsp70 mechanism for hindering NF-κB activation has also been described since Hsp70 stabilized phosphorylated and ubiquitinated IκBα. This interaction prevented the degradation of IκBα without affecting the IKK complex (Weiss et al., 2007). Additionally, a separate report suggests a third mechanism where Hsp70 directly upregulates IκBα expression to suppress NF-κB activation (Wong, Ryan, Menendez, & Wispe, 1999).

Taken together, these studies outline a highly redundant system where iHsp70 can prevent NF-κB activation by manipulating IκB. Further, two studies also suggest that Hsp70 can also prevent NF-κB activation without altering IκBα concentrations. Overexpression of Hsp70 in C6 rat glioma cells did not prevent LPS-induced IκBα degradation, despite suppression of p65 NF-κB subunit nuclear localization. The authors speculated that in certain situations, Hsp70 may instead prevent NF-κB nuclear localization by either competing for the nuclear pore or binding NF-κB after its initial dissociation from IκB (Feinstein et al., 1997). A later study replicated this finding using liposomal delivery of Hsp70 (Meldrum, 2003) and coimmunoprecipitation of Hsp70 with the p65 subunit of NF-κB, which supports the viability of this hypothesis (Chen et al., 2005). Clearly, this body of work demonstrates that NF-κB can be intimately regulated by iHsp70. However, as activation of the NF-κB pathway in sensory neurons has been shown to ameliorate DPN (Saleh, Roy Chowdhury, et al., 2013; Saleh, Schapansky, et al., 2013), it may be counterproductive to invoke therapies which may upregulate both the NF-κB pathway and iHsp70 in neurons.

4.4 Insulin Sensitivity, JNK, and c-Jun

In Type 2 diabetes, iHsp70 may be able to improve insulin resistance and blood glucose levels. Mice fed a high-fat diet show both insulin resistance and glucose intolerance compared to normal chow-fed animals. Transgenic animals overexpressing Hsp70 show no insulin resistance and markedly improved glucose tolerance when fed the high-fat diet (Henstridge, Bruce, et al., 2014; Henstridge, Whitham, & Febbraio, 2014). The actions of Hsp70 overexpression were attributed to decreased phosphorylation of c-jun N-terminal kinase (JNK), a known regulator of insulin sensitivity. Heat treatment and pharmacological induction of Hsp70 also inhibited JNK phosphorylation (Chung et al., 2008; Henstridge, Bruce, et al., 2014). Similarly, Hsp70 overexpression and heat shock suppressed JNK1 activity in NIH 3T3 cells after ultraviolet light stress. JNK1 and Hsp70 can physically interact and this prevented JNK1 phosphorylation and suppressed c-jun transactivating activity (Park, Lee, Huh, Seo, & Choi, 2001). However, JNK regulation is not the only mechanism by which iHsp70 can modulate targets downstream of JNK.

C-jun is emerging as a key negative regulator of myelination (Arthur-Farraj et al., 2012; Hutton et al., 2011; Parkinson et al., 2008) and altering its expression via modulation of iHsp70 may offer an attractive approach to treat the segmental demyelination that occurs in human DPN, as well as in various inherited neuropathies. In myelinated DRG explants, induction of Hsp70 by treatment with the C-terminal Hsp90 modulator KU-32 inhibited c-jun/phospho-c-jun expression and decreased the extent of neuregulin-induced demyelination (Fig. 2A and C–E). Though KU-32 did not decrease the phosphorylation of JNK and ERK, inhibiting the

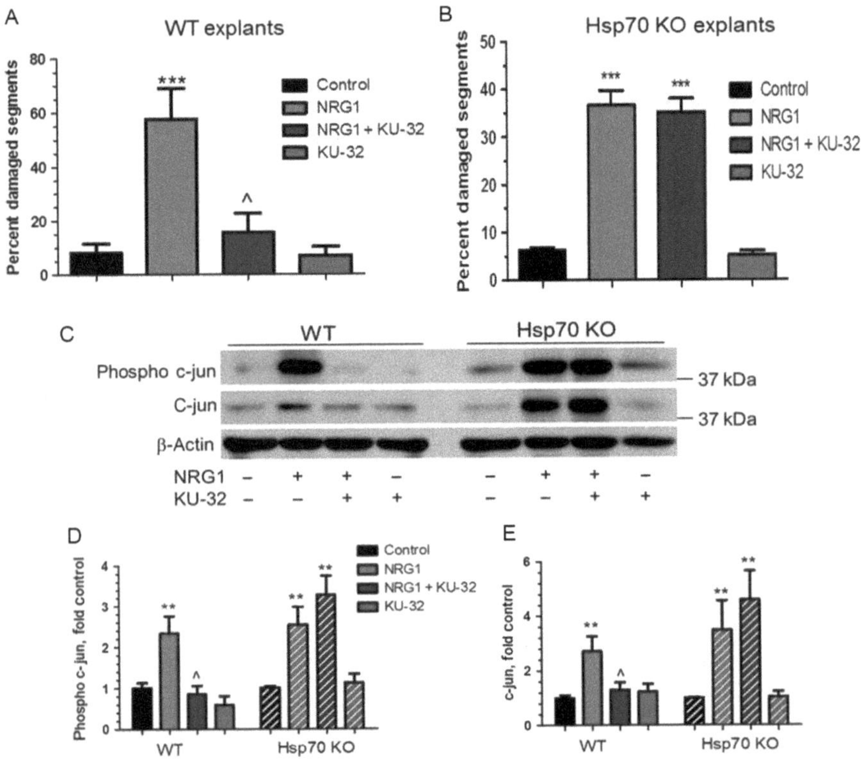

Fig. 2 Modulating Hsp70 with KU-32 decreased neuregulin-induced c-jun expression and demyelination in an Hsp70-dependent manner. Myelinated DRG explants were prepared from wild-type or Hsp70 KO mice. The cultures were treated with vehicle or 1 μM KU-32 overnight prior to inducing demyelination with 200 ng/mL neuregulin 1. Cultures were stained for myelin basic protein and the degeneration of the myelinated segments was quantified (A and B). ***$p < 0.001$ vs control; $\hat{p} < 0.01$ vs NRG1. Other cultures were prepared for immunoblot analysis of c-jun and the effects of the treatments on c-jun and phospho-c-jun expression quantified (C–E). **$p < 0.01$ vs control in WT and Hsp70 KO; $\hat{p} < 0.05$ vs NRG1 in WT only. *Modified with permission from Li, C., Ma, J., Zhao, H., Blagg, B. S., & Dobrowsky, R. T. (2012). Induction of heat shock protein 70 (Hsp70) prevents neuregulin-induced demyelination by enhancing the proteasomal clearance of c-jun.* ASN Neuro, 4, *e00102. ASN Neuro published by Portland Press.*

proteasome with MG132 prevented the decrease in c-jun expression. These data support that iHsp70 facilitates the proteasomal degradation of c-jun to attenuate demyelination (Fig. 2) (Li et al., 2012). Importantly, the necessity of iHsp70 in preventing the increase in c-jun expression and demyelination was validated using myelinated DRG explants prepared from Hsp70 KO mice. Although neuregulin treatment promoted a robust induction of c-jun in these cultures, treatment with KU-32 was unable to block this increase and prevent demyelination, as was observed in myelinated neuronal cultures prepared from wild-type mice (Fig. 2B–E). Similarly, adenovirus-mediated reexpression of Hsp70 in the Hsp70 KO neurons was sufficient to decrease c-jun induction following neuregulin treatment. Since transcriptomic analysis of mRNA from sural nerve of patients with progressive DPN showed an increase in a gene subnetwork centered on c-jun (Hur et al., 2011), it is tempting to speculate that iHsp70 modulation of c-jun signaling may benefit later stages of progressive DPN. Unfortunately, diabetic rodent models do not recapitulate the extensive segmental demyelination that is observed in human DPN, but ongoing work using animal models of demyelinating neuropathies is examining whether decreasing c-jun expression by modulating iHsp70 may improve myelinated fiber function.

5. DPN AND MODULATING Hsp70

Clearly, a large body of evidence supports the concept that modulating Hsp70 may have both beneficial and deleterious effects (Fig. 3). Indeed overexpression of iHsp70 is associated with various forms of cancer and its downregulation can increase sensitivity to chemotherapeutics (Lianos et al., 2015). eHsp70 may contribute to the development of experimental autoimmune encephalitis, an animal model of multiple sclerosis (Mansilla et al., 2014). Furthermore, elevated eHsp70 levels may contribute to inflammation in obese, Type 2 diabetic patients, ketoacidotic Type 1 diabetics (Oglesbee, Herdman, Passmore, & Hoffman, 2005; Rodrigues-Krause et al., 2012) and play a role in the progression of diabetes based on its ability to induce β-cell dysfunction and death (Krause et al., 2014). On the other hand, increasing iHsp70 expression in skeletal muscle improved insulin resistance and decreased metabolic parameters of Type 2 diabetes (Chung et al., 2008; Henstridge, Bruce, et al., 2014). The chaperone balance hypothesis gives context to these observations by proposing that a high ratio of [eHsp70]:[iHsp70] is associated with promoting chronic inflammatory responses, while a low [eHsp70]:[iHsp70] ratio is more antiinflammatory

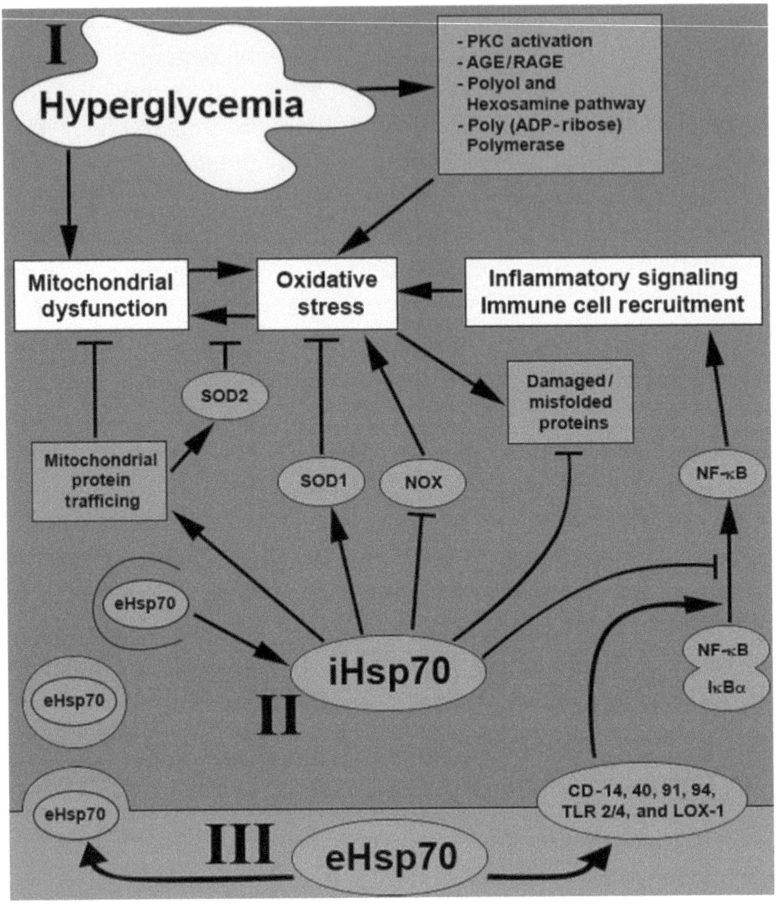

Fig. 3 Summary of potential impact of iHsp70 and eHsp70 on hyperglycemia-mediated dysfunction. (I) Hyperglycemia manipulates multiple metabolic pathways to induce both intra- and extracellular stresses to the cell. These events converge at three points affecting mitochondrial dysfunction, oxidative stress, and proinflammatory signaling to facilitate the progression of DPN. (II) iHsp70 inhibits hyperglycemia-induced dysfunctions through multiple mechanisms including mitochondrial protein shuttling, up/downregulation of anti/prooxidant proteins, respectively, repair or degradation of damaged and misfolded protein, and inhibition of proinflammatory signaling via NF-κB. (III) eHsp70 may play a complicated role since if internalized, it can bolster iHsp70 stores and actions while external binding to one of the multiple purported receptors can antagonize the action of iHsp70 by initiating or exacerbating NF-κB-mediated inflammatory signaling.

(Krause et al., 2015). If this hypothesis proves to be true, then attempts to alter the progression of DPN by modulating Hsp70 may benefit from examining this relationship. This begets the question, what are our options for modulating Hsp70?

Though genetic approaches to downregulate or overexpress Hsp70 provide an avenue to modulate its activity in animal models of disease, these approaches have not been translated to humans. Similarly, no selective small-molecule inhibitors (or activators) of inducible Hsp70 exist that can be translated to the clinic (Evans, Chang, & Gestwicki, 2010; Lianos et al., 2015). This difficulty is partially attributed to the fact that five cellular Hsp70 family members share 86–99% homology (Daugaard, Rohde, & Jaattela, 2007). However, there are viable small-molecule therapies that can indirectly modulate Hsp70 levels by targeting Hsp90 or HSF1.

Hsp90 is required for the folding of many proteins into their final conformation and is the master regulator of the heat shock response. Numerous compounds have been developed that target either the N- or C-terminal of the protein to alter its enzymatic activity or interaction with cochaperones (Miyata, Nakamoto, & Neckers, 2013; Peterson & Blagg, 2009; Walton-Diaz et al., 2013). Since many Hsp90 client proteins are also oncoproteins, N-terminal inhibitors may serve as chemotherapeutics by promoting client protein degradation. As chemotherapeutics, one attribute of N-terminal Hsp90 inhibitors is their selectivity for inhibiting Hsp90 and promoting client protein degradation in cancerous vs noncancerous cells (Chiosis et al., 2003; Kamal et al., 2003; Luo, Rodina, & Chiosis, 2008). However, induction of client protein degradation and cytotoxicity often occurs at drug concentrations that also induce the heat shock response. Unfortunately, an increase in chaperone expression can antagonize the compound's cytotoxic potential by facilitating oncoprotein refolding, narrowing the drug's effective therapeutic window (Peterson & Blagg, 2009). On the other hand, an increase in chaperone synthesis may be useful for treating neurodegenerative diseases since N-terminal Hsp90 inhibitors have been shown to limit or decrease tau aggregation in models of Alzheimer's disease (Dickey et al., 2007; Luo et al., 2007), huntingtin protein in motor neurons (Waza et al., 2005), α-synuclein in Parkinson's disease (Ebrahimi-Fakhari, Saidi, & Wahlster, 2013), and peripheral myelin protein 22 in Schwann cells from a mouse model of Charcot Marie Tooth 1A (Chittoor-Vinod, Lee, Judge, & Notterpek, 2015). However, it is still necessary to dissociate the degradation of client proteins from stimulation of the heat shock response since client protein degradation could lessen the neuroprotective effectiveness of the drugs.

Over the past 5 years we have identified novologues as a class of C-terminal Hsp90 inhibitors derived from novobiocin that induce Hsp70 in the absence of significant client protein degradation (Kusuma et al.,

2012; Urban et al., 2010). KU-32 and KU-596 are two such compounds that were mentioned above and these drugs improved multiple clinical indices of DPN in models of both Type 1 and Type 2 diabetes, in an Hsp70-dependent manner (Ma et al., 2014, 2015; Urban et al., 2010, 2012). Similarly, an N-terminal Hsp90 inhibitor has proven beneficial in treating diabetic nephropathy and the physiologic improvement in renal function may also be Hsp70 dependent (Lazaro et al., 2015). As neither of these diabetic complications has an etiology that centers on formation of any one protein aggregate, clearly modulating Hsp70 may prove efficacious toward treating complex metabolic complications of diabetes. Although all these compounds were shown to increase iHsp70, their effects on eHsp70 levels were not examined. Consistent with many of the studies described in Section 4, novologues diminished the level of oxidative stress, improved mitochondrial bioenergetics, and decreased an array of genes in DRG that were associated with inflammation (Ma et al., 2014, 2015; Urban et al., 2012; Zhang et al., 2012). The improvement of DPN and decrease in inflammation by KU-596 therapy was also associated with a transcriptomic profile that was predicted to inhibit the NF-κB pathway in DRG of diabetic wild-type mice but not diabetic Hsp70 KO mice (Ma et al., 2015). These actions would be inconsistent with enhancing systemic levels of eHsp70, but it is possible that a more localized exchange of eHsp70 between Schwann cells and sensory neurons may contribute to the dependence of drug effectiveness on Hsp70.

It is interesting that increasing Hsp70 in skeletal muscle using the HSF1 activator, BGP-15, improved fasting glucose and insulin signaling in obese mice (Henstridge, Bruce, et al., 2014; Henstridge, Whitham, et al., 2014). However, novologues improved DPN in an Hsp70-dependent manner without correcting any metabolic parameters in models of both Type 1 and Type 2 diabetes. Similarly, the N-terminal Hsp90 inhibitor and geldanamycin derivative (DMAG) improved diabetic nephropathy without effecting metabolic measures of diabetes: weight loss, blood glucose, and HbA1c (Lazaro et al., 2015). The reason for the difference between these compounds is curious, but may lie in BGP-15 being a more direct activator of HSF1. It is encouraging that no adverse effects were reported from a short 28-day clinical trial in patients given 200 or 400 mg BGP-15 daily, which may be consistent with minimal drug effect on increasing eHsp70 activity (Literáti-Nagy et al., 2009). As treatment of diabetic complications such as DPN will require prolonged drug administration, it will be important to assess any effects on eHsp70 as novel modulators of Hsp70 move forward clinically. Fortunately, eHsp70 should be easy to monitor in a clinical trial and this will provide some insight into whether this pool of the protein correlates with adverse, presumably inflammatory, side effects.

6. CONCLUDING REMARKS

In summary, results in animal models and small clinical trials clearly provide compelling proof of concept that modulating molecular chaperones and particularly iHsp70 provides a viable approach to improve insulin resistance and diabetic complications. Although we are not putting forth the premise that directly targeting proteins and pathways that are linked to the pathophysiological progression of DPN is a less than fruitful approach to disease management, it seems that altering any one pathway may be insufficient to treat the human disease. A similar fate may befall an approach that only modulates iHsp70 to treat DPN. However, since modulating Hsp70 improves oxidative stress, mitochondrial function, and to some extent inflammation, increasing the ability of neurons and Schwann cells to tolerate ongoing diabetic stress may facilitate the action of additional agents or improve the ability of glycemic control to slow the rate of onset and diminish the severity of DPN, especially in Type 2 diabetic patients.

ACKNOWLEDGMENTS

This work was supported by Grant DK095911 from the National Institute of Diabetes, Digestive, and Kidney Diseases to R.T.D. and a postdoctoral fellowship from the Pharmaceutical Research and Manufacturers of America Foundation to S.M.E.

REFERENCES

Afolayan, A. J., Teng, R. J., Eis, A., Rana, U., Broniowska, K. A., Corbett, J. A., et al. (2014). Inducible Hsp70 regulates superoxide dismutase-2 and mitochondrial oxidative stress in the endothelial cells from developing lungs. *American Journal of Physiology. Lung Cellular and Molecular Physiology*, *306*, L351–L360.

Agou, F., Ye, F., Goffinont, S., Courtois, G., Yamaoka, S., Israel, A., et al. (2002). Nemo trimerizes through its coiled-coil C-terminal domain. *Journal of Biological Chemistry*, *277*, 17464–17475.

Arthur-Farraj, P. J., Latouche, M., Wilton, D. K., Quintes, S., Chabrol, E., Banerjee, A., et al. (2012). C-jun reprograms Schwann cells of injured nerves to generate a repair cell essential for regeneration. *Neuron*, *75*, 633–647.

Asea, A., Kraeft, S. K., Kurt-Jones, E. A., Stevenson, M. A., Chen, L. B., Finberg, R. W., et al. (2000). Hsp70 stimulates cytokine production through a CD14-dependent pathway, demonstrating its dual role as a chaperone and cytokine. *Nature Medicine*, *6*, 435–442.

Asea, A., Rehli, M., Kabingu, E., Boch, J. A., Bare, O., Auron, P. E., et al. (2002). Novel signal transduction pathway utilized by extracellular Hsp70: Role of toll-like receptor (TLR) 2 and TLR4. *Journal of Biological Chemistry*, *277*, 15028–15034.

Baseler, W. A., Dabkowski, E. R., Williamson, C. L., Croston, T. L., Thapa, D., Powell, M. J., et al. (2011). Proteomic alterations of distinct mitochondrial subpopulations in the Type 1 diabetic heart: Contribution of protein import dysfunction. *American Journal of Physiology. Regulatory, Integrative and Comparative Physiology*, *300*, R186–R200.

Basu, S., Binder, R. J., Ramalingam, T., & Srivastava, P. K. (2001). CD91 is a common receptor for heat shock proteins GRP96, Hsp90, Hsp70, and calreticulin. *Immunity*, *14*, 303–313.

Bausero, M. A., Gastpar, R., Multhoff, G., & Asea, A. (2005). Alternative mechanism by which IFN-g enhances tumor recognition: Active release of heat shock protein 72. *Journal of Immunology*, *175*, 2900–2912.

Becker, T. (2002). CD40, an extracellular receptor for binding and uptake of Hsp70-peptide complexes. *Journal of Cell Biology*, *158*, 1277–1285.

Becker, J., & Craig, E. A. (1994). Heat-shock proteins as molecular chaperones. *European Journal of Biochemistry*, *219*, 11–23.

Beckmann, R. P., Lovett, M., & Welch, W. J. (1992). Examining the function and regulation of Hsp70 in cells subjected to metabolic stress. *Journal of Cell Biology*, *117*, 1137–1150.

Bellmann, K., Jaattela, M., Wissing, D., Burkart, V., & Kolb, H. (1996). Heat shock protein Hsp70 overexpression confers resistance against nitric oxide. *FEBS Letters*, *391*, 185–188.

Bellmann, K., Wenz, A., Radons, J., Burkart, V., Kleemann, R., & Kolb, H. (1995). Heat shock induces resistance in rat pancreatic islet cells against nitric oxide, oxygen radicals and streptozotocin toxicity in vitro. *Journal of Clinical Investigation*, *95*, 2840–2845.

Benarroch, E. E. (2011). Heat shock proteins: Multiple neuroprotective functions and implications for neurologic disease. *Neurology*, *76*, 660–667.

Bocanegra, V., Manucha, W., Pena, M. R., Cacciamani, V., & Valles, P. G. (2010). Caveolin-1 and Hsp70 interaction in microdissected proximal tubules from spontaneously hypertensive rats as an effect of losartan. *Journal of Hypertension*, *28*, 143–155.

Broome, C. S., Kayani, A. C., Palomero, J., Dillmann, W. H., Mestril, R., Jackson, M. J., et al. (2006). Effect of lifelong overexpression of Hsp70 in skeletal muscle on age-related oxidative stress and adaptation after nondamaging contractile activity. *FASEB Journal*, *20*, 1549–1551.

Brownlee, M. (2001). Biochemistry and molecular cell biology of diabetic complications. *Nature*, *414*, 813–820.

Cadenas, E., & Davies, K. J. (2000). Mitochondrial free radical generation, oxidative stress, and aging. *Free Radical Biology and Medicine*, *29*, 222–230.

Callaghan, B. C., Cheng, H. T., Stables, C. L., Smith, A. L., & Feldman, E. L. (2012). Diabetic neuropathy: Clinical manifestations and current treatments. *Lancet Neurology*, *11*, 521–534.

Campisi, J., Leem, T. H., & Fleshner, M. (2003). Stress-induced extracellular Hsp72 is a functionally significant danger signal to the immune system. *Cell Stress & Chaperones*, *8*, 272–286.

Candas, D., & Li, J. J. (2014). MnSOD in oxidative stress response-potential regulation via mitochondrial protein influx. *Antioxidants & Redox Signaling*, *20*, 1599–1617.

Chan, J. Y., Ou, C. C., Wang, L. L., & Chan, S. H. (2004). Heat shock protein 70 confers cardiovascular protection during endotoxemia via inhibition of Nuclear Factor-KappaB activation and inducible nitric oxide synthase expression in the rostral ventrolateral medulla. *Circulation*, *110*, 3560–3566.

Chase, M. A., Wheeler, D. S., Lierl, K. M., Hughes, V. S., Wong, H. R., & Page, K. (2007). Hsp72 induces inflammation and regulates cytokine production in airway epithelium through a TLR- and NF-kB-dependent mechanism. *Journal of Immunology, 179*, 6318–6324.

Chen, H.-W., Kuo, H.-T., Wang, S.-J., Lu, T.-S., & Yang, R.-C. (2005). In vivo heat shock protein assembles with septic liver NF-kB/I-kB complex regulating NF-kB activity. *Shock, 24*, 232–238.

Chiosis, G., Huezo, H., Rosen, N., Mimnaugh, E., Whitesell, L., & Neckers, L. (2003). 17AAG: Low target binding affinity and potent cell activity—Finding an explanation. *Molecular Cancer Therapeutics, 2*, 123–129.

Chittoor-Vinod, V. G., Lee, S., Judge, S. M., & Notterpek, L. (2015). Inducible Hsp70 is critical in preventing the aggregation and enhancing the processing of PMP22. *ASN Neuro, 7*, 1–17.

Choi, S., Park, K. A., Lee, H. J., Park, M. S., Lee, J. H., Park, K. C., et al. (2005). Expression of Cu/Zn SOD protein is suppressed in Hsp 70.1 knockout mice. *Journal of Biochemistry and Molecular Biology, 38*, 111–114.

Chowdhury, S. K., Smith, D. R., & Fernyhough, P. (2013). The role of aberrant mitochondrial bioenergetics in diabetic neuropathy. *Neurobiology of Disease, 51*, 56–65.

Chung, J., Nguyen, A. K., Henstridge, D. C., Holmes, A. G., Chan, M. H., Mesa, J. L., et al. (2008). Hsp72 protects against obesity-induced insulin resistance. *Proceedings of the National Academy of Sciences of the United States of America, 105*, 1739–1744.

Daugaard, M., Rohde, M., & Jaattela, M. (2007). The heat shock protein 70 family: Highly homologous proteins with overlapping and distinct functions. *FEBS Letters, 581*, 3702–3710.

De Maio, A. (2011). Extracellular heat shock proteins, cellular export vesicles, and the stress observation system: A form of communication during injury, infection, and cell damage. It is never known how far a controversial finding will go! Dedicated to Ferruccio Ritossa. *Cell Stress & Chaperones, 16*, 235–249.

Delneste, Y., Magistrelli, G., Gauchat, J., Haeuw, J., Aubry, J., Nakamura, K., et al. (2002). Involvement of LOX-1 in dendritic cell-mediated antigen cross-presentation. *Immunity, 17*, 353–362.

Deshaies, R. J., Koch, B. D., Werner-Washburne, M., Craig, E. A., & Schekman, R. (1988). A subfamily of stress proteins facilitates translocation of secretory and mitochondrial precursor polypeptides. *Nature, 332*, 800–805.

Devi, T. S., Lee, I., Huttemann, M., Kumar, A., Nantwi, K. D., & Singh, L. P. (2012). TXNIP links innate host defense mechanisms to oxidative stress and inflammation in retinal Muller glia under chronic hyperglycemia: Implications for diabetic retinopathy. *Experimental Diabetes Research, 2012*. 438238.

Dickey, C. A., Kamal, A., Lundgren, K., Klosak, N., Bailey, R. M., Dunmore, J., et al. (2007). The high-affinity Hsp90-Chip complex recognizes and selectively degrades phosphorylated tau client proteins. *Journal of Clinical Investigation, 117*, 648–658.

Dokladny, K., Lobb, R., Wharton, W., Ma, T. Y., & Moseley, P. L. (2010). LPS-induced cytokine levels are repressed by elevated expression of Hsp70 in rats: Possible role of NF-kappaB. *Cell Stress & Chaperones, 15*, 153–163.

Dou, F., Netzer, W. J., Tanemura, K., Li, F., Hartl, F. U., Takashima, A., et al. (2003). Chaperones increase association of tau protein with microtubules. *Proceedings of the National Academy of Sciences of the United States of America, 100*, 721–726.

Dvoriantchikova, G., Santos, A. R., Saeed, A. M., Dvoriantchikova, X., & Ivanov, D. (2014). Putative role of protein kinase C in neurotoxic inflammation mediated by extracellular heat shock protein 70 after ischemia-reperfusion. *Journal of Neuroinflammation, 11*, 81.

Dybdahl, B. (2001). Inflammatory response after open heart surgery: Release of heat-shock protein 70 and signaling through Toll-like receptor-4. *Circulation, 105*, 685–690.

Dybdahl, B., Slordahl, S. A., Waage, A., Kierulf, P., Espevik, T., & Sundan, A. (2005). Myocardial ischaemia and the inflammatory response: Release of heat shock protein 70 after myocardial infarction. *Heart*, *91*, 299–304.

Ebrahimi-Fakhari, D., Saidi, L. J., & Wahlster, L. (2013). Molecular chaperones and protein folding as therapeutic targets in Parkinson's disease and other synucleinopathies. *Acta Neuropathologica Communications*, *1*, 79.

Edwards, J. L., Vincent, A. M., Cheng, H. T., & Feldman, E. L. (2008). Diabetic neuropathy: Mechanisms to management. *Pharmacology and Therapeutics*, *120*, 1–34.

Evans, C. G., Chang, L., & Gestwicki, J. E. (2010). Heat shock protein 70 (Hsp70) as an emerging drug target. *Journal of Medicinal Chemistry*, *53*, 4585–4602.

Evans, C. G., Wisen, S., & Gestwicki, J. E. (2006). Heat shock proteins 70 and 90 inhibit early stages of amyloid beta-(1-42) aggregation in vitro. *Journal of Biological Chemistry*, *281*, 33182–33191.

Evdonin, A. L., Martynova, M. G., Bystrova, O. A., Guzhova, I. V., Margulis, B. A., & Medvedeva, N. D. (2006). The release of Hsp70 from A431 carcinoma cells is mediated by secretory-like granules. *European Journal of Cell Biology*, *85*, 443–455.

Farmer, K. L., Li, C., & Dobrowsky, R. T. (2012). Diabetic peripheral neuropathy: Should a chaperone accompany our therapeutic approach? *Pharmacological Reviews*, *64*, 880–900.

Feinstein, D. L., Galea, E., Aquino, D. A., Li, G. C., Xu, H., & Reis, D. J. (1996). Heat shock protein 70 suppresses astroglial-inducible nitric-oxide synthase expression by decreasing NFkappaB activation. *Journal of Biological Chemistry*, *271*, 17724–17732.

Feinstein, D. L., Galea, E., & Reis, D. J. (1997). Suppression of glial nitric oxide synthase induction by heat shock: Effects on proteolytic degradation of IkappaB-alpha. *Nitric Oxide*, *1*, 167–176.

Garcia-Alcala, H., Santos Vichido, C. I., Islas Macedo, S., Genestier-Tamborero, C. N., Minutti-Palacios, M., Hirales Tamez, O., et al. (2015). Treatment with alpha-lipoic acid over 16 weeks in Type 2 diabetic patients with symptomatic polyneuropathy who responded to initial 4-week high-dose loading. *Journal of Diabetes Research*, *2015*. 189857.

Gastpar, R., Gehrmann, M., Bausero, M. A., Asea, A., Gross, C., Schroeder, J. A., et al. (2005). Heat shock protein 70 surface-positive tumor exosomes stimulate migratory and cytolytic activity of natural killer cells. *Cancer Research*, *65*, 5238–5247.

Gifondorwa, D. J., Robinson, M. B., Hayes, C. D., Taylor, A. R., Prevette, D. M., Oppenheim, R. W., et al. (2007). Exogenous delivery of heat shock protein 70 increases lifespan in a mouse model of amyotrophic lateral sclerosis. *Journal of Neuroscience*, *27*, 13173–13180.

Gil Lorenzo, A. F., Bocanegra, V., Benardon, M. E., Cacciamani, V., & Valles, P. G. (2014). Hsp70 regulation on NOX4/p22Phox and cytoskeletal integrity as an effect of losartan in vascular smooth muscle cells. *Cell Stress & Chaperones*, *19*, 115–134.

Goeckeler, J. L., & Brodsky, J. L. (2010). Molecular chaperones and substrate ubiquitination control the efficiency of endoplasmic reticulum-associated degradation. *Diabetes Obesity & Metabolism*, *12*(Suppl. 2), 32–38.

Gross, C., Hansch, D., Gastpar, R., & Multhoff, G. (2003). Interaction of heat shock protein 70 peptide with NK cells involves the NK receptor CD94. *Biological Chemistry*, *384*, 267–279.

Guzhova, I., Kislyakova, K., Moskaliova, O., Fridlanskaya, I., Tytell, M., Cheetham, M., et al. (2001). In vitro studies show that Hsp70 can be released by glia and that exogenous Hsp70 can enhance neuronal stress tolerance. *Brain Research*, *914*, 66–73.

Haas, I. G., & Wabl, M. (1983). Immunoglobulin heavy chain binding protein. *Nature*, *306*, 387–389.

Henstridge, D. C., Bruce, C. R., Drew, B. G., Tory, K., Kolonics, A., Estevez, E., et al. (2014). Activating Hsp72 in rodent skeletal muscle increases mitochondrial number and oxidative capacity and decreases insulin resistance. *Diabetes*, *63*, 1881–1894.

Henstridge, D. C., Whitham, M., & Febbraio, M. A. (2014). Chaperoning to the metabolic party: The emerging therapeutic role of heat-shock proteins in obesity and Type 2 diabetes. *Molecular Metabolism*, *3*, 781–793.

Hightower, L. E., & Guidon, P. T., Jr. (1989). Selective release from cultured mammalian cells of heat-shock (stress) proteins that resemble glia-axon transfer proteins. *Journal of Cellular Physiology*, *138*, 257–266.

Horn, P., Kalz, A., Lim, C. L., Pyne, D., Saunders, P., Mackinnon, L., et al. (2007). Exercise-recruited NK cells display exercise-associated eHsp-70. *Exercise Immunology Review*, *13*, 100–111.

Huang, C., Wang, J., Chen, Z., Wang, Y., & Zhang, W. (2013). 2-Phenylethynesulfonamide prevents induction of pro-inflammatory factors and attenuates LPS-induced liver injury by targeting NHE1-Hsp70 complex in mice. *PloS One*, *8*, e67582.

Hunter-Lavin, C., Davies, E. L., Bacelar, M. M., Marshall, M. J., Andrew, S. M., & Williams, J. H. (2004). Hsp70 release from peripheral blood mononuclear cells. *Biochemical and Biophysical Research Communications*, *324*, 511–517.

Hur, J., Sullivan, K. A., Pande, M., Hong, Y., Sima, A. A. F., Jagadish, H. V., et al. (2011). The identification of gene expression profiles associated with progression of human diabetic neuropathy. *Brain*, *134*, 3222–3235.

Hutton, E. J., Carty, L., Laurá, M., Houlden, H., Lunn, M. P. T., Brandner, S., et al. (2011). C-jun expression in human neuropathies: A pilot study. *Journal of the Peripheral Nervous System*, *16*, 295–303.

Ianaro, A., Lalenti, A., Maffia, P., Di Meglio, P., Di Rosa, M., & Santoro, M. G. (2003). Anti-inflammatory activity of 15-deoxy-delta 12,14-PGJ2 and 2-cyclopenten-1-one: Role of the heat shock response. *Molecular Pharmacology*, *64*, 85–93.

Jaattela, M., & Wissing, D. (1993). Heat-shock proteins protect cells from monocyte cytotoxicity: Possible mechanism of self-protection. *Journal of Experimental Medicine*, *177*, 231–236.

Johnson, J. D., & Fleshner, M. (2006). Releasing signals, secretory pathways, and immune function of endogenous extracellular heat shock protein 72. *Journal of Leukocyte Biology*, *79*, 425–434.

Jones, Q., Voegeli, T. S., Li, G., Chen, Y., & William Currie, R. (2011). Heat shock proteins protect against ischemia and inflammation through multiple mechanisms. *Inflammation & Allergy Drug Targets*, *10*, 247–259.

Kalmar, B., Lu, C. H., & Greensmith, L. (2014). The role of heat shock proteins in amyotrophic lateral sclerosis: The therapeutic potential of arimoclomol. *Pharmacology and Therapeutics*, *141*, 40–54.

Kamal, A., Thao, L., Sensintaffar, J., Zhang, L., Boehm, M. F., Fritz, L. C., et al. (2003). A high-affinity conformation of Hsp90 confers tumour selectivity on Hsp90 inhibitors. *Nature*, *425*, 407–410.

Kim, Y. E., Hipp, M. S., Bracher, A., Hayer-Hartl, M., & Hartl, F. U. (2013). Molecular chaperone functions in protein folding and proteostasis. *Annual Review of Biochemistry*, *82*, 323–355.

Kondo, T., Murakami, K., Honkaniemi, J., Sharp, F. R., Epstein, C. J., & Chan, P. H. (1996). Expression of Hsp70 mRNA is induced in the brain of transgenic mice overexpressing human CuZn-superoxide dismutase following transient global cerebral ischemia. *Brain Research*, *737*, 321–326.

Kovalchin, J. T., Wang, R., Wagh, M. S., Azoulay, J., Sanders, M., & Chandawarkar, R. Y. (2006). In vivo delivery of heat shock protein 70 accelerates wound healing by up-regulating macrophage-mediated phagocytosis. *Wound Repair and Regeneration*, *14*, 129–137.

Krause, M., Heck, T. G., Bittencourt, A., Scomazzon, S. P., Newsholme, P., Curi, R., et al. (2015). The chaperone balance hypothesis: The importance of the extracellular to

intracellular Hsp70 ratio to inflammation-driven Type 2 diabetes, the effect of exercise, and the implications for clinical management. *Mediators of Inflammation, 2015*. 249205.
Krause, M., Keane, K., Rodrigues-Krause, J., Crognale, D., Egan, B., De Vito, G., et al. (2014). Elevated levels of extracellular heat-shock protein 72 (eHsp72) are positively correlated with insulin resistance in vivo and cause pancreatic beta-cell dysfunction and death in vitro. *Clinical Science (London), 126*, 739–752.
Krause, M. S., Oliveira, L. P., Jr., Silveira, E. M., Vianna, D. R., Rossato, J. S., Almeida, B. S., et al. (2007). MRP1/GS-X pump ATPase expression: Is this the explanation for the cytoprotection of the heart against oxidative stress-induced redox imbalance in comparison to skeletal muscle cells? *Cell Biochemistry and Function, 25*, 23–32.
Krimmer, T., Rassow, J., Kunau, W. H., Voos, W., & Pfanner, N. (2000). Mitochondrial protein import motor: The ATPase domain of matrix Hsp70 is crucial for binding to TIM44, while the peptide binding domain and the carboxy-terminal segment play a stimulatory role. *Molecular and Cellular Biology, 20*, 5879–5887.
Kurucz, I., Morva, A., Vaag, A., Eriksson, K. F., Huang, X., Groop, L., et al. (2002). Decreased expression of heat shock protein 72 in skeletal muscle of patients with Type 2 diabetes correlates with insulin resistance. *Diabetes, 51*, 1102–1109.
Kusuma, B. R., Zhang, L., Sundstrom, T., Peterson, L. B., Dobrowsky, R. T., & Blagg, B. S. J. (2012). Synthesis and evaluation of novologues as C-terminal Hsp90 inhibitors with cytoprotective activity against sensory neuron glucotoxicity. *Journal of Medicinal Chemistry, 55*, 5797–5812.
Lancaster, G. I., & Febbraio, M. A. (2005). Exosome-dependent trafficking of Hsp70: A novel secretory pathway for cellular stress proteins. *Journal of Biological Chemistry, 280*, 23349–23355.
Lane, T., Flam, B., Lockey, R., & Kolliputi, N. (2013). Txnip shuttling: Missing link between oxidative stress and inflammasome activation. *Frontiers in Physiology, 4*, 50.
Lazaro, I., Oguiza, A., Recio, C., Mallavia, B., Madrigal-Matute, J., Blanco, J., et al. (2015). Targeting Hsp90 ameliorates nephropathy and atherosclerosis through suppression of NF-kappaB and STAT signaling pathways in diabetic mice. *Diabetes, 64*, 3600–3613.
Lee, J., & Ozcan, U. (2014). Unfolded protein response signaling and metabolic diseases. *Journal of Biological Chemistry, 289*, 1203–1211.
Lehner, T., Bergmeier, L. A., Wang, Y., Tao, L., Sing, M., Spallek, R., et al. (2000). Heat shock proteins generate beta-chemokines which function as innate adjuvants enhancing adaptive immunity. *European Journal of Immunology, 30*, 594–603.
Leustek, T., Dalie, B., Amir-Shapira, D., Brot, N., & Weissbach, H. (1989). A member of the Hsp70 family is localized in mitochondria and resembles Escherichia coli DNAK. *Proceedings of the National Academy of Sciences of the United States of America, 86*, 7805–7808.
Li, C., Ma, J., Zhao, H., Blagg, B. S., & Dobrowsky, R. T. (2012). Induction of heat shock protein 70 (Hsp70) prevents neuregulin-induced demyelination by enhancing the proteasomal clearance of c-jun. *ASN Neuro, 4*, e00102.
Lianos, G. D., Alexiou, G. A., Mangano, A., Mangano, A., Rausei, S., Boni, L., et al. (2015). The role of heat shock proteins in cancer. *Cancer Letters, 360*, 114–118.
Literáti-Nagy, B., Kulcsár, E., Literáti-Nagy, Z., Buday, B., Péterfai, É., Horváth, T., et al. (2009). Improvement of insulin sensitivity by a novel drug, BGP-15, in insulin-resistant patients: A proof of concept randomized double-blind clinical trial. *Hormone and Metabolic Research, 41*, 374–380.
Luo, W., Dou, F., Rodina, A., Chip, S., Kim, J., Zhao, Q., et al. (2007). Roles of heat-shock protein 90 in maintaining and facilitating the neurodegenerative phenotype in tauopathies. *Proceedings of the National Academy of Sciences of the United States of America, 104*, 9511–9516.
Luo, W., Rodina, A., & Chiosis, G. (2008). Heat shock protein 90: Translation from cancer to Alzheimer's disease treatment? *BMC Neuroscience, 9*(Suppl. 2), S7.

Ma, J., Farmer, K. L., Pan, P., Urban, M. J., Zhao, H., Blagg, B. S., et al. (2014). Heat shock protein 70 is necessary to improve mitochondrial bioenergetics and reverse diabetic sensory neuropathy following KU-32 therapy. *Journal of Pharmacology and Experimental Therapeutics*, *348*, 281–292.

Ma, J., Pan, P., Anyika, M., Blagg, B. S., & Dobrowsky, R. T. (2015). Modulating molecular chaperones improves mitochondrial bioenergetics and decreases the inflammatory transcriptome in diabetic sensory neurons. *ACS Chemical Neuroscience*, *6*, 1637–1648.

Madden, L. A., Sandstrom, M. E., Lovell, R. J., & McNaughton, L. (2008). Inducible heat shock protein 70 and its role in preconditioning and exercise. *Amino Acids*, *34*, 511–516.

Mansilla, M. J., Costa, C., Eixarch, H., Tepavcevic, V., Castillo, M., Martin, R., et al. (2014). Hsp70 regulates immune response in experimental autoimmune encephalomyelitis. *PloS One*, *9*, e105737.

Meldrum, K. K. (2003). Liposomal delivery of heat shock protein 72 into renal tubular cells blocks nuclear factor-kappaB activation, tumor necrosis factor-alpha production, and subsequent ischemia-induced apoptosis. *Circulation Research*, *92*, 293–299.

Miyata, Y., Nakamoto, H., & Neckers, L. (2013). The therapeutic target Hsp90 and cancer hallmarks. *Current Pharmaceutical Design*, *19*, 347–365.

Muchowski, P. J., & Wacker, J. L. (2005). Modulation of neurodegeneration by molecular chaperones. *Nature Reviews. Neuroscience*, *6*, 11–22.

Multhoff, G. (2007). Heat shock protein 70 (Hsp70): Membrane location, export and immunological relevance. *Methods*, *43*, 229–237.

Murakami, H., Pain, D., & Blobel, G. (1988). 70-kd Heat shock-related protein is one of at least two distinct cytosolic factors stimulating protein import into mitochondria. *Journal of Cell Biology*, *107*, 2051–2057.

Neef, D. W., Jaeger, A. M., & Thiele, D. J. (2011). Heat shock transcription factor 1 as a therapeutic target in neurodegenerative diseases. *Nature Reviews. Drug Discovery*, *10*, 930–944.

Obrosova, I. G., Drel, V. R., Pacher, P., Ilnytska, O., Wang, Z. Q., Stevens, M. J., et al. (2005). Oxidative-nitrosative stress and poly(ADP-ribose) polymerase (PARP) activation in experimental diabetic neuropathy: The relation is revisited. *Diabetes*, *54*, 3435–3441.

Oglesbee, M. J., Herdman, A. V., Passmore, G. G., & Hoffman, W. H. (2005). Diabetic ketoacidosis increases extracellular levels of the major inducible 70-kDa heat shock protein. *Clinical Biochemistry*, *38*, 900–904.

Ortega, E., Giraldo, E., Hinchado, M. D., Martinez, M., Ibanez, S., Cidoncha, A., et al. (2006). Role of Hsp72 and norepinephrine in the moderate exercise-induced stimulation of neutrophils' microbicide capacity. *European Journal of Applied Physiology*, *98*, 250–255.

Ortega, E., Hinchado, M. D., Martin-Cordero, L., & Asea, A. (2009). The effect of stress-inducible extracellular Hsp72 on human neutrophil chemotaxis: A role during acute intense exercise. *Stress*, *12*, 240–249.

Park, H. S., Lee, J. S., Huh, S. H., Seo, J. S., & Choi, E. J. (2001). Hsp72 functions as a natural inhibitory protein of c-jun N-terminal kinase. *EMBO Journal*, *20*, 446–456.

Parkinson, D. B., Bhaskaran, A., Arthur-Farraj, P., Noon, L. A., Woodhoo, A., Lloyd, A. C., et al. (2008). C-jun is a negative regulator of myelination. *Journal of Cell Biology*, *181*, 625–637.

Peterson, L. B., & Blagg, B. S. (2009). To fold or not to fold: Modulation and consequences of Hsp90 inhibition. *Future Medicinal Chemistry*, *1*, 267–283.

Polla, B. S., Kantengwa, S., Francois, D., Salvioli, S., Franceschi, C., Marsac, C., et al. (1996). Mitochondria are selective targets for the protective effects of heat shock against oxidative injury. *Proceedings of the National Academy of Sciences of the United States of America*, *93*, 6458–6463.

Qiao, Y., Liu, B., & Li, Z. (2008). Activation of NK cells by extracellular heat shock protein 70 through induction of NKG2D ligands on dendritic cells. *Cancer Immunity*, *8*, 12.

Ran, R., Lu, A., Zhang, L., Tang, Y., Zhu, H., Xu, H., et al. (2004). Hsp70 promotes TNF-mediated apoptosis by binding IKK gamma and impairing NF-kappaB survival signaling. *Genes and Development*, *18*, 1466–1481.

Richard, V., Kaeffer, N., & Thuillez, C. (1996). Delayed protection of the ischemic heart—From pathophysiology to therapeutic applications. *Fundamental and Clinical Pharmacology*, *10*, 409–415.

Robinson, M. B., Tidwell, J. L., Gould, T., Taylor, A. R., Newbern, J. M., Graves, J., et al. (2005). Extracellular heat shock protein 70: A critical component for motoneuron survival. *Journal of Neuroscience*, *25*, 9735–9745.

Rodrigues-Krause, J., Krause, M., O'Hagan, C., De Vito, G., Boreham, C., Murphy, C., et al. (2012). Divergence of intracellular and extracellular Hsp72 in Type 2 diabetes: Does fat matter? *Cell Stress & Chaperones*, *17*, 293–302.

Saibil, H. (2013). Chaperone machines for protein folding, unfolding and disaggregation. *Nature Reviews. Molecular Cell Biology*, *14*, 630–642.

Saleh, A., Roy Chowdhury, S. K., Smith, D. R., Balakrishnan, S., Tessler, L., Martens, C., et al. (2013). Ciliary neurotrophic factor activates NF-kappaB to enhance mitochondrial bioenergetics and prevent neuropathy in sensory neurons of streptozotocin-induced diabetic rodents. *Neuropharmacology*, *65*, 65–73.

Saleh, A., Schapansky, J., Smith, D. R., Young, N., Odero, G. L., Aulston, B., et al. (2013). Normalization of NF-kappaB activity in dorsal root ganglia neurons cultured from diabetic rats reverses neuropathy-linked markers of cellular pathology. *Experimental Neurology*, *241*, 169–178.

Schmidt, O., Pfanner, N., & Meisinger, C. (2010). Mitochondrial protein import: From proteomics to functional mechanisms. *Nature Reviews Molecular Cell Biology*, *11*, 655–667.

Schulze, P. C., Yoshioka, J., Takahashi, T., He, Z., King, G. L., & Lee, R. T. (2004). Hyperglycemia promotes oxidative stress through inhibition of thioredoxin function by thioredoxin-interacting protein. *Journal of Biological Chemistry*, *279*, 30369–30374.

Senf, S. M., Howard, T. M., Ahn, B., Ferreira, L. F., & Judge, A. R. (2013). Loss of the inducible Hsp70 delays the inflammatory response to skeletal muscle injury and severely impairs muscle regeneration. *PloS One*, *8*, e62687.

Singh, L. P. (2013). Thioredoxin interacting protein (TXNIP) and pathogenesis of diabetic retinopathy. *Journal of Clinical & Experimental Ophthalmology*, *4*. http://dx.doi.org/10.4172/2155-9570.1000287.

Stavniichuk, R., Drel, V. R., Shevalye, H., Maksimchyk, Y., Kuchmerovska, T. M., Nadler, J. L., et al. (2011). Baicalein alleviates diabetic peripheral neuropathy through inhibition of oxidative–nitrosative stress and p38 MAPK activation. *Experimental Neurology*, *230*, 106–113.

Stetler, R. A., Gan, Y., Zhang, W., Liou, A. K., Gao, Y., Cao, G., et al. (2010). Heat shock proteins: Cellular and molecular mechanisms in the central nervous system. *Progress in Neurobiology*, *92*, 184–211.

Tan, W., Pasinelli, P., & Trotti, D. (2014). Role of mitochondria in mutant SOD1 linked amyotrophic lateral sclerosis. *Biochimica et Biophysica Acta*, *1842*, 1295–1301.

Theriault, J. R., Mambula, S. S., Sawamura, T., Stevenson, M. A., & Calderwood, S. K. (2005). Extracellular Hsp70 binding to surface receptors present on antigen presenting cells and endothelial/epithelial cells. *FEBS Letters*, *579*, 1951–1960.

Tidwell, J. L., Houenou, L. J., & Tytell, M. (2004). Administration of Hsp70 in vivo inhibits motor and sensory neuron degeneration. *Cell Stress & Chaperones*, *9*, 88–98.

Tomlinson, D. R., & Gardiner, N. J. (2008). Glucose neurotoxicity. *Nature Reviews. Neuroscience*, *9*, 36–45.

Turturici, G., Sconzo, G., & Geraci, F. (2011). Hsp70 and its molecular role in nervous system diseases. *Biochemistry Research International, 2011*. 618127.

Tytell, M., Greenberg, S. G., & Lasek, R. J. (1986). Heat shock-like protein is transferred from glia to axon. *Brain Research, 363*, 161–164.

Urban, M. J., Li, C., Yu, C., Lu, Y., Krise, J. M., McIntosh, M. P., et al. (2010). Inhibiting heat shock protein 90 reverses sensory hypoalgesia in diabetic mice. *ASN Neuro, 2*, 189–199.

Urban, M. J., Pan, P., Farmer, K. L., Zhao, H., Blagg, B. S., & Dobrowsky, R. T. (2012). Modulating molecular chaperones improves sensory fiber recovery and mitochondrial function in diabetic peripheral neuropathy. *Experimental Neurology, 235*, 388–396.

Vareniuk, I., Pavlov, I. A., Drel, V. R., Lyzogubov, V. V., Ilnytska, O., Bell, S. R., et al. (2007). Nitrosative stress and peripheral diabetic neuropathy in leptin-deficient (ob/ob) mice. *Experimental Neurology, 205*, 425–436.

Vihervaara, A., & Sistonen, L. (2014). HSF1 at a glance. *Journal of Cell Science, 127*, 261–266.

Walton-Diaz, A., Khan, S., Bourboulia, D., Trepel, J. B., Neckers, L., & Mollapour, M. (2013). Contributions of co-chaperones and post-translational modifications towards Hsp90 drug sensitivity. *Future Medicinal Chemistry, 5*, 1059–1071.

Wang, Y., Kelly, C. G., Karttunen, J. T., Whittall, T., Lehner, P. J., Duncan, L., et al. (2001). CD40 is a cellular receptor mediating mycobacterial heat shock protein 70 stimulation of CC-chemokines. *Immunity, 15*, 971–983.

Waza, M., Adachi, H., Katsuno, M., Minamiyama, M., Sang, C., Tanaka, F., et al. (2005). 17-AAG, an Hsp90 inhibitor, ameliorates polyglutamine-mediated motor neuron degeneration. *Nature Medicine, 11*, 1088–1095.

Wei, W., Liu, Q., Tan, Y., Liu, L., Li, X., & Cai, L. (2009). Oxidative stress, diabetes, and diabetic complications. *Hemoglobin, 33*, 370–377.

Weiss, Y. G., Bromberg, Z., Raj, N., Raphael, J., Goloubinoff, P., Ben-Neriah, Y., et al. (2007). Enhanced heat shock protein 70 expression alters proteasomal degradation of IkB kinase in experimental acute respiratory distress syndrome. *Critical Care Medicine, 35*, 2128–2138.

Wellen, K. E. (2005). Inflammation, stress, and diabetes. *Journal of Clinical Investigation, 115*, 1111–1119.

Whitham, M., & Fortes, M. B. (2008). Heat shock protein 72: Release and biological significance during exercise. *Frontiers in Bioscience, 13*, 1328–1339.

Whiting, D. R., Guariguata, L., Weil, C., & Shaw, J. (2011). IDF Diabetes Atlas: Global estimates of the prevalence of diabetes for 2011 and 2030. *Diabetes Research and Clinical Practice, 94*, 311–321.

Wiedemann, N., Frazier, A. E., & Pfanner, N. (2004). The protein import machinery of mitochondria. *Journal of Biological Chemistry, 279*, 14473–14476.

Wong, H. R., Ryan, M. A., Menendez, I. Y., & Wispe, J. R. (1999). Heat shock activates the I-kappaBalpha promoter and increases I-kappaBalpha mRNA expression. *Cell Stress & Chaperones, 4*, 1–7.

Wong, H. R., Ryan, M., & Wispe, J. R. (1997). The heat shock response inhibits inducible nitric oxide synthase gene expression by blocking I kappa-B degradation and NF-kappa B nuclear translocation. *Biochemical and Biophysical Research Communications, 231*, 257–263.

Wright, E., Jr., Scism-Bacon, J. L., & Glass, L. C. (2006). Oxidative stress in type 2 diabetes: The role of fasting and postprandial glycaemia. *International Journal of Clinical Practice, 60*, 308–314.

Wu, N., Zheng, B., Shaywitz, A., Dagon, Y., Tower, C., Bellinger, G., et al. (2013). AMPK-dependent degradation of TXNIP upon energy stress leads to enhanced glucose uptake via Glut1. *Molecular Cell, 49*, 1167–1175.

Yang, X. M., Baxter, G. F., Heads, R. J., Yellon, D. M., Downey, J. M., & Cohen, M. V. (1996). Infarct limitation of the second window of protection in a conscious rabbit model. *Cardiovascular Research*, *31*, 777–783.

Young, J. C., Hoogenraad, N. J., & Hartl, F. U. (2003). Molecular chaperones Hsp90 and Hsp70 deliver preproteins to the mitochondrial import receptor Tom70. *Cell*, *112*, 41–50.

Zhang, L., Zhao, H., Blagg, B. S., & Dobrowsky, R. T. (2012). A C-terminal heat shock protein 90 inhibitor decreases hyperglycemia-induced oxidative stress and improves mitochondrial bioenergetics in sensory neurons. *Journal of Proteome Research*, *11*, 129–137.

Ziegler, D., Ametov, A., Barinov, A., Dyck, P. J., Gurieva, I., Low, P. A., et al. (2006). Oral treatment with α-lipoic acid improves symptomatic diabetic polyneuropathy. *Diabetes Care*, *29*, 2365–2370.

Ziegler, D., Low, P. A., Litchy, W. J., Boulton, A. J., Vinik, A. I., Freeman, R., et al. (2011). Efficacy and safety of antioxidant treatment with alpha-lipoic acid over 4 years in diabetic polyneuropathy: The Nathan 1 trial. *Diabetes Care*, *34*, 2054–2060.

CHAPTER ELEVEN

Painful Diabetic Neuropathy: Prevention or Suppression?

S.M. Todorovic[1]
School of Medicine, University of Colorado Anschutz Medical Campus, Aurora, CO, United States
[1]Corresponding author: e-mail address: slobodan.todorovic@ucdenver.edu

Contents

Abstract

Pain-sensing sensory neurons (nociceptors) of the dorsal root ganglia (DRG) and dorsal horn (DH) can become sensitized (hyperexcitable) in response to pathological conditions such as diabetes, which in turn may lead to the development of painful peripheral diabetic neuropathy (PDN). Because of incomplete knowledge about the mechanisms underlying painful PDN, current treatment for painful PDN has been limited to somewhat nonspecific systemic drugs that have significant side effects or potential for abuse. Recent studies have established that several ion channels in DRG and DH neurons are dysregulated and make a previously unrecognized contribution to sensitization of pain responses by enhancing excitability of nociceptors in animal models of type 1 and type 2 PDN. Furthermore, it has been reported that targeting posttranslational modification of nociceptive ion channels such as glycosylation and methylglyoxal metabolism can completely reverse mechanical and thermal hyperalgesia in diabetic animals with PDN in vivo. Understanding details of posttranslational regulation of nociceptive channel activity may facilitate development of novel therapies for treatment of painful PDN. We argue that pharmacological targeting of the specific pathogenic mechanism rather than of the channel per se may cause fewer side effects and reduce the potential for drug abuse in patients with diabetes.

International Review of Neurobiology, Volume 127
ISSN 0074-7742
http://dx.doi.org/10.1016/bs.irn.2016.03.005

1. INTRODUCTION

Diabetes mellitus is a chronic, multifactorial disease characterized by hyperglycemia. The World Health Organization estimates that 220 million people worldwide currently suffer from diabetes. Diabetes mellitus type 1 (formerly called insulin-dependent diabetes or juvenile diabetes) is a form of diabetes that results from autoimmune destruction of insulin-producing β-cells in the pancreas. In contrast, type 2 diabetes (formerly called noninsulin-dependent or adult-onset diabetes), the most common form of diabetes, results from insulin resistance and is most often associated with obesity. Peripheral diabetic neuropathy (PDN) is the most frequent complication of diabetes, and it has been estimated that 60–70% of diabetic patients will develop PDN symptoms with prevalence increasing with the duration of diabetes mellitus (Centers for Disease Control and Prevention, 2011). The prevalence of painful PDN is often underestimated, but a recent large community-based study with more than 15,000 patients with diabetes reported painful symptoms in about one-third of all participants (Abbott, Malik, van Ross, Kulkarni, & Boulton, 2011). These patients can suffer from both stimulus-evoked painful episodes, including hyperalgesia (exaggerated pain experience upon presentation of noxious stimuli) and/or allodynia (normally nonnoxious stimuli that may evoke pain sensation), as well as spontaneous pain in the form of burning or tingling. Interestingly, these painful pathologies in diabetics usually occur in parallel with degeneration of peripheral nerves (Abbott et al., 2011). Eventually, these painful symptoms usually subside as the disabling pain is replaced by complete loss of sensation (insensate PDN). Both intractable pain and loss of sensation have significant adverse effects on quality of life measures. Unfortunately, current treatment options are unable to reverse these symptoms and are often associated with serious side effects. For example, $Ca_V2.2$ (N-type) subtypes of high-voltage-activated (HVA) calcium channels and their regulatory subunit α2δ are considered a major cellular target for the anticonvulsants gabapentin and pregabalin, which are commonly used to relieve diabetes-induced pain in clinics. However, more than 50% of patients using gabapentin or pregabalin experience side effects, such as excessive sedation, ataxia, dizziness, euphoria, and weight gain, all of which limit their clinical use (Edwards, Vincent, Cheng, & Feldman, 2008). Although opioids are partially effective for treatment of painful PDN, they are not suitable for chronic use since their long-term use is associated with serious side effects including constipation, urinary retention, impaired cognitive function, respiratory depression, and issues associated with tolerance and addiction. Furthermore, topical applications of therapeutic creams with lidocaine and capsaicin are inconsistently effective in patients with painful PDN and cause unpleasant side effects like numbness and burning, respectively. This underscores the fact that new mechanism-specific therapeutic treatments must be considered that could be more effective and possibly better tolerated in diabetic patients.

Pain is often a result of the activation of neurons (nociceptors) that are activated following injury of peripheral tissue such as the skin and internal organs. This form of pain, called nociceptive pain, is self-protective and typically evokes actions aimed at limiting or avoiding further tissue damage. Nociceptive pain is typically responsive to treatments with conventional painkillers, such as nonsteroidal antiinflammatory drugs or opiates. In contrast, neuropathic pain results from primary dysfunction of peripheral nociceptive, as well as nonnociceptive nerves and/or components of the central nervous system (central pain). Neuropathic pain typically outlasts the initial stage of injury and frequently leads to debilitating disorders that are not amenable to conventionally available drug therapies. It is generally accepted that pathologies associated with neuropathic pain result from increased excitability of peripheral nerves and/or pain pathways in central nervous system (CNS). For example, nociceptors can become inappropriately sensitized by various mechanisms in response to local or systemic metabolic conditions or peripheral tissue injury associated with diabetes. Multiple pathogenic mechanisms, such as the formation of intracellular advanced glycation end products (AGE), release of inflammatory cytokines, increased glucose metabolism by the polyol pathway, and oxidative stress, may contribute to the impaired function of sensory neurons in animals with PDN. However, preclinical and clinical studies aiming to target several of these mechanisms while simultaneously ensuring proper blood glucose control have been unable to provide definite and complete pain relief (Edwards et al., 2008; Gooch & Podwall, 2004). This incomplete pain reversal strongly suggests that other molecular mechanisms may be involved in hyperglycemia-induced alterations of function of nociceptive sensory neurons.

2. NOCICEPTIVE ION CHANNELS CONTROL NEURONAL EXCITABILITY

Several studies conducted in recent years have reported on the plasticity of various ion channels expressed in nociceptive sensory neurons, also known as dorsal root ganglion (DRG) neurons, which play a critical role in modulating overall cellular excitability (reviewed by Calcutt, 2013; Todorovic & Jevtovic-Todorovic, 2011; Todorovic & Jevtovic-Todorovic, 2014). These findings are both important and relevant, since increased excitability of sensory neurons is believed to contribute directly to the development and maintenance of painful symptoms, including hyperalgesia, allodynia, and spontaneous pain. Studies have shown that the Cav3.2 isoform of the T-type voltage-gated calcium channel (T-channel) is heavily expressed in DRG cells and the dorsal horn (DH) of the spinal cord and plays a distinct role in supporting pathological pain in animal models of both type 1- and type 2-induced PDN (Jacus, Uebele, Renger, & Todorovic, 2012; Jagodic et al., 2007; Latham et al., 2009). T-type channels were first discovered in preparations of acutely dissociated sensory neurons of small size (Carbone & Lux, 1984). The characteristic biophysical features, including activation by small membrane depolarization, relatively fast and voltage-dependent inactivation kinetics, as well as slow deactivation kinetics upon channel closing following maximal activation, make them ideally suited to control cellular excitability of sensory neurons. Relatively soon after their initial discovery, it was indeed demonstrated that T-channels control subthreshold excitability of DRG cells by lowering the threshold for action potential generation in a subpopulation of acutely dissociated rat DRG cells of medium size (White, Lovinger, & Weight, 1989). By convention, it is accepted that most of the smaller diameter DRG cells (cell soma < 31 μm) are putative nociceptors, while medium-size DRG cells (cell soma 32–45 μm) can be either nociceptive or nonnociceptive sensory neurons. Most acutely dissociated small DRG cells expressing T-currents are sensitive to the algogene compound from hot peppers, capsaicin, and have the electrical properties of classically described nociceptors with wide action potentials (Nelson et al., 2007). T-channels are also present in a subpopulation of medium-sized acutely dissociated sensory neurons that are either putative nociceptors, since they bear the IB_4 (isolectin B_4) antigen (Jagodic et al., 2007), or that subserve touch sensation mediated by specific DRG cell type called D-hair mechanoreceptors (Dubreuil et al., 2004; Shin, Martinez-Salgado, Heppenstall, & Lewin, 2003). Furthermore, the $Ca_V3.2$ isoform of T-channels is a critical regulator of cellular excitability of nociceptive DRG neurons termed "T-rich" cells that are also responsive to capsaicin and are IB_4 positive (Nelson, Joksovic, Perez-Reyes, & Todorovic, 2005). Prominent $Ca_V3.2$ T-currents in "T-rich" and medium-size DRG cells generate visible after-depolarizing potentials and high-frequency burst firing along with small depolarizations of the membrane potential. In addition, amplification of subthreshold membrane depolarizations via

$Ca_V3.2$ T-channels typically decreases the threshold for action potential firing in DRG cells that express T-channels.

3. ALTERATIONS OF VOLTAGE-GATED AND LIGAND-GATED ION CHANNELS IN SENSORY NEURONS IN ANIMAL MODELS OF TYPE 1 DIABETES

Several studies have demonstrated upregulation of $Ca_V3.2$ currents in both small- and medium-size DRG cells in animal models of painful PDN in type 1 diabetes. For example, in diabetic Bio-Bred/Worchester (BB/W) rats, a model for autoimmune-mediated type 1 diabetes, upregulation of T-channel current density of about threefold was observed in a mixed population of small- and medium-size DRG cells after 8 months in a diabetic state (Hall, Sima, & Wiley, 1995). The current–voltage relationship in these cells was not different between control, healthy rats, and experimental diabetic rats. Similarly, studies with streptozotocin (STZ)-induced diabetic neuropathy, which used electrophysiological recordings from acutely isolated IB_4-positive medium- and small-size DRG cells, found that T-current amplitude was upregulated about twofold (Cao, Byun, Chen, & Pan, 2011; Jagodic et al., 2007; Khomula et al., 2013). In addition, these three studies have shown that biophysical properties of T-channels were altered in DRG cells from diabetic animals, with a significant depolarizing shift in steady-state inactivation curves. As a result of this shift, more T-channels were available at physiological membrane potentials, and this increased propensity for burst firing in DRG cells from diabetic rats as compared with healthy ones (Jagodic et al., 2007; Khomula et al., 2013).

Additional studies have documented upregulation of pronociceptive ion channels such as voltage-gated sodium channels, particularly the $Na_V1.7$ and $Na_V1.8$ isoforms (Bierhaus et al., 2012; Hoeijmakers, Faber, Merkies, & Waxman, 2014), purinergic P2X3 receptors (Xu, Li, Liu, & Huang, 2011), and the vanilloid family of ligand-gated channels, especially transient receptor potential vanilloid 1, TRPV1 (Brederson, Kym, & Szallasi, 2013), within sensory neurons in animal models of diabetes. Given the prominent role of TRP channels in sensory transduction, it is reasonable to assume that dysregulation of capsaicin-gated TRPV1 channels may also contribute to abnormal pain signaling in different subpopulations of DRG cells from rats with STZ-induced PDN. For example, Hong and Wiley found using immunohistochemistry that membrane expression of TRPV1 protein in rats was increased in myelinated A-type fiber DRG cells, while it was decreased in small unmyelinated C-type nociceptive DRG cells as a result of diabetes (Hong & Wiley, 2005). In contrast, a study by Pabbidi and colleagues demonstrated increased expression of TRPV1 protein and greater whole-cell current in small DRG cells in both STZ-induced and transgene-mediated type 1 painful PDN in mice (Pabbidi, Cao, Parihar, Pauza, & Premkumar, 2008; Pabbidi, Yu, et al., 2008). Consistent with increased function of TRPV1 in DRG cells expressing T-currents, capsaicin-gated currents were increased in medium-size (Jagodic et al., 2007) and small DRG cells (Khomula et al., 2013; Shankarappa, Piedras-Rentería, & Stubbs, 2011) from STZ-induced diabetic rats with painful PDN. This observation underscores the importance of possible cross talk between nociceptive ligand and voltage-gated ion channels that control calcium signaling in DRG cells. Increased activity of both channel types in concert could contribute to cellular hyperexcitability and consequently to hyperalgesia and allodynia in PDN. Pabbidi and colleagues reported that STZ may have had a direct effect on expression of TRPV1 channels in DRG cells that was independent of hyperglycemia and hence may result from the direct toxicity of this compound (Pabbidi, Cao, et al., 2008; Pabbidi, Yu, et al., 2008). However, present evidence demonstrates that this is not the case with DRG $Ca_V3.2$ T-channels. A causal relationship between hyperglycemia in STZ-induced PDN in rats and expression of T-type currents in DRG cell was effectively documented since upregulation of T-currents closely followed development of hyperglycemia and was completely reversed in parallel with diabetic hyperalgesia reversal upon chronic insulin treatments (Messinger et al., 2009). An interesting report has documented that forced exercise in diabetic STZ-treated rats prevented the depolarizing shift in steady-state inactivation of T-currents in small DRG cells in parallel with improvement of pain symptoms of PDN in vivo (Shankarappa et al., 2011). This is important given that these findings could potentially be translated in human medicine. Furthermore, in vivo knockdown of the $Ca_V3.2$ isoform of T-channels in DRG cells using antisense oligonucleotides reversed upregulation of T-currents in small DRG cells and diabetic hyperalgesia in STZ-treated rats (Messinger et al., 2009; Obradovic et al., 2014). Finally, as proof of concept, it was reported that when compared to wild-type littermates, diabetic STZ-treated knockout mice lacking the $Ca_V3.2$ isoform of T-channels failed to develop thermal and mechanical hyperalgesia despite the prominent hyperglycemia (Latham et al., 2009). Taken together, these studies have firmly established that the $Ca_V3.2$ isoform of T-channels in DRG cells is required for the development and maintenance of painful PDN in rats and mouse models of type 1 diabetes.

4. ALTERATIONS OF $Ca_V3.2$ T-TYPE CHANNELS IN NOCICEPTIVE SENSORY NEURONS IN ANIMAL MODELS OF TYPE 2 DIABETES WITH PAINFUL PDN

Morbid obesity may be accompanied by type 2 diabetes and painful PDN that is manifested by mechanical/thermal allodynia and hyperalgesia. Importantly, clinical studies have documented that painful PDN is at least twofold more common in patients with type 2 PDN than in patients with type 1 PDN (Edwards et al., 2008). Hence, the function of $Ca_V3.2$ T-channels in DRG cells and its possible relationship with abnormal pain perception in PDN was investigated in leptin-deficient morbidly obese ob/ob mice (Latham et al., 2009). It was found that ob/ob mice developed significant mechanical and thermal pain hypersensitivity early in life that coincided with the onset of hyperglycemia (maximal at the age of 10–16 weeks) and was reversed with timely administered insulin therapy that was initiated before diabetes fully developed. These disturbances were accompanied by significant biophysical and biochemical modulation of T-channels in DRG neurons as indicated by about a twofold increase in the amplitude of T-currents and about a threefold increase in expression of channel mRNA. The most prevalent isoform of T-channels expressed in nociceptive DRG cells, $Ca_v3.2$, was the most strongly affected. Moreover, systemic injections of (3β,5α,17β)-17-hydroxyestrane-3-carbonitrile (ECN), a neuroactive steroid compound and selective T-channel antagonist, provided dose-dependent alleviation of neuropathic thermal and mechanical hypersensitivity in diabetic ob/ob mice. This study indicates that selective pharmacological antagonism of T-channels can potentially be an important novel therapeutic approach for the management of pain associated with PDN. Thus, it appears that in both type 1 and type 2 painful PDN, metabolic abnormalities leading to hyperglycemia may be causative events that eventually lead to upregulation of DRG T-current, increased cellular excitability of nociceptors, and consequently hyperalgesia and allodynia.

5. POSTTRANSLATIONAL MODIFICATION OF PRONOCICEPTIVE ION CHANNELS IN PDN

Aberrant regulation of function of nociceptive ion channels in sensory neurons via posttranslational modification and ensuing hyperexcitability may be factors contributing to painful symptoms of PDN. One of the most prevalent forms of posttranslational modification of proteins is glycosylation, which involves the attachment of sugar molecules to a protein. Glycosylation regulates protein conformation and activity, is involved in protein protection from proteolytic degradation, and plays an important role in protein trafficking and secretion (Moremen, Tiemeyer, & Naim, 2012). However, very little is known about the possible involvement of glycosylation in the sensitization of nociceptive responses in the setting of diabetic hyperglycemia. Unsurprisingly, it has been reported that diabetes results in a nearly threefold increase in peripheral nerve glycosylation as measured in tissue homogenates from diabetic rats and dogs (Vlassara, Brownlee, & Cerami, 1981). Hence, it is plausible that aberrant and/or excessive glycosylation of certain susceptible proteins in peripheral nerves may play a role in the pathogenesis of PDN. Unfortunately, the identity of these glycosylated proteins and their precise role in the pathogenesis of diabetes-induced nerve dysfunction remains elusive.

Protein glycosylation frequently involves extracellular asparagine residues. Thus, it is plausible to propose that as glucose levels rise in sensory neurons, as a consequence of diabetic hyperglycemia, excessive glycosylation may become a maladaptive process and lead to dysfunction of certain ion channels that control cellular excitability. Prevailing literature has focused on protein glycation in a diabetic environment, a process that typically involves intracellular accumulation of AGE that alter protein function by modifying intracellular arginine and lysine residues (Jack & Wright, 2012; Obrosova, 2009). However, decreasing production of AGE in vivo has inconsistently and incompletely reversed pain symptoms in animal models of diabetes and in patients with PDN (Edwards et al., 2008). Hence, it is possible that other parallel signaling mechanisms, such as glycosylation, may be involved in hyperglycemia-induced alterations of nociceptive ion channel function in sensory neurons.

Upregulation of $Ca_V3.2$ (but not $Ca_V3.1$ or $Ca_V3.3$) transcripts in DRG tissues from ob/ob mice can account at least partially for increased T-current density in small DRG cells, but it does not account for kinetic alterations of T-currents (Latham et al., 2009). Thus, it was considered that these channels in DRGs from diabetic ob/ob mice must be additionally modified by other mechanisms including glycosylation. This was recently addressed by using patch-clamp recordings from small DRG and recombinant human embryonic kidney (HEK-293) cells, immunohistochemistry, biochemical studies, as well as in vivo pain studies using ob/ob mice (Orestes et al., 2013). First, it was found that, in addition to increased current density, T-currents in small DRG cells exhibit marked speeding of macroscopic current kinetics in diabetic ob/ob mice. The glycosylation inhibitor neuraminidase completely reversed upregulation and kinetic alterations of $Ca_V3.2$ in small DRG cells from diabetic ob/ob mice, while it had a much smaller effect on T-currents in healthy wild-type mice. Intraplantar (i.pl.) injections of neuraminidase directly into peripheral receptive fields of diabetic ob/ob mice were able to reverse completely the mechanical and thermal hyperalgesia. To test whether the antihyperalgesic effect of neuraminidase in ob/ob mice is related to its effects on T-currents, the selective T-channel blocker ECN was injected and completely reversed thermal and mechanical hyperalgesia in ob/ob mice, while subsequent i.pl. injections of neuraminidase in paws did not further decrease pain thresholds. Hence, these studies indicate that glycosylation inhibitors like neuraminidase may be used to normalize sensory transmission in DRG cells and peripheral nerves in animals with painful PDN.

Two recent independent studies have reported that glycosylation of critical extracellular asparagine residues of $Ca_V3.2$ channels causes kinetic alterations, upregulation of T-currents, as well as an increase in membrane expression of T-channels (Orestes et al., 2013; Weiss, Black, Bladen, Chen, & Zamponi, 2013). Molecular studies using site-directed mutagenesis identified conserved extracellular asparagine residues, most notably N192 and N1466, as important regulators of $Ca_V3.2$ current kinetics and channel membrane expression, respectively. The involvement of N192 was supported by patch-clamp recordings that demonstrated slower current kinetics in an N192Q $Ca_V3.2$ mutant, with apparently normal membrane expression. Interestingly, T-currents in HEK-293 cells transfected with an N1466Q $Ca_V3.2$ mutant could not be consistently recorded, and imaging studies showed minimal membrane expression of this mutant (Orestes et al., 2013). Thus, different glycosylation sites in $Ca_V3.2$ channels may have distinct functional roles. Surprisingly, authors of both studies could not record T-currents from N271Q Cav3.2 channels despite their presence in the cell membrane. It remains possible that N271Q channels trafficked to the membrane are nonfunctional. In contrast to the finding of Orestes et al., Weiss and colleagues found that asparagine N192 serves as a regulator of $Ca_V3.2$ channel membrane expression and asparagine N1466 as a regulator of channel kinetics (Weiss et al., 2013). It is possible that different levels of glucose in cell culture could have contributed to the different findings between the studies. Glycosylated products have larger molecular mass, meaning that glycosylation or the enzymatic removal of these groups from the channel protein will induce a shift in the migration of proteins bands resolved by gel-shift analysis and visualized by Western blotting. The study by Orestes and colleagues took advantage of this and directly demonstrated, using gel-shift analysis, that recombinant $Ca_V3.2$ channels are, indeed, glycosylated within domain I of the channel protein. However, levels of glycosylation of native $Ca_V3.2$ channels in sensory neurons in diabetic animals remain to be determined in future studies. Finally, neither of these two studies examined the role of neuraminidase or other agents that can modulate glycosylation upon membrane excitability of nociceptors in animals with PDN.

Several other in vitro studies have reported that glycosylation may modulate properties of other pronociceptive voltage-gated ion channels. For example, it was reported that in embryonic DRG neurons, neuraminidase alters steady-state inactivation of voltage-gated sodium channels (Tyrrell, Renganathan, Dib-Hajj, & Waxman, 2001). This study did not find any effects of neuraminidase on voltage-gated sodium channels in small DRG cells from healthy adult animals. However, it remains to be determined in future studies if neuraminidase could modulate voltage-gated sodium channels in nociceptive DRG cells from diabetic animals. Future extensive electrophysiological studies could also be expanded to involve examination of other ligand-gated channels (eg, TRPV1) that are crucial for the control of cellular excitability of DRG cells from diabetic animals that might also be modulated by glycosylation (Cohen, 2006; Jing et al., 2012; Pertusa, Madrid, Morenilla-Palao, Belmonte, & Viana, 2012).

In addition to glycosylation of nociceptive ion channels, posttranslational modification of the voltage-gated sodium channels, specifically Nav1.8 isoform and transient receptor potential channel A1 (TRPA1), via methylglyoxal, an endogenous reactive metabolite of glucose, has been demonstrated in DRG neurons in animals with STZ-induced diabetes (Andersson et al., 2013; Bierhaus et al., 2012; Eberhardt et al., 2012). High plasma concentrations of methylglyoxal correlate well with occurrence of painful symptoms in patients with PDN (Bierhaus et al., 2012). Furthermore, methylglyoxal depolarizes putative nociceptive DRG neurons and induces hyperexcitability of these cells thus facilitating action potential firing and in vivo administration of methylglyoxal evoked thermal and mechanical hyperalgesia. In contrast to glycosylation, above-mentioned studies have demonstrated that this form of posttranslational modification of Nav1.8 channels and TRPA1 channels is via modification of intracellular lysine and arginine-binding sites (Bierhaus et al., 2012; Eberhardt et al., 2012).

Finally, upregulation of HVA calcium currents in sensory neurons in STZ-treated diabetic rats via G-proteins may underlie their diminished responsiveness to opioids (Hall, Liu, Sima, & Wiley, 2001; Hall, Sima, & Wiley, 1996).

6. DIMINISHED INHIBITORY DRIVE IN THE SPINAL CORD MAY ALSO CONTRIBUTE TO PAINFUL PDN

Activation of nociceptors can trigger a lasting increase in the excitability and synaptic efficacy of DH neurons in a phenomenon termed central sensitization, which in turn contributes to many pain disorders in humans including: neuropathic pain, fibromyalgia, and osteoarthritis (Woolf, 2011). This can be manifested as prolonged and/or exaggerated response to painful stimuli (hyperalgesia), a reduction in threshold for pain (allodynia), and/or receptive field expansion that enables input from noninjured tissues to produce pain (secondary hyperalgesia). Hence, drugs like gabapentin and pregabalin that reduce central facilitation of DH neurons by diminishing glutamate-mediated excitatory drive, via binding to the $\alpha2\delta$ ancillary subunit of voltage-gated calcium channels, are among the most effective treatments for hyperalgesia and allodynia in PDN. Unfortunately, in many patients (about 50%) these drugs induce sedation, making them unsuitable for long-term therapy (Gooch & Podwall, 2004). Furthermore, it is well documented that promotion of inhibitory drive mediated via γ-aminobutyric acid (GABA) activation in the DH to activate inhibitory inputs is also beneficial in alleviating neuropathic pain symptoms in various animal models of peripheral nerve injury (Sivilotti & Woolf, 1994; Woolf, 2011). Surprisingly, the possibility that spinal GABAergic system may be involved in painful PDN has not been extensively studied. In a very interesting recent study, Lee-Kubli and Calcutt (2014) have found diabetic rats have impaired spinal $GABA_A$ receptor inhibitory function arising from reduced spinal potassium chloride cotransporter, which collectively leads to spinal disinhibition of pain responses. Hence, designing therapies that aim to restore impaired GABAergic function in spinal cord may be a novel approach to the treatment of painful PDN.

7. CONCLUSIONS

It is likely that multiple signaling pathways targeting nociceptive ion channels may work in concert to promote hyperexcitability of sensory neurons and contribute to hyperalgesia, allodynia, and spontaneous pain in patients with PDN. This chapter, and a recent commentary (Todorovic, 2015), summarize how different pronociceptive ion channels expressed in peripheral nerve terminals in skin, cell somas in DRG, and central terminals of sensory neurons may work together with spinal GABA-mediated dysfunction to increase sensitization of pain responses under diabetic conditions.

It is hoped that future efforts in targeting metabolic pathways such as glycosylation and/or methylglyoxal formation may lead to novel and safer pain therapies in patients with diabetes. We posit that a therapeutic strategy to suppress pain by normalizing nociceptive channel function will be advantageous to that of direct channel blockade because, by correcting the pathology of PDN at its source, this approach should avoid the multitude of unintended side effects that are associated with current therapies. We envision that specific targeting of dysregulated biochemical pathways that regulate function of nociceptive ion channels may be developed into a novel, effective, and safe mechanistic-based therapeutic approach. This may be achieved by specifically targeting peripheral axons of nociceptors using therapeutic ointments and creams and/or applying drugs that reverse cellular hyperexcitability directly onto DRGs using fluoroscopy-guided epidural injections in pain clinics. Alternatively, systemically administered drugs may be used, although this will likely increase the risk of side effects. Most patients with painful PDN ask for medical help well after painful symptoms have developed. It is an attractive alternative approach to attempt to identify the population at risk for painful PDN using screening test for glycosylated and methylglyoxal-modified proteins in patients with PDN. This would allow physicians to initiate treatments before sensitization of peripheral and central pain responses is fully developed. To our knowledge, this topic is controversial, and a recent clinical study failed to find correlation between serum methylglyoxal levels and symptoms of peripheral and cardiovascular autonomic neuropathy (Hansen et al., 2015). However, studies like this remain an important area of future investigations.

ACKNOWLEDGMENTS

Our research is supported by American Diabetes Association National Award for Basic Research 7-09-BS-190 (to S.M.T.) and research funds from the Department of Anesthesiology at the University of Virginia, Charlottesville, VA, as well as Department of Anesthesiology at the University of Colorado Anschutz Medical Campus, Aurora, CO.

REFERENCES

Abbott, C. A., Malik, R. A., van Ross, E. R. E., Kulkarni, J., & Boulton, A. J. M. (2011). Prevalence and characteristics of painful diabetic neuropathy in a large community-based diabetic population in the UK. *Diabetes Care, 34*, 2220–2224.

Andersson, D. A., Gentry, C., Light, E., Vastani, N., Vallortigara, J., Bierhaus, A., et al. (2013). Methylglyoxal evokes pain by stimulating TRPA1. *PloS One, 8*(10), e77986.

Bierhaus, A., Fleming, T., Stoyanov, S., Leffler, A., Babes, A., Neacsu, C., et al. (2012). Methylglyoxal modification of Nav1.8 facilitates nociceptive neuron firing and causes hyperalgesia in diabetic neuropathy. *Nature Medicine, 18*(6), 926–933.

Brederson, J. D., Kym, P. R., & Szallasi, A. (2013). Targeting TRP channels for pain relief. *European Journal of Pharmacology, 716*(1–3), 61–76.

Calcutt, N. A. (2013). Location, location, location? Is the pain of diabetic neuropathy generated by hyperactive sensory neurons? *Diabetes, 62*, 3658–3660.

Cao, X. H., Byun, H. S., Chen, S. R., & Pan, H. L. (2011). Diabetic neuropathy enhances voltage-activated Ca2+ channel activity and its control by M4 muscarinic receptors in primary sensory neurons. *Journal of Neurochemistry, 119*(3), 594–603.

Carbone, E., & Lux, H. D. (1984). A low-voltage activated, fully inactivating Ca^{2+} channel in vertebrate sensory neurons. *Nature, 310*, 501–502.

Centers for Disease Control and Prevention. (2011). *National diabetes fact sheet: national estimates and general information on diabetes and prediabetes in the United States, 2011*. Atlanta, GA: US Department of Health and Human Services, Center for Disease Control and Prevention. http://www.cdc.gov/diabetes/pubs/pdf/ndfs_2011.pdf.

Cohen, D. M. (2006). Regulation of TRP channels by N-linked glycosylation. *Seminars in Cell & Developmental Biology, 17*(6), 630–637.

Dubreuil, A. S., Boukhaddaoui, H., Desmadryl, G., Martinez-Salgado, C., Moshourab, R., Lewin, G. R., et al. (2004). Role of T-type calcium current in identified d-hair mechanoreceptor neurons studied *in vitro*. *The Journal of Neuroscience, 24*, 8480–8484.

Eberhardt, M. J., Filipovic, M. R., Leffler, A., de la Roche, J., Kistner, K., Fischer, M. J., et al. (2012). Methylglyoxal activates nociceptors through transient receptor potential channel A1 (TRPA1): A possible mechanism of metabolic neuropathies. *The Journal of Biological Chemistry, 287*(34), 28291–28306.

Edwards, J. L., Vincent, A. M., Cheng, H. T., & Feldman, E. L. (2008). Diabetic neuropathy: Mechanisms to management. *Pharmacology & Therapeutics, 120*, 1–34.

Gooch, C., & Podwall, D. (2004). The diabetic neuropathies. *The Neurologist, 10*, 311–322.

Hall, K. E., Liu, J., Sima, A. A., & Wiley, J. W. (2001). Impaired inhibitory G-protein function contributes to increased calcium currents in rats with diabetic neuropathy. *Journal of Neurophysiology, 86*(2), 760–770.

Hall, K. E., Sima, A. A., & Wiley, J. W. (1995). Voltage-dependent calcium currents are enhanced in dorsal root ganglion neurons from the Bio Bred/Worchester diabetic rat. *Journal of Physiology (London), 486*(Pt. 2), 313–322.

Hall, K. E., Sima, A. A., & Wiley, J. W. (1996). Opiate-mediated inhibition of calcium signaling is decreased in dorsal root ganglion neurons from the diabetic BB/W rat. *The Journal of Clinical Investigation, 97*(5), 1165–1172.

Hansen, C. S., Jensen, T. M., Jensen, J. S., Nawroth, P., Fleming, T., Witte, D. R., et al. (2015). The role of serum methylglyoxal on diabetic peripheral and cardiovascular autonomic neuropathy: The ADDITION Denmark study. *Diabetic Medicine, 32*(6), 778–785.

Hoeijmakers, J. G., Faber, C. G., Merkies, I. S., & Waxman, S. G. (2014). Channelopathies, painful neuropathy, and diabetes: Which way does the causal arrow point? *Trends in Molecular Medicine, 20*(10), 544–550.

Hong, S., & Wiley, J. W. (2005). Early painful diabetic neuropathy is associated with differential changes in the expression and function of vanilloid receptor 1. *The Journal of Biological Chemistry, 280*(1), 618–627.

Jack, M., & Wright, D. (2012). Role of advanced glycation end products and glyoxalase I in diabetic peripheral sensory neuropathy. *Translational Research, 159*(5), 355–365.

Jacus, M. O., Uebele, V. N., Renger, J. J., & Todorovic, S. M. (2012). Presynaptic $Ca_V3.2$ channels regulate excitatory neurotransmission in nociceptive dorsal horn neurons. *The Journal of Neuroscience, 32*(27), 9374–9382.

Jagodic, M. M., Pathirathna, S., Nelson, M. T., Mancuso, S., Joksovic, P. M., Rosenberg, E. R., et al. (2007). Cell-specific alterations of T-type calcium current in painful diabetic neuropathy enhance excitability of sensory neurons. *The Journal of Neuroscience, 27*(12), 3305–3316.

Jing, L., Chu, X. P., Jiang, Y. Q., Collier, D. M., Wang, B., Jiang, Q., et al. (2012). N-glycosylation of acid-sensing ion channel 1a regulates its trafficking and acidosis-induced spine remodeling. *The Journal of Neuroscience, 32*(12), 4080–4091.

Khomula, E. V., Viatchenko-Karpinski, V. Y., Borisyuk, A. L., Duzhyy, D. E., Belan, P. V., & Voitenko, N. V. (2013). Specific functioning of Cav3.2 T-type calcium and TRPV1 channels under different types of STZ-diabetic neuropathy. *Biochimica et Biophysica Acta, 1832*(5), 636–649.

Latham, J. R., Pathirathna, S., Jagodic, M. M., Choe, W. J., Levin, M. E., Nelson, M. T., et al. (2009). Selective T-type calcium channel blockade alleviates hyperalgesia in ob/ob mice. *Diabetes, 58*(11), 2656–2665.

Lee-Kubli, C. A., & Calcutt, N. A. (2014). Altered rate-dependent depression of the spinal H-reflex as an indicator of spinal disinhibition in models of neuropathic pain. *Pain, 155*(2), 250–260.

Messinger, R. B., Naik, A. K., Jagodic, M. M., Nelson, M. T., Lee, W. Y., Choe, W. J., et al. (2009). In-vivo silencing of the Cav3.2 T-type calcium channels in sensory neurons alleviates hyperalgesia in rats with streptozocin-induced diabetic neuropathy. *Pain, 145*(1–2), 184–195.

Moremen, K. W., Tiemeyer, M., & Naim, A. V. (2012). Vertebrate protein glycosylation: Diversity, synthesis and function. *Nature Reviews. Molecular Cell Biology, 13*, 448–462.

Nelson, M. T., Joksovic, P. M., Perez-Reyes, E., & Todorovic, S. M. (2005). The endogenous redox agent L-cysteine induces T-type Ca^{2+} channel-dependent sensitization of a novel subpopulation of rat peripheral nociceptors. *The Journal of Neuroscience, 25*, 8766–8775.

Nelson, M. T., Woo, J., Kang, H.-W., Barrett, P. Q., Vitko, J., Perez-Reyes, E., et al. (2007). Reducing agents sensitize C-type nociceptors by relieving high-affinity zinc inhibition of T-type calcium channels. *The Journal of Neuroscience, 27*(31), 8250–8260.

Obradovic, A. L. j., Hwang, S. M., Scarpa, J., Hong, S. J., Todorovic, S. M., & Jevtovic-Todorovic, V. (2014). CaV3.2 T-type calcium channels in peripheral sensory neurons are important for mibefradil-induced reversal of hyperalgesia and allodynia in rats with painful diabetic neuropathy. *PLoS One, 9*(4), e9.

Obrosova, I. G. (2009). Diabetic painful and insensate neuropathy: Pathogenesis and potential treatments. *Neurotherapeutics, 6*(4), 638–647.

Orestes, P., Osuru, H. P., McIntire, W. E., Jacus, M. O., Salajegheh, R., Jagodic, M. M., et al. (2013). Reversal of neuropathic pain in diabetes by targeting glycosylation of $Ca_V3.2$ T-type channels. *Diabetes, 62*(11), 3828–3838.

Pabbidi, R. M., Cao, D. S., Parihar, A., Pauza, M. E., & Premkumar, L. S. (2008). Direct role of streptozotocin in inducing thermal hyperalgesia by enhanced expression of transient receptor potential vanilloid 1 in sensory neurons. *Molecular Pharmacology, 73*(3), 995–1004.

Pabbidi, R. M., Yu, S. Q., Peng, S., Khardori, R., Pauza, M. E., & Premkumar, L. S. (2008). Influence of TRPV1 on diabetes-induced alterations in thermal pain sensitivity. *Molecular Pain, 4*, 9.

Pertusa, M., Madrid, R., Morenilla-Palao, C., Belmonte, C., & Viana, F. (2012). N-glycosylation of TRPM8 ion channels modulates temperature sensitivity of cold thermoreceptor neurons. *The Journal of Biological Chemistry, 287*(22), 18218–18229.

Shankarappa, S. A., Piedras-Rentería, E. S., & Stubbs, E. B., Jr. (2011). Forced-exercise delays neuropathic pain in experimental diabetes: Effects on voltage-activated calcium channels. *Journal of Neurochemistry*, *118*(2), 224–236.

Shin, J. B., Martinez-Salgado, C., Heppenstall, P. A., & Lewin, G. R. (2003). A T-type calcium channel required for normal function of a mammalian mechanoreceptor. *Nature Neuroscience*, *6*(7), 724–730.

Sivilotti, L., & Woolf, C. J. (1994). The contribution of GABAA and glycine receptors to central sensitization: Disinhibition and touch-evoked allodynia in the spinal cord. *Journal of Neurophysiology*, *72*, 169–179.

Todorovic, S. M. (2015). Is diabetic pain caused by dysregulated ion channels in sensory neurons? *Diabetes*, *64*(12), 3987–3989.

Todorovic, S. M., & Jevtovic-Todorovic, V. (2011). T-type voltage-gated calcium channels as targets for development of novel pain therapies. *British Journal of Pharmacology*, *163*(3), 484–495.

Todorovic, S. M., & Jevtovic-Todorovic, V. (2014). Targeting of CaV3.2 T-type calcium channels in peripheral sensory neurons for the treatment of painful diabetic neuropathy. *Pflügers Archiv*, *466*(4), 701–706.

Tyrrell, L., Renganathan, M., Dib-Hajj, S. D., & Waxman, S. G. (2001). Glycosylation alters steady-state inactivation of sodium channel Nav1.9/NaN in dorsal root ganglion neurons and is developmentally regulated. *The Journal of Neuroscience*, *21*(24), 9629–9637.

Vlassara, H., Brownlee, M., & Cerami, A. (1981). Nonenzymatic glycosylation of peripheral nerve protein in diabetes mellitus. *Proceedings of the National Academy of Sciences of the United States of America*, *78*(8), 5190–5192.

Weiss, N., Black, S. A. G., Bladen, C., Chen, L., & Zamponi, G. W. (2013). Surface expression and function of $Ca_V3.2$ T-type calcium channels are controlled by asparagines-linked glycosylation. *Pflugers Archiv—European Journal of Physiology*, *465*(8), 1159–1170.

White, G., Lovinger, D. M., & Weight, F. F. (1989). Transient low-threshold Ca^{2+} current triggers burst firing through an afterdepolarizing potential in an adult mammalian neuron. *Proceedings of the National Academy of Sciences of the United States of America*, *86*, 6802–6806.

Woolf, C. J. (2011). Central sensitization: Implications for the diagnosis and treatment of pain. *Pain*, *152*(3 Suppl.), S2–S15.

Xu, G. Y., Li, G., Liu, N., & Huang, L. Y. (2011). Mechanisms underlying purinergic P2X3 receptor-mediated mechanical allodynia induced in diabetic rats. *Molecular Pain*, *7*, 60.

SECTION IV

Translating Science into Medicine

CHAPTER TWELVE

New Directions in Diabetic Neuropathy: Evolution or Extinction?

P. Fernyhough*,†,1, N.A. Calcutt‡
*University of Manitoba, Winnipeg, MB, Canada
†St. Boniface Hospital Research Centre, Winnipeg, MB, Canada
‡University of California, San Diego, La Jolla, CA, United States
1Corresponding author: e-mail address: pfernyhough@sbrc.ca

Contents

The earlier chapters "Neuropathy in the DCCT/EDIC" by Pop-Busui and "The Perfect Clinical Trial" by Bril explained the value of long clinical trials (5 or more years) to enable any therapeutic effect against progressive diabetic neuropathy to be detected. These calculations were based primarily on findings from the Diabetes Control and Complications Trial (DCCT), where intensive insulin therapy significantly protected nerve conduction velocity (NCV) in type 1 diabetic patients (DCCT, 1993, 2002). Because of the slow degradation of NCV at a rate that has historically been of approximately 0.5–1.0 m/s per annum, a trial of this duration was required to see separation between untreated and treated groups given the study group sizes and data variability. NCV clearly provides an objective measure that gives information about the current status of large myelinated nerve fiber dysfunction and is predictive of its progression. Thus, it is a strongly validated measure of large fiber neuropathy that can detect effective therapeutic intervention.

International Review of Neurobiology, Volume 127
ISSN 0074-7742
http://dx.doi.org/10.1016/bs.irn.2016.03.009

1. NCV AS AN ENDPOINT IN ANIMAL MODELS

Drugs entered into clinical trial against diabetic neuropathy have, to date, been selected primarily based upon their ability to prevent NCV slowing in rodent models of diabetes. A potential problem with this approach is the applicability of such preclinical efficacy studies to the human setting. The NCV deficit in animal models usually develops within the first month of onset of type 1 diabetes (Eliasson, 1964; Greene, De Jesus, & Winegrad, 1975). The acute disruption of metabolism causes rapid changes in endoneurial blood flow and other parameters that negatively impact nerve conduction such as Schwann cell function, nodal function, and axonal function (Hounsom & Tomlinson, 1997; Sima, 2004; Yagihashi, Yamagishi, & Wada, 2007). These acute alterations in cell biology and function are preventable and rapidly reversible by many interventions and occur months before any structural deterioration of large myelinated nerve fibers is detectable. It remains to be unequivocally proven that these early deficits in nerve function are causal factors in subsequent structural deterioration of large myelinated nerve fibers. The other proboscidean in this particular parlour is that structural deterioration of large nerve fibers is a very subtle affair in diabetic rodents. In STZ-diabetic rats, where the majority of preclinical drug studies have been performed, there is minimal overt degeneration of large myelinated nerve fibers, no demyelination, and little axonal pathology (Kalichman, Dines, Bobik, & Mizisin, 1998; Zochodne, Verge, Cheng, Sun, & Johnston, 2001). The only established and widely reproduced sign of a degenerative process is impaired maturation of axonal diameter and later axonal dwindling (Jakobsen, 1976a, 1976b). As discussed earlier in this volume, alternative models that develop a large-fiber neuropathology that reflects the human disease, such as cats, are available but have not yet been widely used (Mizisin et al., 2007; Mizisin, Shelton, Burgers, Powell, & Cuddon, 2002). At present, there is little evidence to suggest that positive effects on NCV deficits in animal models of diabetic neuropathy provide any true mechanistic insight into the degenerative process occurring in human diabetic neuropathy or have predictive value for efficacy when translated to clinical use. It may well be that the preclinical pipeline that has fed clinical trials has always been broken, or at least is not correctly aligned.

2. TREATMENT PARADIGMS

The majority of preclinical studies have used a prevention paradigm. Drugs are applied at the beginning of the study and forestall development of deficits in NCV. In some longer-term studies this approach has also prevented onset of reduced axonal caliber of large myelinated fibers (Calcutt et al., 2003, 1999; Kato, Mizuno, Makino, Suzuki, & Yagihashi, 2000). However, the ability of an intervention to correct established structural deterioration of large myelinated fibers has rarely, if ever, been evaluated. It is therefore not surprising that drugs have shown minimal effects in phase 2 and phase 3 clinical trials with patients exhibiting established neuropathy that is likely accompanied by large-fiber degeneration and loss. Indeed it is plausible that the statistically significant, but sadly unimpressive, 1.0–1.5 m/s improvements in NCV reported in recent well-designed and executed clinical trials (Bril, Hirose, Tomioka, & Buchanan, 2009; Wahren, Foyt, Daniels, & Arezzo, 2016) represent the maximal effect that can be achieved against large-fiber dysfunction when the bulk of the NCV deficit in humans may reflect established structural pathology that is unresponsive to the intervention. To date, the drugs under investigation have not demonstrated efficacy against pertinent indices of neuropathy in appropriate animal models.

3. INSANITY: DOING THE SAME THING OVER AND OVER AGAIN AND EXPECTING A DIFFERENT RESULT

After 30+ years of failed clinical trials, we owe it to those with diabetic neuropathy, and those likely to face it in the future, to review current approaches to developing therapies for diabetic neuropathy. Fortunately, we may not have to completely reinvent the wheel, just adjust our approaches, biases, and inertia. The measurement of NCV is a valid and useful endpoint in clinical trials for diabetic neuropathy. However, it need not be the central component upon which any decision on success or failure is made. Indeed, the perceived failure of a recent clinical trial of C-peptide that could not separate improvement of NCV in treated subjects vs those treated with placebo but did report statistically significant improvement of vibration perception (Wahren et al., 2016) highlights the absurdity of focusing on NCV as the dominant biomarker for neuropathy while ignoring the patient. As the authors of the above study noted, measuring large-fiber NCV may not even reflect other more relevant aspects of large-fiber sensory function. The chapters "Alternative Quantitative Tools in the Assessment of Diabetic Peripheral and Autonomic Neuropathy" by Vinik and "Wherefore Art Thou O Treatment for Diabetic Neuropathy?" by Malik describe interesting and, for us, compelling arguments for using measures of small-fiber neuropathy, both sensory and autonomic, that allow us to reassess our model of the preclinical drug screen as a precursor to clinical trials. Perhaps the most pertinent development in recent years is that studies in rodent models of diabetes are now using an array of small-fiber endpoints that replicate the clinical condition to test ideas on etiology and screen new therapies (Beiswenger, Calcutt, & Mizisin, 2008; Christianson, Riekhof, & Wright, 2003; Christianson, Ryals, Johnson, Dobrowsky, & Wright, 2007; Davidson, Coppey, Holmes, & Yorek, 2012a, 2012b). Loss of epidermal fibers, loss of sweat gland innervation, loss of thermal (hot and cold) sensation, and loss of corneal nerve fiber density all occur in both diabetic animals and humans (Arezzo, Schaumburg, & Laudadio, 1986; Breiner, Lovblom, Perkins, & Bril, 2014; Hossain, Sachdev, & Malik, 2005; Kennedy, Wendelschafer-Crabb, & Johnson, 1996; Liu et al., 2015; Malik et al., 2003; Zinman, Bril, & Perkins, 2004). Other advantages of working with small-fiber endpoints arise from reports that onset of structural change can be detected early in the course of disease in both diabetic animals and humans (Azmi et al., 2015; Christianson et al., 2003, 2007; Mehra et al., 2007; Petropoulos et al., 2015; Quattrini et al., 2007; Smith, Ramachandran, Tripp, & Singleton, 2001). Corneal nerve fiber loss, which can be detected by confocal microscopy, can be viewed noninvasively so as to allow frequent iterative assessment before and after therapeutic interventions in diabetic rodents and humans (Chen et al., 2013; Tavakoli et al., 2013). It should also be remembered that the aspects of diabetic neuropathy that most trouble patients reflect sensory and autonomic dysfunction. Once fully validated, the introduction of measures of small-fiber nerve pathology and dysfunction as primary goals and endpoints in clinical trials of treatments for diabetic neuropathy may allow shorter, less costly, trials that have better chances of gaining regulatory approval and improving the lives of patients.

REFERENCES

Arezzo, J. C., Schaumburg, H. H., & Laudadio, C. (1986). Thermal sensitivity tester. Device for quantitative assessment of thermal sense in diabetic neuropathy. *Diabetes*, *35*, 590–592.

Azmi, S., Ferdousi, M., Petropoulos, I. N., Ponirakis, G., Alam, U., Fadavi, H., et al. (2015). Corneal confocal microscopy identifies small-fiber neuropathy in subjects with impaired glucose tolerance who develop type 2 diabetes. *Diabetes Care*, *38*, 1502–1508.

Beiswenger, K. K., Calcutt, N. A., & Mizisin, A. P. (2008). Dissociation of thermal hypoalgesia and epidermal denervation in streptozotocin-diabetic mice. *Neuroscience Letters*, *442*, 267–272.

Breiner, A., Lovblom, L. E., Perkins, B. A., & Bril, V. (2014). Does the prevailing hypothesis that small-fiber dysfunction precedes large-fiber dysfunction apply to type 1 diabetic patients? *Diabetes Care*, *37*, 1418–1424.

Bril, V., Hirose, T., Tomioka, S., & Buchanan, R. (2009). Ranirestat for the management of diabetic sensorimotor polyneuropathy. *Diabetes Care, 32*, 1256–1260.

Calcutt, N. A., Allendoerfer, K. L., Mizisin, A. P., Middlemas, A., Freshwater, J. D., Burgers, M., et al. (2003). Therapeutic efficacy of sonic hedgehog protein in experimental diabetic neuropathy. *The Journal of Clinical Investigation, 111*, 507–514.

Calcutt, N. A., Campana, W. M., Eskeland, N. L., Mohiuddin, L., Dines, K. C., Mizisin, A. P., et al. (1999). Prosaposin gene expression and the efficacy of a prosaposin-derived peptide in preventing structural and functional disorders of peripheral nerve in diabetic rats. *Journal of Neuropathology and Experimental Neurology, 58*, 628–636.

Chen, D. K., Frizzi, K. E., Guernsey, L. S., Ladt, K., Mizisin, A. P., & Calcutt, N. A. (2013). Repeated monitoring of corneal nerves by confocal microscopy as an index of peripheral neuropathy in type-1 diabetic rodents and the effects of topical insulin. *Journal of the Peripheral Nervous System, 18*, 306–315.

Christianson, J. A., Riekhof, J. T., & Wright, D. E. (2003). Restorative effects of neurotrophin treatment on diabetes-induced cutaneous axon loss in mice. *Experimental Neurology, 179*, 188–199.

Christianson, J. A., Ryals, J. M., Johnson, M. S., Dobrowsky, R. T., & Wright, D. E. (2007). Neurotrophic modulation of myelinated cutaneous innervation and mechanical sensory loss in diabetic mice. *Neuroscience, 145*, 303–313.

Davidson, E. P., Coppey, L. J., Holmes, A., & Yorek, M. A. (2012a). Changes in corneal innervation and sensitivity and acetylcholine-mediated vascular relaxation of the posterior ciliary artery in a type 2 diabetic rat. *Investigative Ophthalmology & Visual Science, 53*, 1182–1187.

Davidson, E. P., Coppey, L. J., & Yorek, M. A. (2012b). Early loss of innervation of cornea epithelium in streptozotocin-induced type 1 diabetic rats: Improvement with ilepatril treatment. *Investigative Ophthalmology & Visual Science, 53*, 8067–8074.

DCCT. (1993). The effect of intensive treatment of diabetes on the development and progression of long-term complications in insulin-dependent diabetes mellitus. The Diabetes Control and Complications Trial Research Group. *The New England Journal of Medicine, 329*, 977–986.

DCCT. (2002). Effect of intensive therapy on the microvascular complications of type 1 diabetes mellitus. *JAMA, 287*, 2563–2569.

Eliasson, S. G. (1964). Nerve conduction changes in experimental diabetes. *The Journal of Clinical Investigation, 43*, 2353–2358.

Greene, D. A., De Jesus, P. V., Jr., & Winegrad, A. I. (1975). Effects of insulin and dietary myoinositol on impaired peripheral motor nerve conduction velocity in acute streptozotocin diabetes. *The Journal of Clinical Investigation, 55*, 1326–1336.

Hossain, P., Sachdev, A., & Malik, R. A. (2005). Early detection of diabetic peripheral neuropathy with corneal confocal microscopy. *Lancet, 366*, 1340–1343.

Hounsom, L., & Tomlinson, D. R. (1997). Does neuropathy develop in animal models? *Clinical Neuroscience, 4*, 380–389.

Jakobsen, J. (1976a). Axonal dwindling in early experimental diabetes. I. A study of cross sectioned nerves. *Diabetologia, 12*, 539–546.

Jakobsen, J. (1976b). Axonal dwindling in early experimental diabetes. II. A study of isolated nerve fibres. *Diabetologia, 12*, 547–553.

Kalichman, M. W., Dines, K. C., Bobik, M., & Mizisin, A. P. (1998). Nerve conduction velocity, laser Doppler flow, and axonal caliber in galactose and streptozotocin diabetes. *Brain Research, 810*, 130–137.

Kato, N., Mizuno, K., Makino, M., Suzuki, T., & Yagihashi, S. (2000). Effects of 15-month aldose reductase inhibition with fidarestat on the experimental diabetic neuropathy in rats. *Diabetes Research and Clinical Practice, 50*, 77–85.

Kennedy, W. R., Wendelschafer-Crabb, G., & Johnson, T. (1996). Quantitation of epidermal nerves in diabetic neuropathy. *Neurology, 47*, 1042–1048.

Liu, Y., Billiet, J., Ebenezer, G. J., Pan, B., Hauer, P., Wei, J., et al. (2015). Factors influencing sweat gland innervation in diabetes. *Neurology, 84*, 1652–1659.

Malik, R. A., Kallinikos, P., Abbott, C. A., van Schie, C. H., Morgan, P., Efron, N., et al. (2003). Corneal confocal microscopy: A non-invasive surrogate of nerve fibre damage and repair in diabetic patients. *Diabetologia, 46*, 683–688.

Mehra, S., Tavakoli, M., Kallinikos, P. A., Efron, N., Boulton, A. J., Augustine, T., et al. (2007). Corneal confocal microscopy detects early nerve regeneration after pancreas transplantation in patients with type 1 diabetes. *Diabetes Care, 30*, 2608–2612.

Mizisin, A. P., Nelson, R. W., Sturges, B. K., Vernau, K. M., Lecouteur, R. A., Williams, D. C., et al. (2007). Comparable myelinated nerve pathology in feline and human diabetes mellitus. *Acta Neuropathologica, 113*, 431–442.

Mizisin, A. P., Shelton, G. D., Burgers, M. L., Powell, H. C., & Cuddon, P. A. (2002). Neurological complications associated with spontaneously occurring feline diabetes mellitus. *Journal of Neuropathology and Experimental Neurology, 61*, 872–884.

Petropoulos, I. N., Green, P., Chan, A. W., Alam, U., Fadavi, H., Marshall, A., et al. (2015). Corneal confocal microscopy detects neuropathy in patients with type 1 diabetes without retinopathy or microalbuminuria. *PLoS One, 10*, e0123517.

Quattrini, C., Tavakoli, M., Jeziorska, M., Kallinikos, P., Tesfaye, S., Finnigan, J., et al. (2007). Surrogate markers of small fiber damage in human diabetic neuropathy. *Diabetes, 56*, 2148–2154.

Sima, A. A. (2004). Diabetic neuropathy in type 1 and type 2 diabetes and the effects of C-peptide. *Journal of the Neurological Sciences, 220*, 133–136.

Smith, A. G., Ramachandran, P., Tripp, S., & Singleton, J. R. (2001). Epidermal nerve innervation in impaired glucose tolerance and diabetes-associated neuropathy. *Neurology, 57*, 1701–1704.

Tavakoli, M., Mitu-Pretorian, M., Petropoulos, I. N., Fadavi, H., Asghar, O., Alam, U., et al. (2013). Corneal confocal microscopy detects early nerve regeneration in diabetic neuropathy after simultaneous pancreas and kidney transplantation. *Diabetes, 62*, 254–260.

Wahren, J., Foyt, H., Daniels, M., & Arezzo, J. C. (2016). Long-acting C-peptide and neuropathy in type 1 diabetes: A 12-month clinical trial. *Diabetes Care, 39*, 596–602.

Yagihashi, S., Yamagishi, S., & Wada, R. (2007). Pathology and pathogenetic mechanisms of diabetic neuropathy: Correlation with clinical signs and symptoms. *Diabetes Research and Clinical Practice*, 77(Suppl. 1), S184–S189.

Zinman, L. H., Bril, V., & Perkins, B. A. (2004). Cooling detection thresholds in the assessment of diabetic sensory polyneuropathy: Comparison of CASE IV and Medoc instruments. *Diabetes Care, 27*, 1674–1679.

Zochodne, D. W., Verge, V. M., Cheng, C., Sun, H., & Johnston, J. (2001). Does diabetes target ganglion neurones? Progressive sensory neurone involvement in long-term experimental diabetes. *Brain, 124*, 2319–2334.

CHAPTER THIRTEEN

Alternative Quantitative Tools in the Assessment of Diabetic Peripheral and Autonomic Neuropathy

A.I. Vinik*,[1], C. Casellini*, M.-L. Névoret†
*Eastern Virginia Medical School, Strelitz Diabetes and Neuroendocrine Center, Norfolk, VA, United States
†Impeto Medical Inc., San Diego, CA, United States
[1]Corresponding author: e-mail address: vinikai@evms.edu

Contents

Abstract

Here we review some seldom-discussed presentations of diabetic neuropathy, including large fiber dysfunction and peripheral autonomic dysfunction, emphasizing the impact of sympathetic/parasympathetic imbalance. Diabetic neuropathy is the most common complication of diabetes and contributes additional risks in the aging adult. Loss of sensory perception, loss of muscle strength, and ataxia or incoordination lead to a risk of falling that is 17-fold greater in the older diabetic compared to their young nondiabetic counterparts. A fall is accompanied by lacerations, tears, fractures, and worst of all, traumatic brain injury, from which more than 60% do not recover. Autonomic neuropathy has been hailed as the "Prophet of Doom" for good reason. It is conducive to increased risk of myocardial infarction and sudden death. An imbalance in the

International Review of Neurobiology, Volume 127
ISSN 0074-7742
http://dx.doi.org/10.1016/bs.irn.2016.03.010

autonomic nervous system occurs early in the evolution of diabetes, at a stage when active intervention can abrogate the otherwise relentless progression. In addition to hypotension, many newly recognized syndromes can be attributed to cardiac autonomic neuropathy such as orthostatic tachycardia and bradycardia. Ultimately, this constellation of features of neuropathy conspire to impede activities of daily living, especially in the patient with pain, anxiety, depression, and sleep disorders. The resulting reduction in quality of life may worsen prognosis and should be routinely evaluated and addressed. Early neuropathy detection can only be achieved by assessment of both large and small- nerve fibers. New noninvasive sudomotor function technologies may play an increasing role in identifying early peripheral and autonomic neuropathy, allowing rapid intervention and potentially reversal of small-fiber loss.

1. INTRODUCTION

Neuropathy is the most common and troublesome complication of diabetes mellitus, leading to the greatest morbidity and mortality and resulting in a huge economic burden for diabetes care (Holzer et al., 1998; Vinik, Mitchell, et al., 1995) (Fig. 1). It is the most common form of neuropathy in the developed countries of the world, accounts for more hospitalizations than all the other diabetic complications combined, and is

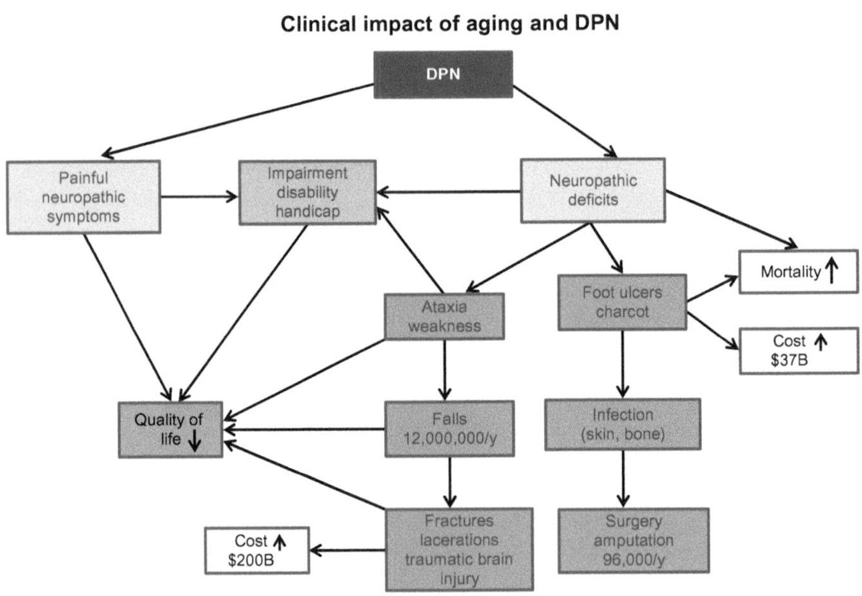

Fig. 1 The clinical and financial implications of aging and diabetic peripheral neuropathy (DPN).

responsible for 50–75% of nontraumatic amputations (Caputo, Cavanagh, Ulbrecht, Gibbons, & Karchmer, 1994; Holzer et al., 1998). Diabetic neuropathy represents a set of clinical syndromes that affect distinct regions of the nervous system, single or combined. It may be silent and go undetected while exercising its ravages or it may present with clinical symptoms and signs that, although nonspecific and insidious, mimic those seen in many other diseases. Diabetic neuropathy is, therefore, diagnosed by exclusion. Unfortunately neither endocrinologists nor nonendocrinologists are routinely trained to recognize the condition and, even when neuropathy is symptomatic, less than one-third of physicians recognize the cause or discuss this with their patients (Herman & Kennedy, 2005).

The true prevalence of diabetic neuropathy is not known and reports vary in their estimates from 10% to 90% in diabetic patients, depending on the criteria and methods used to define neuropathy (Caputo et al., 1994; Dyck et al., 1993; Holzer et al., 1998; Young, Boulton, MacLeod, Williams, & Sonksen, 1993). Twenty-five percent of patients attending a diabetes clinic volunteered symptoms and 50% were found to have neuropathy after a simple clinical test such as the ankle jerk or vibration perception test, while almost 90% tested positive in more sophisticated evaluations of autonomic or peripheral nerve function (Vinik, 1999). Neurologic complications occur equally in type 1 and type 2 diabetes mellitus and in the various forms of acquired diabetes (Dyck et al., 1993). The major morbidity associated with neuropathy is foot ulceration, the precursor of gangrene and limb loss. Neuropathy increases the risk of amputation 1.7-fold, 12-fold if there is deformity (itself a consequence of neuropathy), and 36-fold if there is a history of previous ulceration (Armstrong, Lavery, & Harkless, 1998). In 2010, 73,000 amputations were performed on diabetic patients in the United States, yet up to 75% are preventable (Caputo et al., 1994). Globally there is an amputation every 30 s. Diabetic neuropathy also has a tremendous impact on patients' quality of life (QOL), predominantly by causing weakness, ataxia, and incoordination predisposing to falls and fractures (Vinik et al., 2005). Once autonomic neuropathy sets in, life can become quite dismal and the mortality rate approximates 25–50% within 5–10 years (Levitt, Stansberry, Wychanck, & Vinik, 1996; Rathmann, Ziegler, Jahnke, Haastert, & Gries, 1993).

Diabetes results in a variety of peripheral neuropathy syndromes for which there is no universally accepted classification. These syndromes are generally subdivided into focal/multifocal neuropathies, including diabetic amyotrophy, and symmetric polyneuropathies, including diabetic sensorimotor polyneuropathy (DSPN). The latter is the most common type, affecting about 30% of diabetic patients in hospital care and 25% of those in the community (Sadosky, McDermott, Brandenburg, & Strauss, 2008; Vinik, Ullal, Parson, & Casellini, 2006; Ziegler, Papanas, Vinik, & Shaw, 2014). The Toronto Diabetic Neuropathy Expert Group defined DSPN as a symmetrical, length-dependent sensorimotor polyneuropathy attributable to metabolic, and microvascular alterations as a result of chronic hyperglycemia exposure (diabetes) and cardiovascular risk covariates (Tesfaye et al., 2010). Its onset is generally insidious and without treatment the course is chronic and progressive. Loss of small fiber-mediated sensation results in the loss of thermal and pain perception, whereas large fiber impairment results in loss of touch and vibration perception. Sensory fiber involvement may also result in "positive" symptoms, such as paresthesias and pain, although up to 50% of neuropathic patients are asymptomatic. DSPN can be associated with concurrent involvement of the autonomic nervous system (ANS) that rarely causes severe symptoms (Boulton, Malik, Arezzo, & Sosenko, 2004; Ziegler, 2008) but in its cardiovascular form is associated with at least a threefold increased risk for mortality (Vinik, Maser, & Ziegler, 2010, 2011; Vinik & Ziegler, 2007).

2. DIABETIC PERIPHERAL NEUROPATHY

2.1 Clinical Presentation

Neuropathy is frequently missed in diabetic patients due to absence of symptoms or lack of appropriate diagnostic scrutiny. The clinical presentation varies depending on the effort placed into finding it. A population survey reported that 30% of type 1 and 36–40% of type 2 diabetic patients experienced neuropathic symptoms (Harris, Eastman, & Cowie, 1993). Several studies have also suggested that impaired glucose tolerance (IGT) may lead to polyneuropathy, with IGT reporting rates in patients with chronic idiopathic polyneuropathies of 25–62% (Novella, Inzucchi, & Goldstein, 2001; Singleton, Smith, & Bromberg, 2001a, 2001b; Sumner, Sheth, Griffin, Cornblath, & Polydefkis, 2003; Ziegler et al., 2014). Studies using skin and nerve biopsies have shown a progressive reduction in peripheral nerve fiber density from the time of diagnosis of diabetes or even in earlier prediabetic stages (Dyck, Lais, Karnes, O'Brien, & Rizza, 1986; Pittenger et al., 2004). Sensory symptoms are more prominent than motor symptoms and usually involve the lower limbs and include pain, paresthesiae, hyperesthesiae, deep aching, burning, and sharp stabbing sensations similar to, but less severe than, those described in acute sensory neuropathy. In addition, patients may experience negative symptoms such as numbness in the feet and legs that presage painless foot ulcers and subsequent amputations if the neuropathy is not promptly recognized and treated. Unsteadiness is also frequent, due to abnormal proprioception and muscle sensory function (Cavanagh, Simoneau, & Ulbrecht, 1993; Katoulis et al., 1997). Alternatively, some patients may be completely asymptomatic and signs may be only discovered by a detailed neurological examination.

The natural history of neuropathy syndromes in diabetes separates them into two very distinctive entities, namely those which progress gradually with increasing duration of diabetes and those which remit, usually completely. DSPN and autonomic neuropathies generally progress, while mononeuropathies, radiculopathies, and acute painful neuropathies, although manifesting severe symptoms, are short lived and tend to recover (Watkins, 1993). Progression of neuropathy is related to glycemic control in both type 1 and type 2 diabetes (DCCT Research Group, 1993, 1995). It appears that the most rapid deterioration of nerve function occurs soon after the onset of type 1 diabetes; then within 2–3 years there is a slowing of progression. In type 2 diabetes, slowing of nerve conduction velocity (NCV) may be one of the earliest neuropathic abnormalities and is often present at diagnosis (Ziegler, Cicmir, Mayer, Wiefels, & Gries, 1988).

On physical examination a symmetrical stocking-like distribution of sensory abnormalities in both lower limbs is usually seen. In more severe cases hands may also be involved. All sensory modalities can be affected, particularly vibration, touch, and position perceptions that are indicative of large $A\alpha/\beta$ fiber damage and pain, with abnormal heat and cold temperature perception that is indicative of small thinly myelinated $A\delta$ and unmyelinated C-fiber damage. Deep tendon reflexes may be absent or reduced, especially in the lower extremities. Mild muscle wasting may be seen but severe weakness is rare and should raise the question of a possible nondiabetic etiology (Boulton et al., 2004; Sinnreich, Taylor, & Dyck, 2005). Most patients show a "mixed" neuropathy with both large and small nerve fiber damage and DSPN is also frequently accompanied by autonomic neuropathy, which will be described in more detail below. It is important to remember that all patients with DSPN are at increased risk of neuropathic complications such as foot ulceration and Charcot's neuroarthropathy.

2.2 Diagnosis

The diagnosis of DSPN should rest on the findings of the clinical and neurological examinations, with consideration of the presence of neuropathic symptoms (positive and negative, sensory, and motor) and signs (sensory deficit, allodynia and hyperalgesia, motor weakness, and absence of reflexes). According to the American Academy of Neurology report on the case definition of distal symmetric polyneuropathy, there is good evidence that:

(A) Symptoms alone have poor diagnostic accuracy in predicting the presence of polyneuropathy,

(B) Signs are better predictors than symptoms,

(C) Multiple signs are better predictors than a single sign, and

(D) Relatively simple examinations are as accurate as complex scoring systems (England et al., 2005).

The most recent position statement of the American Diabetes Association recommends that all patients with diabetes be screened for neuropathy at diagnosis in type 2 diabetes and 5 years after diagnosis in type 1 diabetes. Neuropathy screening should be repeated annually and must include sensory examination of the feet and ankle reflexes (American Diabetes Association, 2015). The basic neurological assessment comprises the general medical and neurological history to exclude other potential causes of neuropathy, inspection of the feet, and neurological examination of sensation. As for all patients with distal symmetric polyneuropathy (England et al., 2009), screening laboratory tests may be considered in selected patients with DSPN, with serum B12 with its metabolites and serum protein immunofixation electrophoresis being those with the highest yield of abnormalities (England et al., 2009). The neurological examination should focus on the lower extremities and always include an accurate foot inspection for deformities, ulcers, fungal infection, muscle wasting, hair distribution or loss, and the presence or absence of pulses. Footwear should also be inspected at every visit. Sensory modalities should be assessed using simple handheld devices (Table 1) (Haanpaa et al., 2009). The most sensitive measure is the vibration detection threshold, although sensitivity of 10-g Semmes-Weinstein monofilament to identify feet at risk varies between 86% and 100% (Armstrong, Lavery, Vela, Quebedeaux, & Fleischli, 1998; Kumar et al., 1991). Combinations of more than one test have more than 87% sensitivity in detecting DSPN (Boulton et al., 2005; Vinik, Suwanwalaikorn, et al., 1995). Longitudinal studies show that these simple tests are good predictors of foot ulcer risk (Abbott et al., 2002). Knee and ankle reflexes should also be tested

Table 1 Examination—Bedside Sensory Tests

Sensory Modality	Nerve Fiber	Instrument	Associated Sensory Receptors
Vibration	Aβ (large)	128 Hz Tuning fork	Ruffini corpuscle mechanoreceptors
Pain (pinprick)	C (small)	Neurotips	Nociceptors for pain and warmth
Pressure	Aβ, Aα (large)	1 and 10 g Monofilament	Pacinian corpuscle
Light touch	Aβ, Aα (large)	Wisp of cotton	Meissner's corpuscle
Cold	Aδ (small)	Cold tuning fork	Cold thermoreceptors

(Boulton, Gries, & Jervell, 1998; Boulton et al., 2005) and assessment of joint position and motor power may be indicated.

The following findings should alert the physician to consider causes for DSPN other than diabetes and cause referral for a detailed neurological work-up: (1) pronounced asymmetry of the neurological deficits; (2) predominantly motor deficits, mononeuropathy, or cranial nerve involvement; (3) rapid development or progression of the neuropathic impairments; (4) progression of the neuropathy despite optimal glycemic control; (5) symptoms from the upper limbs; and (6) family history of nondiabetic neuropathy.

2.3 Falls Risk in Older Adults with Diabetes

2.3.1 Falls and Aging

The older adult is often faced with a myriad of health-related issues and falling is a major concern for persons over 65 years of age. The outcomes of a fall extend from short-term health problems (lacerations, fractures, and traumatic brain injury) to long-term consequences (decline in muscle strength, increased fatigue, and heightened fear of falling) that reduce physical activity (Stevens, Corso, Finkelstein, & Miller, 2006). In 2010, the Centers for Disease Control reported that 2.3 million older adults were treated in emergency departments for nonfatal fall-related injuries. Furthermore, falls and fall-related consequences have been reported to be the leading cause of injury-related death and hospitalization in adults over 65 years of age. A staggering one-third of persons this age who fall are likely to suffer one of these adverse events within a given year. Given the range of health issues that can arise following a fall, prevention should be the first course of action.

The key to preventing falls is identification of persons at risk and implementing appropriate interventions. This requires recognizing factors that can lead to increased risk of falling. However, despite our understanding of the seriousness of falls and the high cost of medical care, identifying a few critical factors that are strongly predictive of falls in high-risk populations is challenging, partly because over 400 factors are linked with falls in adult populations (Close, Lord, Menz, & Sherrington, 2005). Something as simple as an individual's perception of threat around them when they move (often referred to as a "fear of falling") can be a significant health issue. Nearly 13 million (36%) of American adults aged 65 years or more have been found to be moderately or very afraid of falling, illustrating that developing a fear of suffering an adverse event is strongly linked with actual falls (Boyd & Stevens, 2009).

2.3.2 Risk of Falling Is Increased for Older Adults with Type 2 Diabetes

Persons with type 2 diabetes are at increased risk of falling compared to healthy adults of a similar age. A combination of age (>65 years) and diabetes further increases the risk of falling 17-fold (Cavanagh, Derr, Ulbrecht, Maser, & Orchard, 1992; Pijpers et al., 2012) due to factors such as sensory neuropathy, declining cognitive function, and use of multiple prescription medications (Close et al., 2005; Peron & Ogbonna, 2015; Schwartz et al., 2002, 2008). The likelihood of suffering a fall increases dramatically with increasing age and/or the emergence of type 2 diabetes, with risk being increased significantly by the presence of diabetes alone (Clark, Lord, & Webster, 1993; Close et al., 2005; Lord, 1996; Lord & Clark, 1996; Pickering et al., 2007; Robinovitch et al., 2000; Sosnoff, Motl, & Morrison, 2013; Sosnoff et al., 2011). Older persons with diabetes must contend with both age-related declines in balance control, muscle strength, walking ability, and proprioception (Morrison, Colberg, Mariano, Parson, & Vinik, 2010; Morrison, Colberg, Parson, & Vinik, 2012; Schwartz et al., 2002, 2008; Wallace et al., 2002) and health-related issues associated with diabetes (Berlie & Garwood, 2010; Maurer, Burcham, & Cheng, 2005; Richardson & Hurvitz, 1995; Tilling, Darawil, & Britton, 2006). Indeed, the additional range of potential risk factors in anyone with diabetes is extensive, covering neuropathy, visual deficits, loss of coordination, cognitive impairment, autonomic dysfunction with orthostatic hypotension, tachycardia, bradycardia, pain, poor lower body function, high-body mass index, cardiovascular syncope, vestibular dysfunction, frontal cortex dysfunction, and use of various medications, all of which may interact and can have an additive effect (Close et al., 2005; Morrison et al., 2010, 2012; Volpato, Leveille, Blaum, Fried, & Guralnik, 2005; Volpato, Maraldi, & Fellin, 2010). Thus, the older person with diabetes typically has a significantly greater risk of suffering a fall when compared to a healthy adult of similar age (Berlie & Garwood, 2010; Maurer et al., 2005; Pijpers et al., 2012).

2.3.3 Neuropathy Is Strongly Linked to Falling

Declining sensory and motor function arising from neuropathy is a major contributing factor to the overall increase in fall risk factor for persons with diabetes (Boucher, Teasdale, Courtemanche, Bard, & Fleury, 1995; Richardson & Hurvitz, 1995; Uccioli et al., 1995; Volpato et al., 2005; Witzke & Vinik, 2005). Consequences extend to decrements in balance and altered walking function (Chiles et al., 2014; Resnick et al., 2002, 2000; Strotmeyer et al., 2008), which are obvious mediators for increased falls risk (Morrison et al., 2010; Pittenger, Mehrabyan, et al., 2005; Richardson & Hurvitz, 1995; Vinik, Strotmeyer, Nakave, & Patel, 2008; Witzke & Vinik, 2005). The ability to optimally control one's balance is essential for mobility, avoidance of disability, and preservation of independence in older people (Colberg et al., 2005). The complexity of the balance system makes localization of the problem difficult since the abnormality may occur in one or more of the sensory sites or in the motor system. A thorough evaluation of the sensory-motor systems affecting balance is required to arrive at a diagnosis and to create a platform to provide a menu for treatment and management choices.

Since neuropathy progression follows a distal-to-proximal gradient, the effects of neuropathy on strength and balance are most evident at the ankles and feet and loss of nerve function can have dramatic implications for both standing and walking tasks (Colberg et al., 2005). For example, diabetic persons with sensory deficits in the feet can exhibit increased postural motion and slower gait speed (Chiles et al., 2014; Resnick et al., 2002, 2000; Strotmeyer et al., 2008) with increased stride time variability (Corriveau et al., 2000; Fioretti, Scocco, Ladislao, Ghetti, & Rabini, 2010; Lafond, Corriveau, & Prince, 2004; Lalli et al., 2013; Resnick et al., 2002; Simoneau, Ulbrecht, Derr, Becher, & Cavanaugh, 1994; Turcot, Allet, Golay, Hoffmeyer, & Armand, 2009; Uccioli et al., 1995). Their impact is further magnified when the task is made more difficult, as when walking or standing on irregular surfaces (Richardson & Hurvitz, 1995; Richardson, Thies, & Ashton-Miller, 2008; Richardson, Thies, DeMott, & Ashton-Miller, 2005). In addition, there is slowing of the reaction time, loss of the ability to prevent progression to a fall after its initiation, and dorsiflexion weakness (Strotmeyer et al., 2009), which increase the susceptibility to tripping on loose rugs, carpets, and minor variations in step height. Although older adults with diabetes are a heterogeneous group, ranging from fit and healthy to frail with many comorbidities and functional disabilities, poorly controlled diabetes with polyuria escalates falls risk simply because of the number of visits to, and urgency to use, the bathroom (Berlie & Garwood, 2010; Maurer et al., 2005; Volpato et al., 2005, 2010).

2.3.4 Can Risk of Falling Be Reversed?

Reversal of falls risk depends first and foremost on screening and identification of the candidate at risk. Fig. 2 shows an algorithm that identifies candidates and matches them to the intervention most likely to succeed. Numerous studies have shown the benefits of various balance/exercise programs in reducing falls risk in healthy older persons (Barnett, Smith, Lord, Williams, & Baumand, 2003; Howe, Rochester, Jackson,

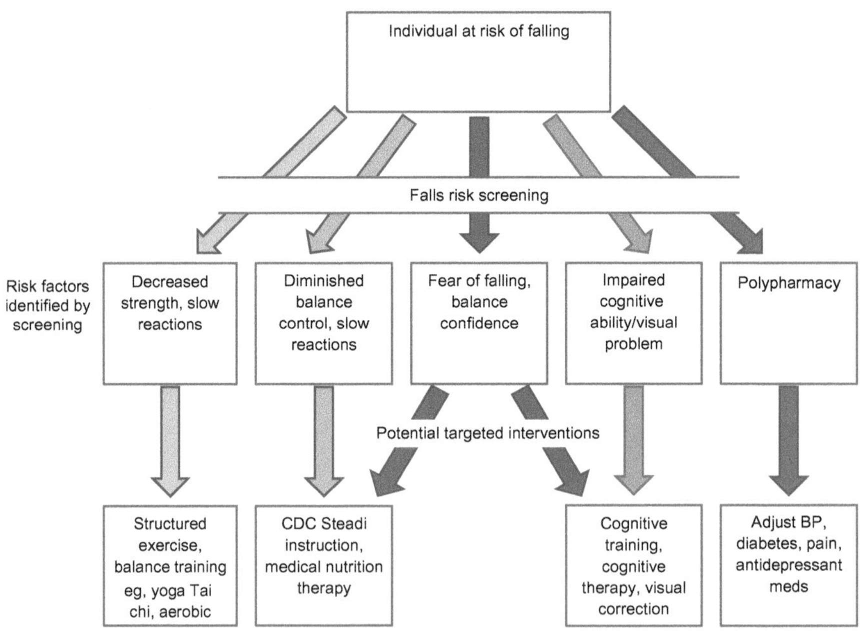

Fig. 2 A clinical algorithm for the evaluation of and targeted intervention for the aging patient at risk of falling.

Banks, & Blair, 2007; Peterson et al., 2009; Robinovitch et al., 2000; Sherrington, Lord, & Herbert, 2003) and studies have reported that targeted interventions can improve balance and walking ability and also reduce falls risk in diabetes (Allet et al., 2009; Morrison et al., 2010, 2012). Structured balance training can lead to improvements in posture and/or gait function (Allet et al., 2010; Morrison et al., 2010, 2012) as well as general gains from physical activity like faster reaction times, improvements in sensory perception and lower limb strength, and better sympathetic/parasympathetic balance (Colberg, 2006; Colberg, Stansberry, McNitt, & Vinik, 2002; Herriott, Colberg, Parson, Nunnold, & Vinik, 2004; Morrison et al., 2010, 2012). Interestingly, recent studies have reported that exercise can also lead to improvements in neuropathy symptoms, including increased nerve fiber branching (Kluding et al., 2012) and improved sensory responses in the lower limbs (Balducci et al., 2006).

3. DIABETIC AUTONOMIC NEUROPATHY

Diabetic autonomic neuropathy (DAN) is among the least recognized and understood complications of diabetes despite its significant negative impact on survival and QOL in people with diabetes (Vinik & Erbas, 2001). It is also a major source of increased cost of caring for the diabetic patient. The two divisions of the ANS—the parasympathetic and the sympathetic nervous systems—work in balanced opposition to control the heart rate, the force of cardiac contraction, the dilatation and constriction of blood vessels, the contraction and relaxation of smooth muscle in the digestive and urogenital systems, and the secretions of glands and pupillary size. Diabetes can cause dysfunction of any or every part of the ANS, leading to a wide range of disorders.

Reported prevalence of DAN can vary. For example, in a community-based study the prevalence of DAN, as defined by one or more abnormal heart rate variability (HRV) test results, was 16.7% (Neil, Thompson, John, McCarthy, & Mann, 1989). If stricter criteria were used (abnormalities present in at least three of six autonomic function tests), the prevalence was 16.8% for individuals with type 1 diabetes and 22.1% for individuals with type 2 diabetes (Ziegler, Gries, Spuler, & Lessmann, 1992).

3.1 Clinical Manifestations

The ubiquitous distribution of the ANS renders virtually all organs susceptible to dysfunction (Table 2) so that clinical manifestations frequently

Table 2 Symptoms and Signs Associated with Diabetic Autonomic Neuropathy

Cardiovascular Autonomic Neuropathy	Gastrointestinal	Urogenital
Resting tachycardia **Abnormal blood pressure regulation** • Nondipping • Reverse dipping **Orthostatic hypotension/ tachycardia/bradycardia** • Lightheadedness • Weakness • Faintness • Dizziness • Visual impairment • Syncope **(All with standing)**	**Gastroparesis** • Nausea • Bloating • Loss of appetite • Early satiety • Postprandial vomiting • Brittle diabetes **Esophageal dysfunction** • Heartburn • Dysphagia for solids **Diabetic diarrhea** • Profuse and watery diarrhea • Fecal incontinence • Alternates with constipation **Constipation**	**Bladder dysfunction** • Frequency • Urgency • Nocturia • Hesitancy • Weak stream • Dribbling • Urinary incontinence • Urinary retention **Male sexual dysfunction** • Erectile dysfunction • Decreased libido • Abnormal ejaculation **Female sexual dysfunction** • Decreased sexual desire • Increased pain during intercourse • Decreased sexual arousal • Inadequate lubrication

occur concurrently, but with inconsistent patterns. A patient diagnosed with diabetes should be suspected of having at least subclinical disturbances of the ANS, which can occur within a year of diagnosis in type 2 diabetic patients and within 2 years in type 1 diabetic patients (Pfeifer et al., 1984). Clinical symptoms generally do not occur until long after the onset of diabetes. The most important etiologic factors that have been associated with DAN are poor glycemic control, diabetes duration, age, female sex, and higher body mass index (BMI). Of patients with symptomatic autonomic dysfunction, 25–50% die within 5–10 years of diagnosis (Ewing, Boland, Neilson, Cho, & Clarke, 1991; Rathmann et al., 1993) and the 5-year mortality rate in patients with DAN is three times higher than in diabetic patients without autonomic involvement

(O'Brien, McFadden, & Corrall, 1991). Leading causes of death in patients with either symptomatic or asymptomatic DAN are heart disease and nephropathy while DAN is also an independent risk factor for stroke.

3.2 Cardiovascular Autonomic Neuropathy

Perhaps one of the most overlooked of all serious complications of diabetes is cardiovascular autonomic neuropathy (CAN) (Maser, Lenhard, & DeCherney, 2000). CAN results from damage to autonomic nerve fibers that innervate the heart and blood vessels that causes abnormalities in heart rate control and vascular dynamics. CAN has been linked to resting tachycardia, postural hypotension, exercise intolerance, enhanced intraoperative or perioperative cardiovascular liability, increased incidence of asymptomatic ischemia, myocardial infarction (MI), and decreased rate of survival after MI. The Diabetes Control and Complications Trial (DCCT) found that 1.65% had abnormal HRV at baseline for less than 5 years duration of diabetes, rising to 6.2% among those with less than 9 years but more than 5 years duration of diabetes, and 12.2% by 9 years or more (Diabetes Control and Complications Trial Research Group, 1998). CAN may be present at diagnosis and prevalence increases with age, duration of diabetes, and poor glycemic control. CAN also cosegregates with distal symmetric polyneuropathy, microangiopathy, and macroangiopathy. Age, diabetes, obesity, and smoking are risk factors for HRV in type 2 diabetes. Hemoglobin A1c, hypertension, distal symmetrical polyneuropathy, retinopathy, and exposure to hyperglycemia were all shown to be risk factors for developing CAN in type 1 diabetes (Witte et al., 2005; Ziegler et al., 2004, 2006).

CAN may be associated with abnormalities in left ventricular (LV) systolic, and particularly diastolic, function in the absence of cardiac disease in diabetic patients. Echocardiographic studies have shown a significant correlation of the severity of CAN with reduced peak diastolic filling rate and with an augmented atrial contribution to diastolic filling as assessed by Doppler echocardiography. CAN is also associated with left ventricular diastolic dysfunction (LVDD) at rest, both in patients with long-term type 2 and type 1 diabetes. LVDD may progress to heart failure, mainly with preserved LV systolic function (diastolic heart failure). The pathophysiology of LVDD includes delayed relaxation, impaired LV filling, and/or increased stiffness. In patients with CAN, vagal impairment can lead to a relative predominance of sympathetic activity in the sympathovagal balance. Sympathetic overactivity stimulates the renin–angiotensin–aldosterone system and increases heart rate, stroke volume, and peripheral vascular resistance, thus contributing to LV dysfunction. Such sympathetic hyperactivity, in combination with regional myocardial sympathetic denervation, has been shown to lead to diminished coronary blood flow reserve and diastolic dysfunction in diabetic patients with early microangiopathy (Dinh et al., 2011; Sacre et al., 2010; Vinik & Erbas, 2013).

3.2.1 Diagnosis of CAN

There are simple bedside tests to diagnose CAN using HRV, responses to breathing, the Valsalva maneuver, and standing. Functional abnormalities and imbalance between the sympathetic and parasympathetic nervous system are discerned with respiratory modulation of different frequency oscillations in HRV.

3.2.2 Cardiovascular Symptoms and Signs

CAN is identified by a number of presentations:

3.2.2.1 Resting Tachycardia

Resting tachycardia and a fixed heart rate are characteristic late findings in diabetic patients with vagal impairment. Resting heart rates of 90–100 bpm and occasional heart rate increments up to 130 bpm occur. A fixed heart rate that is unresponsive to moderate exercise, stress, or sleep indicates almost complete cardiac denervation (Ewing & Clarke, 1986).

3.2.2.2 Exercise Intolerance

Autonomic dysfunction impairs exercise tolerance, reduces response in heart rate and blood pressure, and blunts increases in cardiac output in response to exercise. Diabetic patients who are likely to have CAN should be tested for cardiac stress before undertaking an exercise program. Patients with CAN need to rely on their perceived exertion, not heart rate, to avoid hazardous levels of intensity of exercise (Albers, Krichavsky, & Balady, 2006; Colberg, Swain, & Vinik, 2003; Vinik & Erbas, 2002, Vinik, Erbas, & Pfeifer, 2003).

3.2.2.3 Intraoperative Cardiovascular Liability

Hemodynamic changes occur during surgery for individuals without diabetes. Perioperative cardiovascular morbidity and mortality are increased two- to threefold in patients with diabetes. Compared with nondiabetic subjects, diabetic patients undergoing general anesthesia may experience a greater degree of decline in heart rate and blood pressure during induction of anesthesia and less of an increase after tracheal intubation and extubation. There is also an association between CAN and more severe intraoperative hypothermia that results in decreased drug metabolism and impaired wound healing.

3.2.2.4 Orthostatic Hypotension

Orthostatic hypotension is defined as a fall in blood pressure (>20 mmHg for systolic or >10 mmHg for diastolic) in response to postural change, from supine to standing. In patients with diabetes, orthostatic hypotension is usually due to damage to the efferent sympathetic vasomotor fibers, particularly in the splanchnic vasculature. In addition, there is a decrease in cutaneous, splanchnic, and total vascular resistance. Normally, in response to postural change there is an increase in plasma norepinephrine. For individuals with orthostatic hypotension, there may be a reduction in this response relative to the fall in blood pressure. Diminished cardiac acceleration and cardiac output, particularly in association with exercise, may also be present. Patients with orthostatic hypotension typically present with lightheadedness and presyncopal symptoms. Many patients, however, remain asymptomatic despite significant falls in blood pressure.

3.2.2.5 Silent Myocardial Ischemia/Cardiac Denervation Syndrome

The presence of both symptomatic and asymptomatic coronary artery disease is increased in diabetic patients and subclinical neuropathy is an important cause of silent ischemia in patients with diabetes (Airaksinen & Koistinen, 1992; Marchant, Umachandran, Stevenson, Kopelman, & Timmis, 1993). Features of MI in patients with CAN are silence, cough, nausea and vomiting, dyspnea, tiredness, and ECG changes. Reduced appreciation for ischemic pain can impair timely recognition of myocardial ischemia or infarction, thereby delaying appropriate therapy. Silent ischemia in diabetic patients may either result from CAN, from autonomic dysfunction attributable to coronary artery disease itself, or from both. In the Framingham Study, the rates of unrecognized MI's were 39% in diabetic patients and 22% in nondiabetic subjects, but the difference was not significant (Margolis, Kannel, Feinleib, Dawber, & McNamara, 1973). In a survey from the National Registry of Myocardial Infarction, of 434,877 patients presenting with MI, 33% did not have chest pain. Among those presenting without chest pain, 32% had diabetes vs 25.4% in the group with chest pain (Canto et al., 2000). Thus, patients with CAN warrant careful attention and cardiovascular autonomic function testing may be an important component in the risk assessment of diabetic patients with coronary artery disease.

3.2.2.6 Increased Risk of Mortality

A number of studies have shown a 2.3-fold increased risk of CAN in diabetic patients showing a prolonged QT interval, leading to speculation that CAN might also predispose to malignant ventricular arrhythmias and to sudden death from cardiac arrest caused by torsades de pointes, as in long QT syndrome. Furthermore, symptomatic CAN at 5 years of diabetes predicted mortality at 10 years, even after adjusting for conventional cardiovascular disease (CVD) risk factors. HRV was found to be an independent predictor of all-cause mortality during a period of 9 years (Wirta, Pasternack, Mustonen, & Laippala, 1997). Moreover, impaired autonomic function is associated with increased all-cause and cardiovascular mortality so that CAN in patients already at risk (diabetes, hypertension, or history of CVD) may be especially hazardous (Gerritsen et al., 2001). Results from the ACCORD trial confirmed the association of CAN and mortality as individuals in this trial with baseline CAN were 1.55–2.14 times as likely to die as individuals without CAN. Furthermore, CAN in the presence of peripheral neuropathy was the highest predictor of CVD mortality and combining indexes of autonomic dysfunction have been shown to be associated with the risk of mortality (Lykke, Tarnow, Parving, & Hilsted, 2008; Maser, Mitchell, Vinik, & Freeman, 2003; Ziegler et al., 2008).

3.2.2.7 Increased Mortality After MI

Mortality rates after an MI are higher for diabetic than nondiabetic patients. A simple bedside test that measured 1-min HRV during deep breathing was a good predictor of all-cause mortality for 185 patients (17.8% with diabetes) after a first MI (Stewart, Medow, & Montgomery, 2003).

3.2.2.8 CAN and Sudden Death

Sudden, unexpected deaths occur among subjects with CAN. One potential cause may be severe but asymptomatic ischemia, which can induce lethal arrhythmias. QT prolongation may also predispose individuals to life-threatening cardiac arrhythmias and sudden death. Male patients with impaired HRV had a higher corrected QT prolongation than males without this complication (Veglio, Borra, Stevens, Fuller, & Perin, 1999).

3.2.3 Impact of CAN on Diabetes Management

The question of more intensive therapy and reduction in cardiovascular complications was addressed for people with type 2 diabetes of long duration in three studies: ACCORD (Buse et al., 2007), ADVANCE (Patel et al., 2008), and VADT (Duckworth et al., 2009). Although the three studies utilized different patient cohorts, with varying durations of diabetes and had different treatment regimens, the results indicated that intensive glucose control did not reduce CVD events. Recent analyses of the ACCORD trial examined whether the effects of CAN or self-reported history of neuropathy at baseline could have been a contributor to the higher mortality risk in the intensive glycemic arm (272,273). The paradox was that higher HbA1C was associated with higher hypoglycemic rates within both intensive and standard treatment groups. The HR for hypoglycemia by each 1% higher updated average HbA1C was 1.15 for the intensive and 1.76 for the standard and hypoglycemia doubled the risk of death with intensification of glycemic control. This is in stark contrast with the DCCT study, where rates of hypoglycemia increased with lowering of the blood glucose. It was not the hypoglycemia which killed the patients but rather the predictive capacity of hypoglycemia for sudden death!

3.3 Prevention and Reversibility of Autonomic Neuropathy

Restoration of autonomic balance is possible and has been shown following lifestyle changes, increased physical activity, β-adrenergic blockers, aldose reductase inhibitors, ACE inhibitors, angiotensin receptor blockers, and potent antioxidants such as α-lipoic acid. Obesity in humans has been associated with autonomic dysfunction and increased sympathetic activity (Lambert, Sari, et al., 2010; Lambert, Straznicky, Lambert, Dixon, & Schlaich, 2010; Piestrzeniewicz, Luczak, Lelonek, Wranicz, & Goch, 2008; Straznicky et al., 2009). Furthermore, several studies have shown that weight loss improves HRV measures and autonomic imbalance (Casellini et al., 2015; Karason, Molgaard, Wikstrand, & Sjostrom, 1999; Maser, Lenhard, Irgau, & Wynn, 2007; Nault et al., 2007; Perugini et al., 2010; Ravussin, 2010; Ziegler et al., 2015).

There are exciting new prospects for pathogenesis-oriented intervention to reverse established CAN (Vinik et al., 2011). Strict glycemic control (DCCT Research Group, 1995) and a stepwise progressive management of hyperglycemia, lipids, and blood pressure as well as the use of antioxidants (Ziegler & Gries, 1997) and ACE inhibitors (Athyros et al., 1998) reduce the odds ratio for autonomic neuropathy (Gaede, Vedel, Parving, & Pedersen, 1999). The EDIC study, a longitudinal cohort follow-up study for the DCCT in which patients with type 1 diabetes were randomized to conventional or intensive glycemic control, demonstrated persistent beneficial effects of past glucose control on microvascular complications despite the loss of glycemic separation in type 1 diabetic patients (DCCT EDIC Research Group, 2003). During the EDIC follow-up, CAN progressed in both treatment groups, but the incidence and prevalence of CAN remained lower in the formerly intensive group than in the formerly conventional group, despite similar levels of glycemic control during EDIC. To minimize the development of autonomic neuropathy, intensive glucose control of type 1 diabetes should therefore be started as early as possible (Pop-Busui et al., 2009). However, while glycemic control with a reduction of HbA1c from 9.5 to 8.4 improved HRV in those patients with mild autonomic abnormalities, this was not so in type 1 diabetics with advanced autonomic abnormalities (Burger, Weinrauch, D'Elia, & Aronson, 1999).

The effects of glycemic control in type 2 diabetics are less definite. The VA Cooperative Study showed no difference in the prevalence of autonomic neuropathy after 2 years of intensive glycemic control in type 2 diabetic patients (Azad et al., 1999). In the Steno-2 Study, where people with type 2 diabetes received intensive multifactorial treatment that targeted hyperglycemia, hypertension, dyslipidemia, and microalbuminuria, along with secondary prevention of CVD with aspirin, the approach reduced autonomic dysfunction by 63%. The glucose-lowering agents appeared to have the least effect when compared with antihypertensive treatment, lipid-lowering agents, aspirin, and vitamin–mineral supplements (Gaede et al., 1999). In addition, a survey of evidence from clinical trials shows that early identification of autonomic neuropathy permits timely initiation of therapy with the antioxidant alpha-lipoic acid, which slows or reverses progression of CAN (Ziegler et al., 1997).

Early ACE inhibition or angiotensin receptor blockade improved both DAN and LVDD after 1 year of treatment in asymptomatic patients with long-term diabetes, with the combination being slightly better than monotherapies and auguring well for the patient with established CAN (Didangelos et al., 2006). Fluvastatin treatment also improves cardiac sympathetic neuropathy in the diabetic rat heart, in association with attenuation of increased cardiac oxidative stress (Matsuki et al., 2010). On the other hand, selective COX-2 inactivation confers protection against sympathetic denervation during experimental diabetes by reducing intramyocardial oxidative stress and inflammation (Kellogg, Converso, Wiggin, Stevens, & Pop-Busui, 2009).

Therefore, statins and COX-2 inactivation might help to attenuate cardiac sympathetic dysfunction. It has also been shown that early mortality is a function of loss of beat-to-beat variability with MI. This can be reduced by 33% with acute administration of insulin (Malmberg, Norhammar, Wedel, & Ryden, 1999). Kendall, Rooney, Smets, Salazar Bolding, and Robertson (1997) reported that successful pancreas transplantation improves epinephrine response and normalizes hypoglycemia symptom recognition in patients with long-standing diabetes and established autonomic neuropathy. Burger et al. (1999) showed that a reversible metabolic component of CAN exists in patients with early CAN. There are, therefore, exciting new prospects for pathogenesis-oriented intervention (Vinik et al., 2011).

4. MEASURING DIABETIC NEUROPATHY: ESTABLISHED AND INNOVATIVE APPROACHES

A summary of the different diagnostic tools available to the physician for the evaluation of diabetic neuropathies is provided in Table 3.

4.1 QOL Measures

A number of instruments have been developed and validated to assess QOL. The NeuroQoL measures patients' perceptions of the impact of neuropathy and foot ulcers (Vileikyte et al., 2003). The Norfolk QOL questionnaire is a validated tool addressing specific symptoms and the impact of large, small, and autonomic nerve fiber functions. When tested in 262 subjects differences between DSPN patients and both diabetic and healthy controls were significant ($P < 0.05$) for all item groupings (small fiber, large fiber, and autonomic nerve function; symptoms; and activities of daily living (ADL)). The nerve fiber-specific domains in particular have been shown to correlate with objective measures of nerve function (Vinik et al., 2005; Vinik, Hayes, Oglesby, & Vinik, 2004; Vinik, Paulson, Ford-Molvik, & Vinik, 2008). Total QOL scores correlated with total neuropathy scores. The ADL, total scores, and autonomic scores were also greater in diabetic controls compared to healthy controls ($P < 0.05$), suggesting that diabetes per se impacts some aspects of QOL (Vinik et al., 2005). The tool has been translated into 60 languages and has served as an endpoint in trials of diabetic neuropathy (Boyd, Casellini, Vinik, & Vinik, 2011; Vinik et al., 2014). It captured 6600 patients with previously undiagnosed neuropathy in a screening of 25,000 patients in Romania (Veresiu et al., 2015) and recently served as a tool to demonstrate safety in a study of topical capsaicin (Vinik, Perrot, et al., 2015).

Table 3 Performance Range of Tools for the Diagnosis of Peripheral Neuropathy

Test	Sensitivity	Specificity	Gold Standard Used in Study	Study Subject Population
Large fiber				
10 g Monofilament (Dros, Wewerinke, Bindels, & van Weert, 2009)	41–93%	68–100%	NCS	Unselected diabetic (DM) patients
Vibration perception (Lai, Ahmed, Bollineni, Lewis, & Ramchandren, 2014)	20–26%	88–89%	NCS	Neurology outpatient clinic
Vibration perception (Martin et al., 2010)	75–87%	51–62%	Clinical signs and symptoms; abnormal NCS; or both	Adult type 1 diabetic patients (T1D)
Ankle reflexes (Shehab et al., 2012)	91.5%	67.4%	NCS	Type 2 diabetes patients (T2D)
Ankle reflexes (Taksande, Ansari, Jaikrishnan, & Karwasara, 2011)	72.22–78.95%	43.42–46.77%	NCS	Unselected DM patients
EMG—sural-nerve amplitude (Ebadi, Perkins, Katzberg, Lovblom, & Bril, 2012)	77%	73%	IENFD < 5.4 fibers/mm	Patients with painful polyneuropathy and normal NCS
EMG—sural and peroneal (Gibbons, Freeman, & Veves, 2010)	59–73%	53–78%	Neuropathy Disability Score (NDS) > 2	DM patients and healthy controls
Small fiber				
Pinprick (Neurotip®) (Nather et al., 2011)	50%	50%	N/A—all patients with diabetic foot	Patients with diabetic foot

Pinprick (Neurotip®) (Paisley, Abbott, van Schie, & Boulton, 2002)	91.8%	41%	NDS $\geq$ 6	Unselected DM patients
QST: CDT foot and HP foot (Ebadi et al., 2012)	62–65%	63–65%	IENFD < 5.4 fibers/mm	Patients with painful polyneuropathy and normal NCS
QST: CDT, CP, HDT, HP foot (Gibbons et al., 2010)	67–89%	29–72%	NDS > 2	DM patients and healthy controls
IENFD (Smith, Lessard, Reyna, Doudova, & Singleton, 2014)	63%	63%	Utah early neuropathy scale $\geq$ 4	Healthy controls and suspected distal neuropathy
IENFD (England et al., 2009)	45–90%	95–97%	Unknown: American Academy of Neurology evidence-based review	Healthy controls and established polyneuropathy
LDI flare (Ebadi et al., 2012)	54%	54%	IENFD < 5.4 fibers/mm	Patients with painful polyneuropathy and normal NCS
LDI flare (Vas & Rayman, 2013b)	75–77%	85–90%	NDS $\geq$ 3	Healthy controls and DM patients
Sudoscan (Casellini, Parson, Richardson, Nevoret, & Vinik, 2013)	78%	92%	NIS-LL	Healthy controls and DM patients
Sudoscan (Selvarajah et al., 2015)	87.5%	80.0%	Toronto Consensus DPN definition	Healthy controls and T1D patients

CDT, cooling detection threshold; *CP*, cold pain; *EMG*, electromyography; *HDT*, heat detection threshold; *HP*, heat pain; *LDI flare*, laser Doppler flare imaging; *NIS-LL*, Neuropathy Impairment Score of the Lower Limbs.

4.2 Clinical Assessment Tools

Clinical assessment should be standardized and conducted using validated, sufficiently reproducible, scores for both the severity of symptoms and the degree of neuropathic deficits. Tools include the Michigan Neuropathy Screening Instrument (MNSI) (Feldman et al., 1994); the Neuropathy Symptom Score (NSS) for neuropathic symptoms and the Neuropathy Disability Score (NDS) or the Neuropathy Impairment Score (NIS) for neuropathic deficits. These questionnaires are useful for patient follow-up and to assess response to treatment. The neurological history and examination should be performed initially and then with all subsequent visits. Minimum criteria for the clinical diagnosis of neuropathy according to the NSS and NIS are: (1) moderate signs with or without symptoms or (2) mild signs with moderate symptoms. However, this means that the exclusive presence of neuropathic symptoms without deficits is not sufficient to diagnose DSPN. Therefore, early stages of DSPN or a painful small-fiber neuropathy with or without minimal deficits can only be verified using more sophisticated tests such as quantitative sensory testing (QST). QST enables more accurate assessment of sensory deficits, including those related to small-fiber function, by applying controlled and quantified stimuli, and standardized procedures. Multiple studies have proven the value of QST measures in the detection of subclinical small-fiber neuropathy, the assessment of progression of neuropathy, and the prediction of risk of foot ulceration (Abbott, Vileikyte, Williamson, Carrington, & Boulton, 1998; Dyck, Dyck, Larson, O'Brien, & Velosa, 2000; Vinik, Suwanwalaikorn, et al., 1995; Yarnitsky & Sprecher, 1994). QST standardized measures of vibration and thermal perception thresholds are probably an effective tool in diabetic neuropathy (Shy et al., 2003) and also play an important role in multicenter clinical trials as primary efficacy endpoints.

4.3 Objective Measurements

4.3.1 Nerve Conduction Studies

An atypical pattern of presentation of symptoms or signs, such as the presence of relevant motor deficits, an asymmetrical or proximal distribution, or rapid progression, always requires referral for electrodiagnostic testing. Furthermore, in the presence of such atypical neuropathic signs and symptoms other forms of neuropathy should be sought and excluded. After diagnosis, slowing of NCV generally progresses at a steady rate of approximately 1 m/s/year and the level of impairment is positively correlated with duration of diabetes. Although most studies have documented that symptomatic patients are more likely to have slower NCVs than patients without symptoms, these do not relate with the severity of symptoms. In a long-term follow-up study of type 2 diabetes patients (Partanen et al., 1995), electrophysiologic abnormalities in the lower limb increased from 8% at baseline to 42% after 10 years with a decrease in sensory and motor amplitudes (indicating axonal destruction) being more pronounced than NCV slowing. Using objective measures of sensory function such as the vibration perception threshold test, the rate of decline in function has been reported as 1–2 vibration units/year. However, the most recent clinical studies suggest that there is a decline in this rate of progression. The advent of therapeutic lifestyle change and the use of statins and ACE inhibitors likely contribute to slowed progression of neuropathy and have drastically changed the requirements for placebo-controlled studies (Casellini et al., 2007). It is also important to recognize that diabetic neuropathy is a disorder wherein the prevailing abnormality is loss of nerve fibers and that this is reflected by a reduction in electromyogram amplitudes, not conduction velocity. Therefore, changes in NCV may not be an appropriate means of monitoring progress or deterioration of nerve function. Furthermore, small, unmyelinated nerve fibers are affected early in diabetes and are not reflected in NCV studies. Other methods of measuring neuropathy that do not depend on NCV, such as QST (see earlier), or skin biopsy with quantification of intraepidermal nerve fibers (IENF), are necessary to identify these patients (Pittenger et al., 2004; Pittenger, Simmons, et al., 2005; Sinnreich et al., 2005).

4.3.2 Skin Biopsy

The importance of the skin biopsy as a diagnostic tool for DSPN is increasingly being recognized as it allows investigation of small caliber sensory neurons including somatic unmyelinated IENF, dermal myelinated nerve fibers, and autonomic nerve fibers (Lauria, Bakkers, et al., 2010; Lauria, Hsieh, et al., 2010; Pittenger, Mehrabyan, et al., 2005). Since small fibers are impacted early in the course of diabetic neuropathy (Vinik, Casellini, & Nevoret, 2015), it is imperative to catch onset prior to further progression, when treatment and reversibility may be limited. For diagnostic purposes a 3-mm punch skin biopsy at the distal leg is recommended and quantification of the linear density of IENF in at least three 50 μm thick sections per biopsy. Tissue is viewed by bright-field immunohistochemistry or immunofluorescence using protein gene product 9.5 antibodies (Joint Task Force of the EFNS and the PNS, 2010). Quantification of IENF density appeared more sensitive than sensory nerve conduction or sural-nerve biopsy in diagnosing small-fiber neuropathy. Use of skin biopsies has also led to the recognition of the small nerve fiber syndrome as part of IGT and the metabolic syndrome so that when patients present with the "burning foot or hand syndrome," an evaluation for glucose tolerance and the metabolic syndrome becomes mandatory. Therapeutic life style changes can result in nerve fiber regeneration, reversal of the neuropathy, and alleviation of symptoms (Smith et al., 2006).

4.3.3 Corneal Confocal Microscopy

Corneal confocal microscopy (CCM) is a noninvasive technique used to detect small nerve fiber loss in the cornea, which correlates with both increasing neuropathic severity and reduced IENF in diabetic patients (Quattrini et al., 2007; Tavakoli et al., 2010). This novel technique of real-time mapping permits an area of 3.2 mm^2 to be mapped with a total of 64 theoretically nonoverlapping single 400 μm^2 images (Zhivov, Blum, Guthoff, & Stachs, 2010). In type 1 diabetes subjects who received a pancreas transplantation, CCM was able detect nerve regrowth in the cornea before any improvement of IENF (Tavakoli et al., 2013). However, it is as yet unclear whether improvements in CCM parameters predict improvements in peripheral neuropathy or patient-oriented outcomes such as pain, disability, and QOL. Furthermore, the performance of CCM as an early biomarker of nerve regeneration in patients with type 2 diabetes receiving other interventions has not yet been demonstrated (Shtein & Callaghan, 2013).

4.3.4 Contact Heat Evoked Potentials

Contact heat-evoked potentials (CHEPS) have been studied in healthy controls, newly diagnosed diabetic patients, established diabetic patients, and patients with the metabolic syndrome. It appears that CHEPS is capable of detecting small-fiber neuropathy in the absence of other indices and that CHEPS correlates with quantitative sensory perception and objective tests of small-fiber function such as the cooling detection threshold and cold pain (Parson, Nguyen, Boyd, & Vinik, 2009). CHEPS is therefore a useful diagnostic tool for the evaluation of small nerve fiber function in neuropathic patients (Chen, Niddam, & rendt-Nielsen, 2001; Itskovich, Fei, & Harkins, 2000; Opsommer, Masquelier, & Plaghki, 1999). In the past, laser-evoked potentials were used but left undesirable thermal damage to intact nerve fibers and surrounding healthy tissue. Long-term assessment using this modality is not practical for quality patient care (Bromm & Treede, 1991). CHEPS, with its rapidly heating thermode (70°C/s), allows for repeated assessments without the risk of long-term damage to the area of interest. Studies supporting reproducibility of CHEPS after 6 months (Ruscheweyh, Emptmeyer, Putzer, Kropp, & Marziniak, 2013), especially in areas of interest (Kramer et al., 2012), lend further support for its use in repeated clinical assessments.

Various groups, including ours, have performed comparison studies using CHEPS in neuropathic populations, and CHEPS intrapeak amplitudes (IA) have proven to be a distinguishable marker differentiating neuropaths from nonneuropaths (Atherton et al., 2007; Chao et al., 2010; Parson et al., 2013; Wong & Chung, 2011). CHEPS IA also correlates with other assessments of small nerve fibers, including skin-flare response and IENF density (Atherton et al., 2007; Chao, Hsieh, Tseng, Chang, & Hsieh, 2008; Chao et al., 2010). Assessment of structural changes may miss early nerve dysfunction in the presence of normal IENF density. However, CHEPS provides a sensitive (76.4%) and specific (80.6%) tool in assessing nerve fiber loss in the dorsal foot with IENF as a standard for comparison, while providing a functional evaluation (Casanova-Molla, Grau-Junyent, Morales, & Valls-Sole, 2011).

In a comparison of diabetic and nondiabetic patients, CHEPS IAs of the lumbosacral region, dorsal, and volar forearm negatively correlated with HbA1c (Parson et al., 2013). Diabetic neuropaths also tend to have CHEPS responses of longer latencies and lower amplitudes (Parson et al., 2013) (Chao et al., 2010). Upon closer examination of patients with DSPN, CHEPS responses tend to be higher in those with pain than those without (Parson et al., 2013), which has led to further investigation of the tool for studying pain models and assessing intervention.

The versatility of CHEPS to study function in conjunction with other devices, such as fMRI and magnetoencephalography, further adds to its value in providing a more global evaluation of future therapeutics and broadening our understanding of the mechanism of pain (Gopalakrishnan, Machado, Burgess, & Mosher, 2013; Kramer, Jutzeler, Haefeli, Curt, & Freund, 2015; Shenoy et al., 2011). Thus, there is a great need to create a standard for CHEPS testing and normative values. Endeavors by Dutch (Lagerburg et al., 2015) and Taiwanese groups (Chen et al., 2006) to create such values should further accelerate its validation in diabetic neuropathy and its various subtypes. However, future normative values must be obtained from a more universally diverse cohort to truly create a robust standard with meaningful clinical significance.

4.3.5 LDI Flare

Vasodilation induced by both acetylcholine iontophoresis (ACh Ionto) and the laser Doppler flare technique (LDIFT) can be used to measure small-fiber function. When local anesthetic is applied, axonal transport but not ACh Ionto is inhibited, thus hyperemic flow induced by Ach Ionto is not neurogenic. By contrast, the hyperemia induced by heating the skin to 44°C is blocked by local anesthetic and therefore reflects C-fiber function. Krishnan and Rayman demonstrated that the ratio of hyperemic area to stimulus area was significantly reduced by local anesthesia in the LDIFT [ratio (mean ± SD): 2.33 ± 0.67 with local anesthesia; 6.84 ± 1.33 without local anesthesia; $P < 0.0001$], but not in the ACh Ionto group (2.61 ± 0.57 with local anesthesia; 2.67 ± 1.27 without local anesthesia), confirming that LDIFT measures small-fiber function (Green, Krishnan, & Rayman, 2009). The same authors showed that this measure of C-fiber function was more sensitive than QST using the Case IV device (Krishnan & Rayman, 2004).

A modified laser Doppler imager method (mLDIf) uses 47°C (vs 44°C), producing larger flares than the older method for assessing C-fiber function in foot skin and making it quicker and better suited for clinical use. It has also been validated for assessing C-fiber function in the clinical setting (Vas & Rayman, 2013a). Rayman et al. evaluated a rapid, low-cost, point-of-care nerve conduction device (POCD; NC-stat®|DPNCheck™) for the assessment of DPN and compared it with mLDIf in 162 patients with diabetes and 80 healthy controls. SNCV measured with the POCD correlated significantly with mLDIf in both control and diabetic subjects ($r < 0.90$ and $r = 0.78$, respectively) as did SNAP ($r = 0.88$ and $r = 0.73$, respectively) (Sharma, Vas, & Rayman, 2015). Sharma and Rayman also presented the results of a longitudinal 1-year study of the technique in type 1 and 2 diabetes at Neurodiab 2014 (Sopron, Hungary—September 24, 2014) (Tucker, 2014). At year 1, the LDI-flare size compared to baseline was reduced by 6.8% in the type 1 diabetes patients and by 5.9% in the type 2 patients, but just 0.84% among the healthy controls ($P < 0.001$ for both diabetes groups vs controls). Among the type 1 diabetes patients, the percent reduction was significantly greater among the 56 with microangiopathy compared with the 24 without microangiopathy (8.7% vs 4.6%; $P < 0.0001$). However, there was no significant difference between the 49 type 2 diabetes patients with microangiopathy and the 33 without (4.9% vs 6.0%; $P = 0.06$). The differences in findings between the type 1 and type 2 diabetes patients can be explained on the grounds of differences in the pathogenesis of small-fiber dysfunction between the two groups. As Sharma commented, "While type 2 diabetes is more related to metabolic syndrome, where insulin resistance plays an important part, the same cannot be said of type 1, where glycemic control plays an important role in the genesis of small-fiber neuropathy." There was a highly significant correlation between triglycerides and LDI-flare size in both the controls and the diabetes patients (both types) at baseline ($P=0.008$ and 0.001, respectively) and at year 1 ($P=0.01$ and 0.004, respectively). The change in triglycerides also correlated with the change in LDI-flare size ($P=0.008$ and 0.009, respectively), supporting increasing evidence to suggest that triglycerides play an important role in the pathogenesis of diabetic polyneuropathy. This relationship is so far shown to be independent of glycemic control in various cross-sectional studies. Change in small-fiber function as measured by LDI flare did not correlate with BMI, other lipid indices, or blood pressure. Neuropathy disability scores, sural-nerve amplitude, and sural-nerve conduction velocity did not change over the study year, nor did they correlate with changes in HbA_{1c}, lipids, or blood pressure suggesting that the technique may have distinct advantages in evaluating prediabetic neurovascular dysfunction.

Neurovascular dysfunction studies conducted by ourselves in lean controls, obese controls, relatives of patients with IGT, IGT, and type 2 diabetes show that skin perfusion response to occlusion and raising the ambient temperature from 32°C to 44°C may be the single most sensitive tool for detecting abnormalities in IGT and obesity but even more so in relatives of patients with diabetes (Vinik, Nevoret, Casellini, & Parson, 2013) (Fig. 3). A study comparing new techniques of quantification of small-fiber structure and function with LDI flare would be of great interest.

4.3.6 Assessment of Autonomic Neuropathy

Since many conditions affect the ANS, and autonomic neuropathy is not unique to diabetes, the diagnosis of DAN rests with establishing the diagnosis and excluding other causes (Table 4). The best-studied diagnostic methods, for which there are large databases and evidence to support their use in clinical practice, relate to the evaluation of cardiovascular reflexes. In addition, the evaluation of orthostasis is fairly straightforward and is readily done in clinical practice, as is the establishment of the cause of gastrointestinal symptoms and erectile dysfunction. The combination of cardiovascular autonomic tests with sudomotor function tests may allow a more accurate diagnosis of DAN (England et al., 2009; Vinik et al., 2013). Table 5 presents the diagnostic tests that would be applicable to the diagnosis of CAN. These

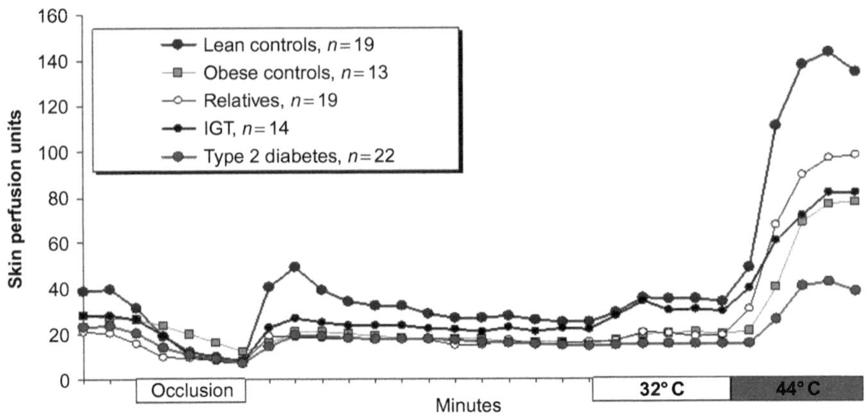

Fig. 3 Abnormalities in heat-mediated vasodilation of hairy skin in obese controls, relatives of subjects with diabetes, impaired glucose tolerance (*IGT*), and type 2 diabetes subjects, as compared with lean controls (Vinik et al., 2013).

Table 4 Differential Diagnosis of Diabetic Autonomic Neuropathy

Clinical Manifestations	Differential Diagnosis
Cardiovascular	Cardiovascular disorders
Resting tachycardia, exercise intolerance	Idiopathic orthostatic hypotension, multiple system atrophy with Parkinsonism, orthostatic tachycardia, hyperadrenergic hypotension
Orthostatic tachycardia and bradycardia syndromes	Shy-Drager syndrome
Cardiac denervation, painless myocardial infarction	Panhypopituitarism
Orthostatic hypotension	Pheochromocytoma
Intraoperative and perioperative cardiovascular instability	Hypovolemia
	Congestive heart disease
	Carcinoid syndrome

tests can be used as a surrogate for the diagnosis of DAN in any system since it is generally rare to find involvement of any other division of the ANS in the absence of cardiovascular autonomic dysfunction. For example, if one entertains the possibility that the patient has erectile dysfunction due to DAN, then prior to embarking upon a sophisticated and expensive evaluation of erectile status, a measure of heart rate and its variability in response to deep breathing would—if normal—exclude the likelihood that the erectile dysfunction is a consequence of disease of the ANS. The cause thereof would have to be sought elsewhere. Similarly it is extremely unusual to find gastroparesis secondary to DAN in a patient with normal cardiovascular autonomic reflexes.

Table 5 Diagnostic Tests of Cardiovascular Autonomic Neuropathy

Test	Method/Parameters
Time and frequency evaluation of resting heart rate beat-to-beat heart rate variation[a] and response to deep breathing	>100 Beats/min is abnormal. With the patient at rest and supine (no overnight coffee or hypoglycemic episodes), breathing 6 breaths/min, heart rate monitored by EKG or ANSCORE device, a difference in heart rate of >15 beats/min is normal and <10 beats/min is abnormal, R-R inspiration/R-R expiration >1.17. All indices of HRV are age dependent[b]. SDNN and RMSSD.
Heart rate response to standing[a]	During continuous EKG monitoring, the R-R interval is measured at beats 15 and 30 after standing. Normally, a tachycardia is followed by reflex bradycardia. The 30:15 ratio is normally >1.03.
Heart rate response to Valsalva maneuver[a]	The subject forcibly exhales into the mouthpiece of a manometer to 40 mmHg for 15 s during EKG monitoring. Healthy subjects develop tachycardia and peripheral vasoconstriction during strain and an overshoot bradycardia and rise in blood pressure with release. The ratio of longest R-R shortest R-R should be >1.2.
Spectral analysis of heart rate variation, very low-frequency power (VLFP 0.003–0.04), and high-frequency power (HFP 0.15–0.40 Hz)	Series of sequential R-R intervals into its various frequent components. It defines two fixed spectral regions for the low-frequency and high-frequency measure.
Systolic blood pressure response to standing	Systolic blood pressure is measured in the supine subject. The patient stands and the systolic blood pressure is measured after 2 min. Normal response is a fall of <10 mmHg, borderline is a fall of 10–29 mmHg, and abnormal is a fall of >30 mmHg with symptoms.
Diastolic blood pressure response to isometric exercise	The subject squeezes a handgrip dynamometer to establish a maximum. Grip is then squeezed at 30% maximum for 5 min. The normal response for diastolic blood pressure is a rise of >16 mmHg in the other arm.

Continued

Table 5 Diagnostic Tests of Cardiovascular Autonomic Neuropathy—cont'd

Test	Method/Parameters
EKG QT/QTc intervals spectral analysis with respiratory frequency	The QTc (corrected QT interval on EKG) should be <440 ms. VLF peak (sympathetic dysfunction) LF peak (sympathetic dysfunction) HF peak (parasympathetic dysfunction) LH/HF ratio (sympathetic imbalance).
Neurovascular flow	Using noninvasive laser Doppler measures of peripheral sympathetic responses to nociception.
Sudorimetry	Use of Sudoscan to evaluate the electrochemical skin conductance response to low voltage stimulation applied to the skin of the volar surfaces of the hands and feet measured as electrical skin conductance (ESC)

[a]These can now be performed quickly (<15 min) in the practitioners' office, with a central reference laboratory providing quality control and normative values. *LF, VLF, HF*—low-, very low-, and high-frequency peaks on spectral analysis. These are now readily available in most cardiologist's practice.
[b]Lowest normal value of E/I ratio: Age 20–24:1.17, 25–29:1.15, 30–34:1.13, 35–30:1.12, 40–44:1.10, 45–49:1.08, 50–54:1.07, 55–59:1.06, 60–64:1.04, 65–69:1.03, and 70–75:1.02.

4.3.7 Sudorimetry

Small unmyelinated C- and Aδ-fibers in the periphery subserve somatic and autonomic nerve functions such as warm, cold, and pain perception as well as innervating sweat glands. The gold standard for their evaluation has been skin biopsy and quantification of IENF density, an invasive procedure (Pittenger et al., 2004; Vinik et al., 2013). On the other hand, the evaluation of sweating has the appeal of quantifiable noninvasive determination of the integrity of the ANS. Sudomotor nerves are thin unmyelinated C-fibers that are primarily cholinergic at the ganglion although, epinephrine, norepinephrine, vasoactive intestinal peptide, atrial natriuretic peptide, calcitonin gene-related polypeptide, galanin, ATP, and substance P have also been localized to periglandular nerves (Freeman & Chapleau, 2013). A variety of techniques have capitalized on quantifying sweating (Illigens & Gibbons, 2009) or the innervation of sweat glands (Gibbons, Illigens, Wang, & Freeman, 2009; Wang & Gibbons, 2013) to capture the impact of disordered regulation on structure and function of sweat glands as surrogates for the underlying autonomic dysfunction.

Sudorimetry, or sudomotor function testing, is the science of measuring the function of sweat gland innervation and is unique among the autonomic function tests in that it evaluates the peripheral sympathetic system but relies principally on cholinergic postganglionic neurotransmission. The Neuropad is an indicator test for the detection of sudomotor dysfunction that assesses plantar sweat production by means of a color change from blue to pink. The patch contains the complex salt anhydrous cobalt-II-chloride. In the presence of water, this salt absorbs water molecules, normally changing its color from blue to pink. If the patch remains completely or partially blue within 10 min, the result is considered abnormal (Papanas et al., 2005).

Recently, a newer sudomotor function technology has become available and adds a new tool at point of care when autonomic testing had been previously restricted to specialized neurological laboratories. The Sudoscan®, an instrument capable of detecting chloride ion flux in response to a very low current, is an objective, noninvasive, and quantitative sudomotor function test with promising sensitivity and specificity in the investigation of small-fiber function in DSPN (Casellini et al., 2013; Gin, Baudoin, Raffaitin, Rigalleau, & Gonzalez, 2011; Smith, Lessard, Reyna, Doudova, & Singleton, 2014; Yajnik, Kantikar, Pande, & Deslypere, 2012) and has shown correlation to the severity of pain symptoms (Casellini et al., 2013; Selvarajah et al., 2015; Smith et al., 2014). The entire evaluation takes only 2 min and can be done in an ambulatory setting. The unique qualities of sudomotor function tests are that they may yield diagnostic information not only of autonomic dysfunction but also enhance the assessment of the small somatosensory nerves (Boger et al., 2012; Thaisetthawatkul, Fernandes Filho, & Herrmann, 2013). An excellent review on sudomotor function tests can be read elsewhere (Illigens & Gibbons, 2009).

Sudoscan™ technology is based on different electrochemical principles (reverse iontophoresis and chronoamperometry) to measure sudomotor function than prior technologies, affording it a much more practical and precise performance profile for routine clinical use with potential as a research tool. The device consists of a desktop computer connected to two pairs of large surface stainless steel electrodes: one pair for application of the palms and the other for the soles. The subject places both hands and feet simultaneously on the designated electrodes and a painless scanning process ensues over 2–3 min, during which a low DC voltage (1–4 V) is incrementally applied to the electrodes, with the left and right electrodes serving alternatively as cathode and anode. At voltages under 10 V, the stratum corneum is electrically insulating whereas the sweat glands consist of a cellular bilayer and therefore can transmit electrically charged ions to the skin surface and the electrodes by reverse iontophoresis. The current of sweat-derived chloride ions generated in response to the incremental voltage applied can be quantified and reflects sweat gland function and, by inference, functional innervation. This chloride ion current is reported as electrochemical skin conductance (ESC)—the inverse of resistance—measured in microSiemens (μS). The palms and soles are assessed due to their very high density of sweat glands and the frequency of small-nerve degeneration. By alternating the left and right extremities as anode and cathode, each extremity is evaluated separately during the 3 min scan, and asymmetry between the extremities can be calculated.

4.3.7.1 Clinical Diagnostic Evaluation Using Sudomotor Function

The clinical diagnostic accuracy of Sudoscan™ has been evaluated for DSPN, peripheral small-fiber neuropathy, and autonomic neuropathy against other standard methods. Casellini et al. (2013) examined 83 type 1 and 2 diabetic patients, both with and without DSPN, and compared them to 210 healthy controls. Diabetic patients with DSPN had significantly worse ESCs of both feet and hands than patients without DSPN or healthy controls. Somewhat surprisingly, ESCs correlated significantly with clinical neuropathy scores of sensory, motor, and reflex function, somatic (QST and NCV studies) and autonomic (quantitative autonomic function testing) measures of DSPN suggesting that the measure captures both somatic and autonomic nerve function. Using a Neuropathy Impairment Score of the Lower Limb (NIS-LL) > 2 as the gold standard for the diagnosis of neuropathy, the sensitivity and specificity of ESC measured in the feet for the diagnosis of DSPN were 78% and 92%, respectively (Fig. 4). The investigators further determined correlations between sudorimetry and somatic and autonomic function and found strong correlations between ESC and both somatic and autonomic functions with sudorimetry scores being able to separate painful from nonpainful DSPN.

The value of ESC has been confirmed by others. Smith et al. (2014) studied 55 patients with suspected idiopathic or diabetes-related small-fiber neuropathy and 42 controls to evaluate the diagnostic performance of ESC. Using an abnormal Utah Early Neuropathy Score (UENS ≥ 4) as the diagnostic gold standard, ESC of feet and hands were significantly lower than those of controls, regardless of the etiology of small-fiber neuropathy. More importantly, ESC and IENF density had similar diagnostic performance in detecting small-fiber neuropathy. Mirroring the findings of Casellini et al.

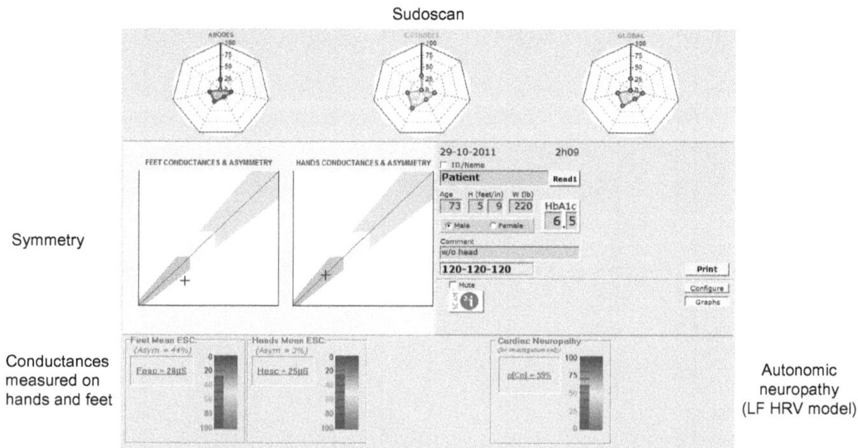

Fig. 4 Display of electrochemical skin conductance (ESC) results for the assessment of sudomotor function. Numeric ESC scores are provided in microSiemens and visually displayed on a scale of normative ranges. Asymmetry represents the percent difference in ESC measures between left and right limbs. The cardiac neuropathy risk score (pCN) is calculated from the measured ESC and patient characteristics.

(2013), ESC in the feet had a very high negative predictive value (83%), and significantly correlated with both symptoms and signs assessed using the MNSI and the Utah early neuropathy symptom scale. ESC measurement may also be a simple tool for early identification of peripheral autonomic neuropathy and useful in screening for subclinical CAN. Some of the major studies completed using ESC measurement in a clinical diagnostic capacity are summarized in Table 6.

4.3.7.2 Validation of ESC Robustness

For ESC measurement to be clinically meaningful, widely applicable thresholds for normal and abnormal sudomotor function must be determined and high reproducibility of results demonstrated. Initially, a large cohort of healthy subjects (432 females, 177 males) indicated that there was almost no drop-off in normal sudomotor function from age 21 to 80 and no sexual dimorphism. However, data from population studies in different countries indicate that normative values for sudomotor function as quantified by ESC varies by ethnicity so that diagnostic thresholds are needed to address these differences (Table 7). Reproducibility of testing has also been performed under various stress scenarios including hyperglycemia, CVD, and pre- and postexercise and coefficients of variation for ESC ranged from 7% to 15% for hands and 4–10% for feet ESC. Comparison between tests

Table 6 Feet ESC Diagnostic Accuracy vs Established Clinical Tools

Study	Diagnostic Variable	Comparison	Sensitivity (%)	Specificity (%)	PPV (%)	NPV (%)
Casellini et al. (2013)	DPN	NIS-LL	78.34	92.38	74.6	93.72
Selvarajah et al. (2015)	DPN	AAN guidelines for DPN	87.5	76.2	N/A	N/A
Selvarajah et al. (2015)	CAN—diabetes	5 Cardiac autonomic reflex tests	65	85	N/A	N/A
Smith et al. (2014)	Distal symmetric polyneuropathy	UENS	77	67	59	83
Yajnik et al. (2012)	DPN	VPT	73	62	N/A	N/A
Eranki et al. (2013)	DPN	VPT $\geq$ 15 V	82	55	N/A	N/A
Yajnik et al. (2013)	CAN—diabetes	3 Cardiac autonomic reflex tests	92	49	N/A	N/A
Ozaki et al. (2011)	Diabetic kidney disease	eGFR, ACR	94	78	81	93

ACR, albumin creatinine ratio; *CAN*, cardiac autonomic neuropathy; *CARTs*, cardiac autonomic reflex tests; *DPN*, diabetic peripheral neuropathy; *eGFR*, estimated glomerular filtration rate; *NIS-LL*, neuropathy impairment score-lower limbs; *PPV*, positive predictive value; *NPV*, negative predictive value; *UENS*, Utah Early Neuropathy Score; *VPT*, vibration perception threshold.

Table 7 Electrochemical Skin Conductance (ESC) Zones for Normal Sudomotor Function (Green), Moderate Sudomotor Dysfunction (Yellow), and Severe Sudomotor Dysfunction (Red) as Defined by Racial Background

Race	Hands ESC Red Zone	Hands ESC Yellow Zone	Hands ESC Green Zone	Feet ESC Red Zone	Feet ESC Yellow Zone	Feet ESC Green Zone
Caucasian	0–0	40–60	60–100	0–50	50–70	70–100
African American Ancestry	0–30	30–50	50–100	0–40	40–60	60–100
Other (Asian, Indian)	0–40	40–60	60–100	0–40	40–60	60–100

All numeric values in microSiemens (μS). Reproducibility testing was performed under various stress scenarios as well as normal conditions including hyperglycemia, cardiovascular disease, and pre- and post-exercise; coefficients of variation ranged from 7% to 15% for hands ESC and 4% to 10% for feet ESC. Comparison between tests performed twice on three different devices demonstrated that for feet ESC the mean coefficient of variation for repeatability was 2.8 ± 1.6% in healthy volunteers (HV) and 6.9 ± 6.3% in patients with type 2 DM (PDT2), while coefficient of variation for reproducibility were 3.1 ± 1.5% and 6.9 ± 6.3%, respectively (submitted for publication).

Data from Casellini, C. M., Parson, H. K., Richardson, M. S., Nevoret, M. L., & Vinik, A. I. (2013). SUDOSCAN, a noninvasive tool for detecting diabetic small fiber neuropathy and autonomic dysfunction. *Diabetes Technology & Therapeutics, 15,* 948–953; Smith, A. G., Lessard, M., Reyna, S., Doudova, M., & Singleton, J. R. (2014). The diagnostic utility of SUDOSCAN for distal symmetric peripheral neuropathy. *Journal of Diabetes and Its Complications, 28,* 511–516; Freedman, B. I., Smith, S. C., Bagwell, B. M., Xu, J., Bowden, D. W., & Divers, J. (2015). Electrochemical skin conductance in diabetic kidney disease. *American Journal of Nephrology 41,* 438–447.

performed twice on three different devices demonstrated that, for feet ESC, the mean coefficient of variation for repeatability was 2.8 ± 1.6% in healthy volunteers and 6.9 ± 6.3% in patients with type 2 diabetes while coefficient of variation for reproducibility were 3.1 ± 1.5% and 6.9 ± 6.3%, respectively (submitted for publication).

4.3.7.3 Sudomotor Function as a Tool to Measure Progression and Regression of Disease

One of the most difficult and frustrating tasks in the study of peripheral neuropathy has been to measure disease progression and response to therapeutic intervention with an easy, objective, and accurate tool. Several research studies have now demonstrated that ESC may be a clinically meaningful tool in measuring neuropathy progression and regression. In a cohort of 52 type 1 and 115 type 2 diabetic patients followed for 1 year with no change in their glycemic control, type 2 patients treated without insulin showed a decrease in hand and foot ESC, while a significant increase was observed in those

Table 8 Change Between Baseline and 12-Month Follow-Up in Weight, Waist, VO_2max, and ESC Risk Score Value in 154 Women Included in a Lifestyle Intervention Program (Raisanen, Eklund, Calvet, & Tuomilehto, 2014)

	Without Follow-Up of Training Level ($n = 72$)		Low Weekly Activity[a] ($n = 62$)		High Weekly Activity[b] ($n = 20$)		
	Mean	**SD**	**Mean**	**SD**	**Mean**	**SD**	p
Change in weight (kg)	−0.9	3.4	−1.6	4.0	−3.3	4.8	NS
Change in waist (cm)	−2.1	4.7	−2.4	4.5	−3.6	5.9	NS
Change in estimated VO_2max (METs)	+0.5	0.9	+0.8	0.9	+1.1	1.2	NS
Change in hand ESC (μS)	+5.0	8.4	+3.0	9.4	+8.4	12.3	0.043
Change in foot ESC (μS)	+5.6	8.9	+4.9	8.9	+10.8	12.8	0.024
Change in ESC risk score (%)	−5.1	5.3	−4.7	6.4	−8.5	6.8	0.027

[a]Less than 150 min of moderate activity and 75 min of high activity.
[b]More than 150 min of moderate activity or 75 min of high activity; moderate activity 3–7 METs, high activity > 7 METs.

receiving insulin. Type 1 patients taking insulin therapy also had a nonsignificant increase in their feet ESC (Calvet, Dupin, Winiecki, & Schwarz, 2013). Raisanen et al. (2014) demonstrated that a 12-month lifestyle intervention could improve small-fiber function in metabolic syndrome. Among 154 female participants with the lowest fitness level at baseline, those performing the highest level of weekly activity showed the greatest improvement in ESC, which was more pronounced than the changes in weight, waist circumference, or VO_2max (Table 8). These, and other ongoing studies, indicate that ESC may provide information on the status of small nerve fiber function that may not otherwise be known to or measurable by the physician.

5. CONCLUDING REMARKS

Diabetic neuropathy is the most common complication of diabetes and contributes additional risks in the aging adult. For eons, the focus has been on pain and loss of sensory perception leading to development of foot ulcers and amputations. There is now recognition that a broader view of neuropathy needs to be taken. Loss of sensory perception, loss of muscle strength, and ataxia or incoordination lead to an increased risk of falling

17-fold greater in the older diabetic than their young nondiabetic counterparts. A fall is accompanied by lacerations, tears, fractures and, worst of all, traumatic brain injury from which more than 60% do not recover. There clearly is an increased need for early recognition of people at risk and to embark on preventive measures. Autonomic neuropathy has been hailed as the "Prophet of Doom" (Vinik et al., 2011) for good reason. It is conducive to increased risk of MI and sudden death. It is therefore critical that we identify people with cardiac autonomic neuropathy and embark on preventive strategies. It also has been recognized that an imbalance in the ANS occurs early in the evolution of diabetes at a stage when active intervention can abrogate the relentless progression. Many new and emerging syndromes can be attributed to CAN such as orthostatic tachycardia and bradycardia in addition to hypotension. While pain is a common and debilitating accompaniment of neuropathy, providers have failed to recognize the importance of the accompanying sleep deprivation, anxiety, and depression which, if unattended, compromise appropriate management of pain. Ultimately the constellation of features of neuropathy conspires to reduce QOL and ADL. Thus, meticulous evaluation of DSPN and DAN in the aging diabetic patient at diagnosis and then on a yearly basis is critical, in addition to falls risk assessment and determination of autonomic integrity. Early neuropathy detection can only be achieved by assessment of small-nerve fibers. New noninvasive sudomotor function technologies may play an increasing role in identifying early peripheral and autonomic neuropathy, allowing rapid intervention and potentially prevention or reversal of small-fiber loss.

REFERENCES

Abbott, C. A., Carrington, A. L., Ashe, H., Bath, S., Every, L. C., Griffiths, J., et al. (2002). The North-West Diabetes Foot Care Study: Incidence of, and risk factors for, new diabetic foot ulceration in a community-based patient cohort. *Diabetic Medicine*, *19*, 377–384.

Abbott, C. A., Vileikyte, L., Williamson, S., Carrington, A. L., & Boulton, A. J. (1998). Multicenter study of the incidence of and predictive risk factors for diabetic neuropathic foot ulceration. *Diabetes Care*, *21*, 1071–1075.

Airaksinen, K. E. J., & Koistinen, M. J. (1992). Association between silent coronary artery disease, diabetes, and autonomic neuropathy: Fact or fallacy? *Diabetes Care*, *15*, 288–292.

Albers, A. R., Krichavsky, M. Z., & Balady, G. J. (2006). Stress testing in patients with diabetes mellitus: Diagnostic and prognostic value. *Circulation*, *113*, 583–592.

Allet, L., Armand, S., de Bie, R. A., Golay, A., Monnin, D., Aminian, K., et al. (2010). The gait and balance of patients with diabetes can be improved: A randomised controlled trial. *Diabetologia*, *53*, 458–466.

Allet, L., Armand, S., de Bie, R. A., Golay, A., Pataky, Z., Aminian, K., et al. (2009). Clinical factors associated with gait alterations in diabetic patients. *Diabetic Medicine*, *26*, 1003–1009.

American Diabetes Association (2015). Standards of medical care in diabetes—2015. *Diabetes Care, 38*(Suppl. S1). http://dx.doi.org/10.2337/dc15-S003.

Armstrong, D. G., Lavery, L. A., & Harkless, L. B. (1998a). Validation of a diabetic wound classification system. The contribution of depth, infection, and ischemia to risk of amputation. *Diabetes Care, 21*, 855–859.

Armstrong, D. G., Lavery, L. A., Vela, S. A., Quebedeaux, T. L., & Fleischli, J. G. (1998b). Choosing a practical screening instrument to identify patients at risk for diabetic foot ulceration. *Archives of Internal Medicine, 158*, 289–292.

Atherton, D. D., Facer, P., Roberts, K. M., Misra, V. P., Chizh, B. A., Bountra, C., et al. (2007). Use of the novel contact heat evoked potential stimulator (CHEPS) for the assessment of small fibre neuropathy: Correlations with skin flare responses and intra-epidermal nerve fibre counts. *BMC Neurology, 7*, 21.

Athyros, V. G., Didangelos, T. P., Karamitsos, D. T., Papageorgiou, A. A., Boudoulas, H., & Kontopoulos, A. G. (1998). Long-term effect of converting enzyme inhibition on circadian sympathetic and parasympathetic modulation in patients with diabetic autonomic neuropathy. *Acta Cardiologica, 53*, 201–209.

Azad, N., Emanuele, N. V., Abraira, C., Henderson, W. G., Colwell, J., Levin, S. R., et al. (1999). The effects of intensive glycemic control on neuropathy in the VA cooperative study on type II diabetes mellitus (VA CSDM). *Journal of Diabetes and Its Complications, 13*, 307–313.

Balducci, S., Iacobellis, G., Parisi, L., Di, B. N., Calandriello, E., Leonetti, F., et al. (2006). Exercise training can modify the natural history of diabetic peripheral neuropathy. *Journal of Diabetes and Its Complications, 20*, 216–223.

Barnett, A., Smith, B., Lord, S. R., Williams, M., & Baumand, A. (2003). Community-based group exercise improves balance and reduces falls in at-risk older people: A randomised controlled trial. *Age and Ageing, 32*, 407–414.

Berlie, H. D., & Garwood, C. L. (2010). Diabetes medications related to an increased risk of falls and fall-related morbidity in the elderly. *The Annals of Pharmacotherapy, 44*, 712–717.

Boger, M. S., Hulgan, T., Haas, D. W., Mitchell, V., Smith, A. G., Singleton, J. R., et al. (2012). Measures of small-fiber neuropathy in HIV infection. *Autonomic Neuroscience, 169*, 56–61.

Boucher, P., Teasdale, N., Courtemanche, R., Bard, C., & Fleury, M. (1995). Postural stability in diabetic polyneuropathy. *Diabetes Care, 18*, 638–645.

Boulton, A. J., Gries, F. A., & Jervell, J. A. (1998). Guidelines for the diagnosis and outpatient management diabetic peripheral neuropathy. *Diabetic Medicine, 15*, 508–514.

Boulton, A. J., Malik, R. A., Arezzo, J. C., & Sosenko, J. M. (2004). Diabetic somatic neuropathies. *Diabetes Care, 27*, 1458–1486.

Boulton, A. J., Vinik, A. I., Arezzo, J. C., Bril, V., Feldman, E. L., Freeman, R., et al. (2005). Diabetic neuropathies: A statement by the american diabetes association. *Diabetes Care, 28*, 956–962.

Boyd, A., Casellini, C., Vinik, E., & Vinik, A. (2011). Quality of life and objective measures of diabetic neuropathy in a prospective placebo controlled trial of Ruboxistaurin and topiramate. *Journal of Diabetes Science and Technology, 5*, 714–722.

Boyd, R., & Stevens, J. A. (2009). Falls and fear of falling: Burden, beliefs and behaviours. *Age and Ageing, 38*, 423–428.

Bromm, B., & Treede, R. D. (1991). Laser-evoked cerebral potentials in the assessment of cutaneous pain sensitivity in normal subjects and patients. *Revue Neurologique (Paris), 147*, 625–643.

Burger, A. J., Weinrauch, L. A., D'Elia, J. A., & Aronson, D. (1999). Effects of glycemic control on heart rate variability in type I diabetic patients with cardiac autonomic neuropathy. *The American Journal of Cardiology, 84*, 687–691.

Buse, J. B., Bigger, J. T., Byington, R. P., Cooper, L. S., Cushman, W. C., Friedewald, W. T., et al. (2007). Action to Control Cardiovascular Risk In Diabetes (ACCORD) trial: Design and methods. *The American Journal of Cardiology*, *99*, 21i–33i.
Calvet, J. H., Dupin, J., Winiecki, H., & Schwarz, P. E. (2013). Assessment of small fiber neuropathy through a quick, simple and non invasive method in a German diabetes outpatient clinic. *Experimental and Clinical Endocrinology & Diabetes*, *121*, 80–83.
Canto, J. G., Shlipak, M. G., Rogers, W. J., Malmgren, J. A., Frederick, P. D., Lambrew, C. T., et al. (2000). Prevalence, clinical characteristics, and mortality among patients with myocardial infarction presenting without chest pain. *JAMA*, *283*, 3223–3229.
Caputo, G. M., Cavanagh, P. R., Ulbrecht, J. S., Gibbons, G. W., & Karchmer, A. W. (1994). Assessment and management of foot disease in patients with diabetes. *The New England Journal of Medicine*, *331*, 854–860.
Casanova-Molla, J., Grau-Junyent, J. M., Morales, M., & Valls-Sole, J. (2011). On the relationship between nociceptive evoked potentials and intraepidermal nerve fiber density in painful sensory polyneuropathies. *Pain*, *152*, 410–418.
Casellini, C. M., Barlow, P. M., Rice, A. L., Casey, M., Simmons, K., Pittenger, G., et al. (2007). A 6-month, randomized, double-masked, placebo-controlled study evaluating the effects of the protein kinase C-{beta} inhibitor Ruboxistaurin on skin microvascular blood flow and other measures of diabetic peripheral neuropathy. *Diabetes Care*, *30*, 896–902.
Casellini, C. M., Parson, H. K., Hodges, K., Edwards, J. F., Lieb, D. C., Wohlgemuth, S. D., et al. (2015). Bariatric surgery restores somatic and autonomic nerve function towards normal in obese patients with and without diabetes. In *AACE 24th annual scientific clinical congress, May 13–17*. Abstract.
Casellini, C. M., Parson, H. K., Richardson, M. S., Nevoret, M. L., & Vinik, A. I. (2013). SUDOSCAN, a noninvasive tool for detecting diabetic small fiber neuropathy and autonomic dysfunction. *Diabetes Technology & Therapeutics*, *15*, 948–953.
Cavanagh, P. R., Derr, J. A., Ulbrecht, J. S., Maser, R. E., & Orchard, T. J. (1992). Problems with gait and posture in neuropathic patients with insulin-dependent diabetes mellitus. *Diabetic Medicine*, *9*, 469–474.
Cavanagh, P. R., Simoneau, G. G., & Ulbrecht, J. S. (1993). Ulceration, unsteadiness, and uncertainty: The biomechanical consequences of diabetes mellitus. *Journal of Biochemistry*, *26*(Suppl. 1), 23–40.
Chao, C. C., Hsieh, S. C., Tseng, M. T., Chang, Y. C., & Hsieh, S. T. (2008). Patterns of contact heat evoked potentials (CHEP) in neuropathy with skin denervation: Correlation of CHEP amplitude with intraepidermal nerve fiber density. *Clinical Neurophysiology*, *119*, 653–661.
Chao, C. C., Tseng, M. T., Lin, Y. J., Yang, W. S., Hsieh, S. C., Lin, Y. H., et al. (2010). Pathophysiology of neuropathic pain in type 2 diabetes: Skin denervation and contact heat-evoked potentials. *Diabetes Care*, *33*, 2654–2659.
Chen, I. A., Hung, S. W., Chen, Y. H., Lim, S. N., Tsai, Y. T., Hsiao, C. L., et al. (2006). Contact heat evoked potentials in normal subjects. *Acta Neurologica Taiwanica*, *15*, 184–191.
Chen, A. C., Niddam, D. M., & rendt-Nielsen, L. (2001). Contact heat evoked potentials as a valid means to study nociceptive pathways in human subjects. *Neuroscience Letters*, *316*, 79–82.
Chiles, N. S., Phillips, C. L., Volpato, S., Bandinelli, S., Ferrucci, L., Guralnik, J. M., et al. (2014). Diabetes, peripheral neuropathy, and lower-extremity function. *Journal of Diabetes and Its Complications*, *28*, 91–95.
Clark, R. D., Lord, S. R., & Webster, I. W. (1993). Clinical parameters associated with falls in an elderly population. *Gerontology*, *39*, 117–123.

Close, J. C., Lord, S. L., Menz, H. B., & Sherrington, C. (2005). What is the role of falls? *Best Practice & Research. Clinical Rheumatology, 19*, 913–935.

Colberg, S. (2006). The impact of exercise on insulin action in type 2 diabetes mellitus: Relationship to prevention and control. *Insulin, 1*, 85–98.

Colberg, S. R., Parson, H. K., Nunnold, T., Holton, D. R., Swain, D. P., & Vinik, A. I. (2005). Change in cutaneous perfusion following 10 weeks of aerobic training in type 2 diabetes. *Journal of Diabetes and Its Complications, 19*, 276–283.

Colberg, S., Stansberry, K., McNitt, P., & Vinik, A. (2002). Chronic exercise is associated with enhanced cutaneous blood flow in type 2 diabetes. *Journal of Diabetes and Its Complications, 16*, 139–145.

Colberg, S., Swain, D., & Vinik, A. (2003). Use of heart rate reserve and rating of perceived exertion to prescribe exercise intensity in diabetic autonomic neuropathy. *Diabetes Care, 26*, 986–990.

Corriveau, H., Prince, F., Hebert, R., Raiche, M., Tessier, D., Maheux, P., et al. (2000). Evaluation of postural stability in elderly with diabetic neuropathy. *Diabetes Care, 23*, 1187–1191.

DCCT EDIC Research Group. (2003). Sustained effect of intensive treatment of type 1 diabetes mellitus on development and progression of diabetic nephropathy. *JAMA, 290*, 2159–2167.

DCCT Research Group. (1993). The effect of intensive treatment of diabetes on the development and progression of long-term complications in insulin dependent diabetes mellitus. *The New England Journal of Medicine, 329*, 977–986.

DCCT Research Group. (1995). The effect of intensive diabetes therapy on the development and progression of neuropathy. *Annals of Internal Medicine, 122*, 561–568.

Diabetes Control and Complications Trial Research Group. (1998). The effect of intensive diabetes therapy on measures of autonomic nervous system function in the Diabetes Control And Complications Trial (DCCT). *Diabetologia, 41*, 416–423.

Didangelos, T. P., Arsos, G. A., Karamitsos, D. T., Athyros, V. G., Georga, S. D., & Karatzas, N. D. (2006). Effect of quinapril or losartan alone and in combination on left ventricular systolic and diastolic functions in asymptomatic patients with diabetic autonomic neuropathy. *Journal of Diabetes and Its Complications, 20*, 1–7.

Dinh, W., Futh, R., Lankisch, M., Bansemir, L., Nickl, W., Scheffold, T., et al. (2011). Cardiovascular autonomic neuropathy contributes to left ventricular diastolic dysfunction in subjects with type 2 diabetes and impaired glucose tolerance undergoing coronary angiography. *Diabetic Medicine, 28*, 311–318.

Dros, J., Wewerinke, A., Bindels, P. J., & van Weert, H. C. (2009). Accuracy of monofilament testing to diagnose peripheral neuropathy: A systematic review. *Annals of Family Medicine, 7*, 555–558.

Duckworth, W., Abraira, C., Moritz, T., Reda, D., Emanuele, N., Reaven, P. D., et al. (2009). Glucose control and vascular complications in veterans with type 2 diabetes. *The New England Journal of Medicine, 360*, 129–139.

Dyck, P. J., Dyck, P. J., Larson, T. S., O'Brien, P. C., & Velosa, J. A. (2000). Patterns of quantitative sensation testing of hypoesthesia and hyperalgesia are predictive of diabetic polyneuropathy: A study of three cohorts. Nerve growth factor study group. *Diabetes Care, 23*, 510–517.

Dyck, P. J., Kratz, K. M., Karnes, J. L., Litchy, W. J., Klein, R., Pach, J. M., et al. (1993). The prevalence by staged severity of various types of diabetic neuropathy, retinopathy, and nephropathy in a population-based cohort: The Rochester Diabetic Neuropathy Study. *Neurology, 43*, 817–824.

Dyck, P. J., Lais, A., Karnes, J. L., O'Brien, P., & Rizza, R. (1986). Fiber loss is primary and multifocal in sural nerves in diabetic polyneuropathy. *Annals of Neurology, 19*, 425–439.

Ebadi, H., Perkins, B. A., Katzberg, H. D., Lovblom, L. E., & Bril, V. (2012). Evaluation of proxy tests for SFSN: Evidence for mixed small and large fiber dysfunction. *PLoS One*, *7*, e42208.
England, J. D., Gronseth, G. S., Franklin, G., Carter, G. T., Kinsella, L. J., Cohen, J. A., et al. (2009). Practice parameter: Evaluation of distal symmetric polyneuropathy: Role of autonomic testing, nerve biopsy, and skin biopsy (an evidence-based review). Report of the American Academy of Neurology, American Association of Neuromuscular and Electrodiagnostic Medicine, and American Academy of Physical Medicine and Rehabilitation. *Neurology*, *72*, 177–184.
England, J. D., Gronseth, G. S., Franklin, G., Miller, R. G., Asbury, A. K., Carter, G. T., et al. (2005). Distal symmetric polyneuropathy: A definition for clinical research: Report of the American Academy of Neurology, the American Association of Electrodiagnostic Medicine, and the American Academy of Physical Medicine and Rehabilitation. *Neurology*, *64*, 199–207.
Eranki, V. G., Santosh, R., Rajitha, K., Pillai, A., Sowmya, P., Dupin, J., et al. (2013). Sudomotor function assessment as a screening tool for microvascular complications in type 2 diabetes. *Diabetes Research in Clinical Practice*, *101*, e11–e13.
Ewing, D. J., Boland, O., Neilson, J. M., Cho, C. G., & Clarke, B. F. (1991). Autonomic neuropathy, QT interval lengthening, and unexpected deaths in male diabetic patient. *Diabetologia*, *34*, 182–185.
Ewing, D. J., & Clarke, B. F. (1986). Diabetic autonomic neuropathy: Present insights and future prospects. *Diabetes Care*, *9*, 648–665.
Feldman, E. L., Stevens, M. J., Thomas, P. K., Brown, M. B., Canal, N., & Greene, D. A. (1994). A practical two-step quantitative clinical and electrophysiological assessment for the diagnosis and staging of diabetic neuropathy. *Diabetes Care*, *17*, 1281–1289.
Fioretti, S., Scocco, M., Ladislao, L., Ghetti, G., & Rabini, R. A. (2010). Identification of peripheral neuropathy in type-2 diabetic subjects by static posturography and linear discriminant analysis. *Gait & Posture*, *32*, 317–320.
Freeman, R., & Chapleau, M. W. (2013). Testing the autonomic nervous system. *Handbook of Clinical Neurology*, *115*, 115–136.
Gaede, P., Vedel, P., Parving, H. H., & Pedersen, O. (1999). Intensified multifactorial intervention in patients with type 2 diabetes mellitus and microalbuminuria: The steno type 2 randomized study. *Lancet*, *353*, 617–622.
Gerritsen, J., Dekker, J. M., TenVoorde, B. J., Kostense, P. J., Heine, R. J., Bouter, L. M., et al. (2001). Impaired autonomic function is associated with increased mortality, especially in subjects with diabetes, hypertension, or a history of cardiovascular disease: The Hoorn study. *Diabetes Care*, *24*, 1793–1798.
Gibbons, C. H., Freeman, R., & Veves, A. (2010). Diabetic neuropathy: A cross-sectional study of the relationships among tests of neurophysiology. *Diabetes Care*, *33*, 2629–2634.
Gibbons, C. H., Illigens, B. M., Wang, N., & Freeman, R. (2009). Quantification of sweat gland innervation: A clinical-pathologic correlation. *Neurology*, *72*, 1479–1486.
Gin, H., Baudoin, R., Raffaitin, C. H., Rigalleau, V., & Gonzalez, C. (2011). Non-invasive and quantitative assessment of sudomotor function for peripheral diabetic neuropathy evaluation. *Diabetes & Metabolism*, *37*, 527–532.
Gopalakrishnan, R., Machado, A. G., Burgess, R. C., & Mosher, J. C. (2013). The use of contact heat evoked potential stimulator (CHEPS) in magnetoencephalography for pain research. *Journal of Neuroscience Methods*, *220*, 55–63.
Green, A. Q., Krishnan, S. T., & Rayman, G. (2009). C-fiber function assessed by the laser Doppler imager flare technique and acetylcholine iontophoresis. *Muscle & Nerve*, *40*, 985–991.

Haanpaa, M. L., Backonja, M. M., Bennett, M. I., Bouhassira, D., Cruccu, G., Hansson, P. T., et al. (2009). Assessment of neuropathic pain in primary care. *The American Journal of Medicine, 122*, S13–S21.

Harris, M., Eastman, R., & Cowie, C. (1993). Symptoms of sensory neuropathy in adults with NIDDM in the U.S. population. *Diabetes Care, 16*, 1446–1452.

Herman, W. H., & Kennedy, L. (2005). Underdiagnosis of peripheral neuropathy in type 2 diabetes. *Diabetes Care, 28*, 1480–1481.

Herriott, M. T., Colberg, S. R., Parson, H. K., Nunnold, T., & Vinik, A. I. (2004). Effects of 8 weeks of flexibility and resistance training in older adults with type 2 diabetes. *Diabetes Care, 27*, 2988–2989.

Holzer, S. E., Camerota, A., Martens, L., Cuerdon, T., Crystal, P., & Zagari, M. (1998). Costs and duration of care for lower extremity ulcers in patients with diabetes. *Clinical Therapeutics, 20*, 169–181.

Howe, T., Rochester, L., Jackson, A., Banks, P., & Blair, V. (2007). Exercise for improving balance in older people. *The Cochrane Database of Systematic Reviews, 17*, CD004963.

Illigens, B. M., & Gibbons, C. H. (2009). Sweat testing to evaluate autonomic function. *Clinical Autonomic Research, 19*, 79–87.

Itskovich, V. V., Fei, D. Y., & Harkins, S. W. (2000). Psychophysiological and psychophysical responses to experimental pain induced by two types of cutaneous thermal stimuli. *The International Journal of Neuroscience, 105*, 63–75.

Joint Task Force of the EFNS and the PNS. (2010). European Federation of Neurological Societies/Peripheral Nerve Society Guideline on the use of skin biopsy in the diagnosis of small fiber neuropathy. Report of a joint task force of the European Federation of Neurological Societies and the Peripheral Nerve Society. *Peripheral Nervous System, 15*, 79–92.

Karason, K., Molgaard, H., Wikstrand, J., & Sjostrom, L. (1999). Heart rate variability in obesity and the effect of weight loss. *The American Journal of Cardiology, 83*, 1242–1247.

Katoulis, E. C., Ebdon-Parry, M., Lanshammar, H., Vileikyte, L., Kulkarni, J., & Boulton, A. J. (1997). Gait abnormalities in diabetic neuropathy. *Diabetes Care, 20*, 1904–1907.

Kellogg, A. P., Converso, K., Wiggin, T., Stevens, M., & Pop-Busui, R. (2009). Effects of cyclooxygenase-2 gene inactivation on cardiac autonomic and left ventricular function in experimental diabetes. *American Journal of Physiology. Heart and Circulatory Physiology, 296*, H453–H461.

Kendall, D. M., Rooney, D. P., Smets, Y. F., Salazar Bolding, L., & Robertson, R. P. (1997). Pancreas transplantation restores epinephrine response and symptom recognition during hypoglycemia in patients with long-standing type I diabetes and autonomic neuropathy. *Diabetes, 46*, 249–257.

Kluding, P. M., Pasnoor, M., Singh, R., Jernigan, S., Farmer, K., Rucker, J., et al. (2012). The effect of exercise on neuropathic symptoms, nerve function, and cutaneous innervation in people with diabetic peripheral neuropathy. *Journal of Diabetes and Its Complications, 26*, 424–429.

Kramer, J. L., Jutzeler, C. R., Haefeli, J., Curt, A., & Freund, P. (2015). Discrepancy between perceived pain and cortical processing: A voxel-based morphometry and contact heat evoked potential study. *Clinical Neurophysiology, 127*, 762–768.

Kramer, J. L., Taylor, P., Haefeli, J., Blum, J., Zariffa, J., Curt, A., et al. (2012). Test-retest reliability of contact heat-evoked potentials from cervical dermatomes. *Journal of Clinical Neurophysiology, 29*, 70–75.

Krishnan, S. T., & Rayman, G. (2004). The LDIflare: A novel test of C-fiber function demonstrates early neuropathy in type 2 diabetes. *Diabetes Care, 27*(12), 2930–2935.

Kumar, S., Fernando, D. J., Veves, A., Knowles, E. A., Young, M. J., & Boulton, A. J. (1991). Semmes-Weinstein monofilaments: A simple, effective and inexpensive screening device for identifying diabetic patients at risk of foot ulceration. *Diabetes Research and Clinical Practice*, *13*, 63–67.

Lafond, D., Corriveau, H., & Prince, F. (2004). Postural control mechanisms during quiet standing in patients with diabetic sensory neuropathy. *Diabetes Care*, *27*, 173–178.

Lagerburg, V., Bakkers, M., Bouwhuis, A., Hoeijmakers, J. G., Smit, A. M., Van Den Berg, S. J., et al. (2015). Contact heat evoked potentials: Normal values and use in small-fiber neuropathy. *Muscle & Nerve*, *51*, 743–749.

Lai, S., Ahmed, U., Bollineni, A., Lewis, R., & Ramchandren, S. (2014). Diagnostic accuracy of qualitative versus quantitative tuning forks: Outcome measure for neuropathy. *Journal of Clinical Neuromuscular Disease*, *15*, 96–101.

Lalli, P., Chan, A., Garven, A., Midha, N., Chan, C., Brady, S., et al. (2013). Increased gait variability in diabetes mellitus patients with neuropathic pain. *Journal of Diabetes and Its Complications*, *27*, 248–254.

Lambert, E., Sari, C. I., Dawood, T., Nguyen, J., McGrane, M., Eikelis, N., et al. (2010). Sympathetic nervous system activity is associated with obesity-induced subclinical organ damage in young adults. *Hypertension*, *56*, 351–358.

Lambert, G. W., Straznicky, N. E., Lambert, E. A., Dixon, J. B., & Schlaich, M. P. (2010). Sympathetic nervous activation in obesity and the metabolic syndrome—Causes, consequences and therapeutic implications. *Pharmacology & Therapeutics*, *126*, 159–172.

Lauria, G., Bakkers, M., Schmitz, C., Lombardi, R., Penza, P., Devigili, G., et al. (2010). Intraepidermal nerve fiber density at the distal leg: A worldwide normative reference study. *Journal of the Peripheral Nervous System*, *15*, 202–207.

Lauria, G., Hsieh, S., Johansson, O., Kennedy, W., Leger, J., Mellgren, S., et al. (2010). European Federation of Neurological Societies/Peripheral Nerve Society Guideline on the use of skin biopsy in the diagnosis of small fiber neuropathy. Report of a joint task force of the European Federation of Neurological Societies and the Peripheral Nerve Society. *European Journal of Neurology*, *17*, 903–912.

Levitt, N. S., Stansberry, K. B., Wychanck, S., & Vinik, A. I. (1996). Natural progression of autonomic neuropathy and autonomic function tests in a cohort of IDDM. *Diabetes Care*, *19*, 751–754.

Lord, S. (1996). Instability and falls in elderly people. In C. Lafont, A. Baroni, & M. Allard et al. (Eds.), *Falls, gait, and balance disorders in the elderly* (pp. 125–139). New York: Springer.

Lord, S. R., & Clark, R. D. (1996). Simple physiological and clinical tests for the accurate prediction of falling in older people. *Gerontology*, *42*, 199–203.

Lykke, J. A., Tarnow, L., Parving, H. H., & Hilsted, J. (2008). A combined abnormality in heart rate variation and QT corrected interval is a strong predictor of cardiovascular death in type 1 diabetes. *Scandinavian Journal of Clinical and Laboratory Investigation*, *68*, 654–659.

Malmberg, K., Norhammar, A., Wedel, H., & Ryden, L. (1999). Glycometabolic state at admission: Important risk marker of mortality in conventionally treated patients with diabetes mellitus and acute myocardial infarction: Long-term results from the Diabetes and Insulin-Glucose Infusion in Acute Myocardial Infarction (DIGAMI) study. *Circulation*, *99*, 2626–2632.

Marchant, B., Umachandran, V., Stevenson, R., Kopelman, P. G., & Timmis, A. D. (1993). Silent myocardial ischemia: Role of subclinical neuropathy in patients with and without diabetes. *Journal of the American College of Cardiology*, *22*, 1433–1437.

Margolis, J. R., Kannel, W. S., Feinleib, M., Dawber, T. R., & McNamara, P. M. (1973). Clinical features of unrecognized myocardial infarction—Silent and symptomatic. Eighteen year follow-up: The Framingham study. *The American Journal of Cardiology*, *32*, 1–7.

Martin, C. L., Waberski, B. H., Pop-Busui, R., Cleary, P. A., Catton, S., Albers, J. W., et al. (2010). Vibration perception threshold as a measure of distal symmetrical peripheral neuropathy in type 1 diabetes: Results from the DCCT/EDIC study. *Diabetes Care, 33*, 2635–2641.

Maser, R., Lenhard, M., & DeCherney, G. (2000). Cardiovascular autonomic neuropathy: The clinical significance of its determination. *Endocrinologist, 10*, 27–33.

Maser, R. E., Lenhard, M. J., Irgau, I., & Wynn, G. M. (2007). Impact of surgically induced weight loss on cardiovascular autonomic function: One-year follow-up. *Obesity (Silver Spring), 15*, 364–369.

Maser, R. E., Mitchell, B. D., Vinik, A. I., & Freeman, R. (2003). The association between cardiovascular autonomic neuropathy and mortality in individuals with diabetes: A meta-analysis. *Diabetes Care, 26*, 1895–1901.

Matsuki, A., Nozawa, T., Igarashi, N., Sobajima, M., Ohori, T., Suzuki, T., et al. (2010). Fluvastatin attenuates diabetes-induced cardiac sympathetic neuropathy in association with a decrease in oxidative stress. *Circulation Journal, 74*, 468–475.

Maurer, M. S., Burcham, J., & Cheng, H. (2005). Diabetes mellitus is associated with an increased risk of falls in elderly residents of a long-term care facility. *The Journals of Gerontology. Series A, Biological Sciences and Medical Sciences, 60*, 1157–1162.

Morrison, S., Colberg, S. R., Mariano, M., Parson, H. K., & Vinik, A. I. (2010). Balance training reduces falls risk in older individuals with type 2 diabetes. *Diabetes Care, 33*, 748–750.

Morrison, S., Colberg, S. R., Parson, H. K., & Vinik, A. I. (2012). Relation between risk of falling and postural sway complexity in diabetes. *Gait & Posture, 35*, 662–668.

Nather, A., Keng, L. W., Aziz, Z., Hj, O. C., Mc, F. B., & Lin, B. (2011). Assessment of sensory neuropathy in patients with diabetic foot problems. *Diabetic Foot & Ankle, 2*.

Nault, I., Nadreau, E., Paquet, C., Brassard, P., Marceau, P., Marceau, S., et al. (2007). Impact of bariatric surgery—Induced weight loss on heart rate variability. *Metabolism, 56*, 1425–1430.

Neil, H. A., Thompson, A. V., John, S., McCarthy, S. T., & Mann, J. I. (1989). Diabetic autonomic neuropathy: The prevalence of impaired heart rate variability in a geographically define population. *Diabetic Medicine, 6*, 20–24.

Novella, S. P., Inzucchi, S. E., & Goldstein, J. M. (2001). The frequency of undiagnosed diabetes and impaired glucose tolerance in patients with idiopathic sensory neuropathy. *Muscle & Nerve, 24*, 1229–1231.

O'Brien, I. A., McFadden, J. P., & Corrall, R. J. (1991). The influence of autonomic neuropathy on mortality in insulin-dependent diabetes. *The Quarterly Journal of Medicine, 79*, 495–502.

Opsommer, E., Masquelier, E., & Plaghki, L. (1999). Determination of nerve conduction velocity of C-fibres in humans from thermal thresholds to contact heat (thermode) and from evoked brain potentials to radiant heat (CO2 laser). *Neurophysiologie Clinique, 29*, 411–422.

Ozaki, R., Cheung, K. K., Wu, E., Kong, A., Yang, X., Lau, E., et al. (2011). A new tool to detect kidney disease in Chinese type 2 diabetes patients: Comparison of EZSCAN with standard screening methods. *Diabetes Technology & Therapeutics, 13*, 937–943.

Paisley, A. N., Abbott, C. A., van Schie, C. H. M., & Boulton, A. J. M. (2002). A comparison of the Neuropen against standard quantitative sensory threshold measures for assessing peripheral nerve function. *Diabetic Medicine, 19*, 400–405.

Papanas, N., Papatheodorou, K., Christakidis, D., Papazoglou, D., Giassakis, G., Piperidou, H., et al. (2005). Evaluation of a new indicator test for sudomotor function (Neuropad) in the diagnosis of peripheral neuropathy in type 2 diabetic patients. *Experimental and Clinical Endocrinology & Diabetes, 113*, 195–198.

Parson, H. K., Nguyen, V. T., Boyd, A. L., & Vinik, A. (2009). CHEPS detects neuropathic changes earlier than traditional clinical measures. *Diabetes*, *58*(Suppl.), 829P.

Parson, H. K., Nguyen, V. T., Orciga, M. A., Boyd, A. L., Casellini, C. M., & Vinik, A. I. (2013). Contact heat-evoked potential stimulation for the evaluation of small nerve fiber function. *Diabetes Technology & Therapeutics*, *15*, 150–157.

Partanen, J., Niskanen, L., Lehtinen, J., Mervaala, E., Siitonen, O., & Uusitupa, M. (1995). Natural history of peripheral neuropathy in patients with non-insulin-dependent diabetes mellitus. *The New England Journal of Medicine*, *333*, 89–94.

Patel, A., MacMahon, S., Chalmers, J., Neal, B., Billot, L., Woodward, M., et al. (2008). Intensive blood glucose control and vascular outcomes in patients with type 2 diabetes. *The New England Journal of Medicine*, *358*, 2560–2572.

Peron, E. P., & Ogbonna, K. C. (2015). Diabetic medications and polypharmacy. *Clinics of Geriatric Medicine*, *31*, 17–27.

Perugini, R. A., Li, Y., Rosenthal, L., Gallagher-Dorval, K., Kelly, J. J., & Czerniach, D. R. (2010). Reduced heart rate variability correlates with insulin resistance but not with measures of obesity in population undergoing laparoscopic Roux-en-Y gastric bypass. *Surgery for Obesity and Related Diseases*, *6*, 237–241.

Peterson, M. J., Giuliani, C., Morey, M. C., Pieper, C. F., Evenson, K. R., Mercer, V., et al. (2009). Physical activity as a preventative factor for frailty: The health, aging, and body composition study. *The Journals of Gerontology. Series A, Biological Sciences and Medical Sciences*, *64*, 61–68.

Pfeifer, M. A., Weinberg, C. R., Cook, D. L., Reenan, A., Halter, J. B., & Ensinck, J. W. (1984). Autonomic neural dysfunction in recently diagnosed diabetic subjects. *Diabetes Care*, *7*, 447–453.

Pickering, R. M., Grimbergen, Y. A., Rigney, U., Ashburn, A., Mazibrada, G., Wood, B., et al. (2007). A meta-analysis of six prospective studies of falling in Parkinson's disease. *Movement Disorders*, *22*, 1892–1900.

Piestrzeniewicz, K., Luczak, K., Lelonek, M., Wranicz, J. K., & Goch, J. H. (2008). Obesity and heart rate variability in men with myocardial infarction. *Cardiology Journal*, *15*, 43–49.

Pijpers, E., Ferreira, I., de Jongh, R. T., Deeg, D. J., Lips, P., Stehouwer, C. D., et al. (2012). Older individuals with diabetes have an increased risk of recurrent falls: Analysis of potential mediating factors: The Longitudinal Ageing Study Amsterdam. *Age and Ageing*, *41*, 358–365.

Pittenger, G., Mehrabyan, A., Simmons, K., Rice, A., Dublin, C., Barlow, P., et al. (2005). Small fiber neuropathy is associated with the metabolic syndrome. *Metabolic Syndrome and Related Disorders*, *3*, 113–121.

Pittenger, G. L., Ray, M., Burcus, N. I., McNulty, P., Basta, B., & Vinik, A. I. (2004). Intraepidermal nerve fibers are indicators of small-fiber neuropathy in both diabetic and nondiabetic patients. *Diabetes Care*, *27*, 1974–1979.

Pittenger, G., Simmons, K., Anandacoomaraswamy, D., Rice, A., Barlow, P., & Vinik, A. (2005). Topiramate improves intraepidermal nerve fiber morphology and quantitative neuropathy measures in diabetic neuropathy patients. *Journal of the Peripheral Nervous System*, *10*(Suppl. 1), 1–108.

Pop-Busui, R., Low, P. A., Waberski, B. H., Martin, C. L., Albers, J. W., Feldman, E. L., et al. (2009). Effects of prior intensive insulin therapy on cardiac autonomic nervous system function in type 1 diabetes mellitus: The Diabetes Control and Complications Trial/Epidemiology of Diabetes Interventions and Complications study (DCCT/EDIC). *Circulation*, *119*, 2886–2893.

Quattrini, C., Tavakoli, M., Jeziorska, M., Kallinikos, P., Tesfaye, S., Finnigan, J., et al. (2007). Surrogate markers of small fiber damage in human diabetic neuropathy. *Diabetes*, *56*, 2148–2154.

Raisanen, A., Eklund, J., Calvet, J.-H., & Tuomilehto, J. (2014). Sudomotor function as a tool for cardiorespiratory fitness level evaluation: Comparison with VO2 max. *International Journal of Environmental Research and Public Health, 11*, 5839–5848.

Rathmann, W., Ziegler, D., Jahnke, M., Haastert, B., & Gries, F. A. (1993). Mortality in diabetic patients with cardiovascular autonomic neuropathy. *Diabetic Medicine, 10*, 820–824.

Ravussin, E. (2010). Impact of six month caloric restriction on autonomic nervous system in healthy, overweight individuals. *Obesity, 18*, 414–416.

Resnick, H. E., Stansberry, K. B., Harris, T. B., Tirivedi, M., Smith, K., Morgan, P., et al. (2002). Diabetes, peripheral neuropathy, and old age disability. *Muscle & Nerve, 25*, 43–50.

Resnick, H., Vinik, A., Schwartz, A., Leveille, S., Brancati, F., Balfour, J., et al. (2000). Independent effects of peripheral nerve dysfunction on lower-extremity physical function in old age. *Diabetes Care, 23*, 1642–1647.

Richardson, J. K., & Hurvitz, E. A. (1995). Peripheral neuropathy: A true risk factor for falls. *The Journals of Gerontology. Series A, Biological Sciences and Medical Sciences, 50*, M211–M215.

Richardson, J., Thies, S., & Ashton-Miller, J. (2008). An exploration of step time variability on smooth and irregular surfaces in older persons with neuropathy. *Clinical biomechanics, 23*, 349–356.

Richardson, J. K., Thies, S. B., DeMott, T. K., & Ashton-Miller, J. A. (2005). Gait analysis in a challenging environment differentiates between fallers and nonfallers among older patients with peripheral neuropathy. *Archives of Physical Medicine and Rehabilitation, 86*, 1539–1544.

Robinovitch, S. N., Hsiao, E. T., Sandler, R., Cortez, J., Liu, Q., & Paiement, G. D. (2000). Prevention of falls and fall-related fractures through biomechanics. *Exercise and Sport Sciences Reviews, 28*, 74–79.

Ruscheweyh, R., Emptmeyer, K., Putzer, D., Kropp, P., & Marziniak, M. (2013). Reproducibility of contact heat evoked potentials (CHEPs) over a 6 months interval. *Clinical Neurophysiology, 124*, 2242–2247.

Sacre, J. W., Franjic, B., Jellis, C. L., Jenkins, C., Coombes, J. S., & Marwick, T. H. (2010). Association of cardiac autonomic neuropathy with subclinical myocardial dysfunction in type 2 diabetes. *JACC. Cardiovascular Imaging, 3*, 1207–1215.

Sadosky, A., McDermott, A. M., Brandenburg, N. A., & Strauss, M. (2008). A review of the epidemiology of painful diabetic peripheral neuropathy, postherpetic neuralgia, and less commonly studied neuropathic pain conditions. *Pain Practice, 8*, 45–56.

Schwartz, A. V., Hillier, T. A., Sellmeyer, D. E., Resnick, H. E., Gregg, E., Ensrud, K. E., et al. (2002). Older women with diabetes have a higher risk of falls: A prospective study. *Diabetes Care, 25*, 1749–1754.

Schwartz, A. V., Vittinghoff, E., Sellmeyer, D. E., Feingold, K. R., de, R. N., Strotmeyer, E. S., et al. (2008). Diabetes-related complications, glycemic control, and falls in older adults. *Diabetes Care, 31*, 391–396.

Selvarajah, D., Cash, T., Davies, J., Sankar, A., Rao, G., Grieg, M., et al. (2015). SUDOSCAN: A simple, rapid, and objective method with potential for screening for diabetic peripheral neuropathy. *PloS One, 10*, e0138224.

Sharma, S., Vas, P. R., & Rayman, G. (2015). Assessment of diabetic neuropathy using a point-of-care nerve conduction device shows significant associations with the LDIFLARE method and clinical neuropathy scoring. *Journal of Diabetes Science and Technology, 9*, 123–131.

Shehab, D. K., Al-Jarallah, K. F., Abraham, M., Mojiminiyi, O. A., Al-Mohamedy, H., & Abdella, N. A. (2012). Back to basics: Ankle reflex in the evaluation of peripheral neuropathy in type 2 diabetes mellitus. *QJM: Monthly Journal of the Association of Physicians*, *105*, 315–320.

Shenoy, R., Roberts, K., Papadaki, A., McRobbie, D., Timmers, M., Meert, T., et al. (2011). Functional MRI brain imaging studies using the contact heat evoked potential stimulator (CHEPS) in a human volunteer topical capsaicin pain model. *Journal of Pain Research*, *4*, 365–371.

Sherrington, C., Lord, S. R., & Herbert, R. D. (2003). A randomised trial of weight-bearing versus non-weight-bearing exercise for improving physical ability in inpatients after hip fracture. *The Australian Journal of Physiotherapy*, *49*, 15–22.

Shtein, R. M., & Callaghan, B. C. (2013). Corneal confocal microscopy as a measure of diabetic neuropathy. *Diabetes*, *62*, 25–26.

Shy, M. E., Frohman, E. M., So, Y., Arezzo, J. C., Cornblath, D. C., Giuliani, M. J., et al. (2003). Quantitative sensory testing: Report on the therapeutic and technology assessment subcommittee of the American Academy of Neurology. *Neurology*, *602*, 898–906.

Simoneau, G. G., Ulbrecht, J., Derr, J. A., Becher, M. B., & Cavanaugh, P. R. (1994). Postural instability in patients with diabetic sensory neuropathy. *Diabetes Care*, *17*, 1411–1421.

Singleton, J. R., Smith, A. G., & Bromberg, M. B. (2001a). Painful sensory polyneuropathy associated with impaired glucose tolerance. *Muscle & Nerve*, *24*, 1225–1228.

Singleton, J. R., Smith, A. G., & Bromberg, M. B. (2001b). Increased prevalence of impaired glucose tolerance in patients with painful sensory neuropathy. *Diabetes Care*, *24*, 1448–1453.

Sinnreich, M., Taylor, B. V., & Dyck, P. J. (2005). Diabetic neuropathies. Classification, clinical features, and pathophysiological basis. *The Neurologist*, *11*, 63–79.

Smith, A. G., Lessard, M., Reyna, S., Doudova, M., & Singleton, J. R. (2014). The diagnostic utility of SUDOSCAN for distal symmetric peripheral neuropathy. *Journal of Diabetes and Its Complications*, *28*, 511–516.

Smith, A. G., Russell, J., Feldman, E. L., Goldstein, J., Peltier, A., Smith, S., et al. (2006). Lifestyle intervention for pre-diabetic neuropathy. *Diabetes Care*, *29*, 1294–1299.

Sosnoff, J., Motl, R., & Morrison, S. (2013). Multiple sclerosis and falls—An evolving tale. *US Neurology*, *9*, 30–34.

Sosnoff, J. J., Socie, M. J., Boes, M. K., Sandroff, B. M., Pula, J. H., Suh, Y., et al. (2011). Mobility, balance and falls in persons with multiple sclerosis. *PloS One*, *6*, e28021.

Stevens, J. A., Corso, P. S., Finkelstein, E. A., & Miller, T. R. (2006). The costs of fatal and non-fatal falls among older adults. *Injury Prevention*, *12*, 290–295.

Stewart, J. M., Medow, M. S., & Montgomery, L. D. (2003). Local vascular responses affecting blood flow in postural tachycardia syndrome. *American Journal of Physiology. Heart and Circulatory Physiology*, *285*, H2749–H2756.

Straznicky, N. E., Lambert, G. W., McGrane, M. T., Masuo, K., Dawood, T., Nestel, P. J., et al. (2009). Weight loss may reverse blunted sympathetic neural responsiveness to glucose ingestion in obese subjects with metabolic syndrome. *Diabetes*, *58*, 1126–1132.

Strotmeyer, E. S., de, R. N., Schwartz, A. V., Faulkner, K. A., Resnick, H. E., Goodpaster, B. H., et al. (2008). The relationship of reduced peripheral nerve function and diabetes with physical performance in older white and black adults: The health, aging, and body composition (health ABC) study. *Diabetes Care*, *31*, 1767–1772.

Strotmeyer, E. S., de, R. N., Schwartz, A. V., Resnick, H. E., Goodpaster, B. H., Faulkner, K. A., et al. (2009). Sensory and motor peripheral nerve function and lower-extremity quadriceps strength: The health, aging and body composition study. *Journal of the American Geriatrics Society*, *57*, 2004–2010.

Sumner, C. J., Sheth, S., Griffin, J. W., Cornblath, D. R., & Polydefkis, M. (2003). The spectrum of neuropathy in diabetes and impaired glucose tolerance. *Neurology, 60*, 108–111.

Taksande, B., Ansari, S., Jaikrishnan, A., & Karwasara, V. (2011). The diagnostic sensitivity, specificity, and reproducibility of the clinical physical examination signs in patients of diabetes mellitus for making diagnosis of peripheral neuropathy. *Journal of Endocrinology and Metabolism, 1*, 21–26.

Tavakoli, M., Mitu-Pretorian, M., Petropoulos, I. N., Fadavi, H., Asghar, O., Alam, U., et al. (2013). Corneal confocal microscopy detects early nerve regeneration in diabetic neuropathy after simultaneous pancreas and kidney transplantation. *Diabetes, 62*, 254–260.

Tavakoli, M., Quattrini, C., Abbott, C., Kallinikos, P., Marshall, A., Finnigan, J., et al. (2010). Corneal confocal microscopy: A novel noninvasive test to diagnose and stratify the severity of human diabetic neuropathy. *Diabetes Care, 33*, 1792–1797.

Tesfaye, S., Boulton, A. J., Dyck, P. J., Freeman, R., Horowitz, M., Kempler, P., et al. (2010). Diabetic neuropathies: Update on definitions, diagnostic criteria, estimation of severity, and treatments. *Diabetes Care, 33*, 2285–2293.

Thaisetthawatkul, P., Fernandes Filho, J. A., & Herrmann, D. N. (2013). Contribution of QSART to the diagnosis of small fiber neuropathy. *Muscle & Nerve, 48*, 883–888.

Tilling, L. M., Darawil, K., & Britton, M. (2006). Falls as a complication of diabetes mellitus in older people. *Journal of Diabetes and Its Complications, 20*, 158–162.

Tucker, M. E. (2014). *Laser-doppler-imager flare detects early diabetic neuropathy.* www.medscape.com/viewarticle/832229.

Turcot, K., Allet, L., Golay, A., Hoffmeyer, P., & Armand, S. (2009). Investigation of standing balance in diabetic patients with and without peripheral neuropathy using accelerometers. *Clinical Biomechanics (Bristol, Avon), 24*, 716–721.

Uccioli, L., Giacomini, P. G., Monticone, G., Magrini, A., Durola, L., Bruno, E., et al. (1995). Body sway in diabetic neuropathy. *Diabetes Care, 18*, 339–344.

Vas, P. R., & Rayman, G. (2013a). Validation of the modified LDIFlare technique: A simple and quick method to assess C-fiber function. *Muscle & Nerve, 47*, 351–356.

Vas, P. R., & Rayman, G. (2013b). The rate of decline in small fibre function assessed using axon reflex-mediated neurogenic vasodilatation and the importance of age related centile values to improve the detection of clinical neuropathy. *PLoS One, 8*, e69920.

Veglio, M., Borra, M., Stevens, L. K., Fuller, J. H., & Perin, P. C. (1999). The relation between QTc interval prolongation and diabetic complications. The EURODIAB IDDM Complications Study Group. *Diabetologia, 42*, 68–75.

Veresiu, A. I., Bondor, C. I., Florea, B., Vinik, E. J., Vinik, A. I., & Gavan, N. A. (2015). Detection of undisclosed neuropathy and assessment of its impact on quality of life: A survey in 25,000 Romanian patients with diabetes. *Journal of Diabetes and Its Complications, 29*, 644–649.

Vileikyte, L., Peyrot, M., Bundy, C., Rubin, R. R., Leventhal, H., Mora, P., et al. (2003). The development and validation of a neuropathy- and foot ulcer-specific quality of life instrument. *Diabetes Care, 26*, 2549–2555.

Vinik, A. (1999). Diabetic neuropathy: Pathogenesis and therapy. *The American Journal of Medicine, 107*(2B), 17S–26S.

Vinik, A., Casellini, C., & Nevoret, M.-L. (2015). Diabetic neuropathies. In L. DeGroot (Ed.), South Dartmouth, MA: MDText.com, Inc.

Vinik, A. I., & Erbas, T. (2001). Recognizing and treating diabetic autonomic neuropathy. *Cleveland Clinic Journal of Medicine, 68*, 928–944.

Vinik, A., & Erbas, T. (2002). Neuropathy. In N. Ruderman, J. T. Devlin, S. Schneider, & A. Kriska (Eds.), *Handbook of exercise in diabetes* (pp. 463–495). Alexandria, VA: American Diabetes Association.

Vinik, A. I., & Erbas, T. (2013). Diabetic autonomic neuropathy. In R. M. Buijs & D. F. Swaab (Eds.), *Handbook of clinical neurology* (3rd ed., pp. 279–294). Amsterdam, The Netherlands: Elsevier.

Vinik, A., Erbas, T., & Pfeifer, M. A. (2003). Diabetic autonomic neuropathy. In D. Porte, Jr, R. S. Sherwin, & A. Baron (Eds.), *Ellenberg & Rifkin's diabetes mellitus* (6th ed., pp. 789–804). New York, NY: McGraw-Hill.

Vinik, E. J., Hayes, R. P., Oglesby, A., Bastyr, E., Barlow, P., Ford-Molvik, S. L., et al. (2005). The development and validation of the Norfolk QOL-DN, a new measure of patients' perception of the effects of diabetes and diabetic neuropathy. *Diabetes Technology & Therapeutics*, *7*, 497–508.

Vinik, E. J., Hayes, C., Oglesby, A., & Vinik, A. I. (2004). Identification of factors in the nerve fiber specific quality of life (QOL-DN) inventory that reflect QOL and health status. *Diabetes*, *53*, A295.

Vinik, A. I., Maser, R. E., & Ziegler, D. (2010). Neuropathy: The crystal ball for cardiovascular disease? *Diabetes Care*, *33*, 1688–1690.

Vinik, A., Maser, R., & Ziegler, D. (2011). Autonomic imbalance: Prophet of doom or scope for hope? *Diabetic Medicine*, *28*, 643–651.

Vinik, A. I., Mitchell, B. D., Leichter, S. B., Wagner, A. L., O'Brian, J. T., & Georges, L. P. (1995). Epidemiology of the complications of diabetes. In R. D. G. Leslie & D. C. Robbins (Eds.), *Diabetes: Clinical science in practice* (pp. 221–287). Cambridge, UK: Cambridge University Press.

Vinik, A. I., Nevoret, M., Casellini, C., & Parson, H. (2013). Neurovascular function and sudorimetry in health and disease. *Current Diabetes Reports*, *13*, 517–532.

Vinik, E., Paulson, J., Ford-Molvik, S., & Vinik, A. (2008). German-translated Norfolk quality of life (QOL-DN) identifies the same factors as the english version of the tool and discriminates different levels of neuropathy severity. *Journal of Diabetes Science and Technology*, *2*, 1075–1086.

Vinik, A. I., Perrot, S., Vinik, E. J., Pazdera, L., Jacobs, H., Stoker, M., et al. (2015). Long-term safety and efficacy of repeated applications of capsaicin 8% patch (QUTENZA™) in painful diabetic peripheral neuropathy: Pace study. AACE ePoster Library. May 13, 2015; 97789. http://aace.multilearning.com/aace/2015/eposter/97789/aaron.vinik.long-term.safety.and.efficacy.of.repeated.applications.of.html?f=p6m3e813o10388 Accessed 24.03.16.

Vinik, A. I., Strotmeyer, E. S., Nakave, A. A., & Patel, C. V. (2008). Diabetic neuropathy in older adults. *Clinics in Geriatric Medicine*, *24*, 407–435. v.

Vinik, A. I., Suwanwalaikorn, S., Stansberry, K. B., Holland, M. T., McNitt, P. M., & Colen, L. E. (1995). Quantitative measurement of cutaneous perception in diabetic neuropathy. *Muscle & Nerve*, *18*, 574–584.

Vinik, A., Ullal, J., Parson, H. K., & Casellini, C. M. (2006). Diabetic neuropathies: Clinical manifestations and current treatment options. *Nature Clinical Practice. Endocrinology & Metabolism*, *2*, 269–281.

Vinik, E. J., Vinik, A. I., Paulson, J. F., Merkies, I. S., Packman, J., Grogan, D. R., et al. (2014). Norfolk QOL-DN: Validation of a patient reported outcome measure in transthyretin familial amyloid polyneuropathy. *Journal of the Peripheral Nervous System*, *19*, 104–114.

Vinik, A. I., & Ziegler, D. (2007). Diabetic cardiovascular autonomic neuropathy. *Circulation*, *115*, 387–397.

Volpato, S., Leveille, S. G., Blaum, C., Fried, L. P., & Guralnik, J. M. (2005). Risk factors for falls in older disabled women with diabetes: The women's health and aging study. *The Journals of Gerontology. Series A, Biological Sciences and Medical Sciences*, *60*, 1539–1545.

Volpato, S., Maraldi, C., & Fellin, R. (2010). Type 2 diabetes and risk for functional decline and disability in older persons. *Current Diabetes Reviews*, *6*, 134–143.

Wallace, C., Reiber, G. E., LeMaster, J., Smith, D. G., Sullivan, K., Hayes, S., et al. (2002). Incidence of falls, risk factors for falls, and fall-related fractures in individuals with diabetes and a prior foot ulcer. *Diabetes Care*, *25*, 1983–1986.

Wang, N., & Gibbons, C. H. (2013). Skin biopsies in the assessment of the autonomic nervous system. *Handbook of Clinical Neurology*, *117*, 371–378.

Watkins, P. J. (1993). Progression of diabetic autonomic neuropathy. *Diabetic Medicine*, *10*(Suppl. 2), 77S–78S.

Wirta, O., Pasternack, A., Mustonen, J., & Laippala, P. (1997). Renal and cardiovascular predictors of 9 year total and sudden cardiac mortality in non-insulin-dependent diabetic subjects. *Nephrology, Dialysis, Transplantation*, *12*, 2612–2617.

Witte, D. R., Tesfaye, S., Chaturvedi, N., Eaton, S. E., Kempler, P., & Fuller, J. H. (2005). Risk factors for cardiac autonomic neuropathy in type 1 diabetes mellitus. *Diabetologia*, *48*, 164–171.

Witzke, K. A., & Vinik, A. I. (2005). Diabetic neuropathy in older adults. *Reviews in Endocrine & Metabolic Disorders*, *6*, 117–127.

Wong, M. C., & Chung, J. W. (2011). Feasibility of contact heat evoked potentials for detection of diabetic neuropathy. *Muscle & Nerve*, *44*, 902–906.

Yajnik, C. S., Kantikar, V. V., Pande, A. J., & Deslypere, J. P. (2012). Quick and simple evaluation of sudomotor function for screening of diabetic neuropathy. *ISRN Endocrinology*, *2012*, 103714.

Yajnik, C. S., Kantikar, V., Pande, A., Deslypere, J. P., Dupin, J., Calvet, J. H., et al. (2013). Screening of cardiovascular autonomic neuropathy in patients with diabetes using noninvasive quick and simple assessment of sudomotor function. *Diabetes & Metabolism*, *39*, 126–131.

Yarnitsky, D., & Sprecher, E. (1994). Thermal testing: Normative data and repeatability for various test algorithms. *Journal of the Neurological Sciences*, *125*, 39–45.

Young, M. J., Boulton, A. J. M., MacLeod, A. F., Williams, D. R. R., & Sonksen, P. H. (1993). A multicenter study of the prevalence of diabetic peripheral neuropathy in the United Kingdom hospital clinic population. *Diabetologia*, *36*, 150–154.

Zhivov, A., Blum, M., Guthoff, R., & Stachs, O. (2010). Real-time mapping of the subepithelial nerve plexus by in vivo confocal laser scanning microscopy. *The British Journal of Ophthalmology*, *94*, 1133–1135.

Ziegler, D. (2008). Painful diabetic neuropathy: Treatment and future aspects. *Diabetes/Metabolism Research and Reviews*, *24*(Suppl. 1), S52–S57.

Ziegler, D., Cicmir, I., Mayer, P., Wiefels, K., & Gries, F. A. (1988). Somatic and autonomic nerve function during the first year after diagnosis of type 1 (insulin-dependent) diabetes. *Diabetes Research*, *7*, 123–127.

Ziegler, D., & Gries, F. A. (1997). Alpha-lipoic acid in the treatment of diabetic peripheral and cardiac autonomic neuropathy. *Diabetes*, *46*(Suppl. 2), S62–S66.

Ziegler, D., Gries, F. A., Spuler, M., & Lessmann, F. (1992). The epidemiology of diabetic neuropathy. Diabetic cardiovascular autonomic neuropathy multicenter study group. *Journal of Diabetes and Its Complications*, *6*, 49–57.

Ziegler, D., Papanas, N., Vinik, A. I., & Shaw, J. E. (2014). Epidemiology of polyneuropathy in diabetes and prediabetes. In D. W. Zochodne & R. A. Malik (Eds.), *Diabetes and The Nervous System: 126 (3rd. series). Handbook of clinical neurology* (pp. 3–22): Elsevier.

Ziegler, D., Schatz, H., Conrad, F., Gries, F. A., Ulrich, H., & Reichel, G. (1997). Effects of treatment with the antioxidant alpha-lipoic acid on cardiac autonomic neuropathy in NIDDM patients. A 4-month randomized controlled multicenter trial (DEKAN study). deutsche kardiale autonome neuropathie. *Diabetes Care*, *20*, 369–373.

Ziegler, D., Strom, A., Nowotny, B., Zahiragic, L., Nowotny, P. J., Carstensen-Kirberg, M., et al. (2015). Effect of Low-energy diets differing in fiber, red meat, and coffee intake on cardiac autonomic function in obese individuals with type 2 diabetes. *Diabetes Care*, *38*, 1750–1757.

Ziegler, D., Zentai, C. P., Perz, S., Rathmann, W., Haastert, B., Doring, A., et al. (2008). Prediction of mortality using measures of cardiac autonomic dysfunction in the diabetic and nondiabetic population: The MONICA/KORA Augsburg Cohort Study. *Diabetes Care*, *31*, 556–561.

Ziegler, D., Zentai, C., Perz, S., Rathmann, W., Haastert, B., Meisinger, C., et al. (2004). Diminished heart rate variability (HRV) and prolonged QTc interval, but not increased QT dispersion (QTD) are predicators of mortality in the diabetic population (Abstract). *Diabetes*, *53*, A57.

Ziegler, D., Zentai, C., Perz, S., Rathmann, W., Haastert, B., Meisinger, C., et al. (2006). Selective contribution of diabetes and other cardiovascular risk factors to cardiac autonomic dysfunction in the general population. *Experimental and Clinical Endocrinology & Diabetes*, *114*, 153–159.

CHAPTER FOURTEEN

Wherefore Art Thou, O Treatment for Diabetic Neuropathy?

R.A. Malik[1]
Weill Cornell Medicine-Qatar, Qatar Foundation, Education City, Doha, Qatar
[1]Corresponding author: e-mail address: rayaz.a.malik@man.ac.uk

Contents

Abstract

As of March 2016, we continue to advocate the diagnosis of diabetic neuropathy using a simple foot examination or monofilament, which identifies only those with severe neuropathy and hence risk of foot ulceration. Given the fact that the 5-year mortality rate of diabetic patients with foot ulceration is worse than that of most common cancers, surely we should be identifying patients at an earlier stage of neuropathy to prevent its progression to a stage with such a high mortality? Of course, we lament that there is no licensed treatment for diabetic neuropathy. Who is to blame? As researchers and carers, we have a duty of care to our patients with diabetic neuropathy. So, we have to look forward not backwards, and move away from our firmly entrenched views on the design

International Review of Neurobiology, Volume 127
ISSN 0074-7742
http://dx.doi.org/10.1016/bs.irn.2016.03.008

and conduct of clinical trials for diabetic neuropathy. Relevant organizations such as Neurodiab, the American Diabetes Association and the Peripheral Nerve Society have to acknowledge that they cannot continue to endorse a bankrupt strategy. The FDA needs an open and self-critical dialogue with these organizations, to give pharmaceutical companies at least a fighting chance to deliver effective new therapies for diabetic neuropathy.

I have not failed, I've just found 10,000 ways that won't work

attr. Thomas Edison

1. THE PROBLEM

As the doctors say of a wasting disease, to start with it is easy to cure but difficult to diagnose; after a time... it becomes easy to diagnose but difficult to cure.

Niccolo Machiavelli

Diabetic neuropathy affects at least 50% of patients with diabetes during their lifetime. The human and economic burden of diabetic neuropathy and its consequences in the form of painful neuropathy, foot ulceration, and amputation are considerable for patients and healthcare systems. This is true in both in the developed (Abbott, Carrington, Ashe, et al., 2002; Boulton, Vileikyte, Ragnarson-Tennvall, et al., 2005) and particularly in the developing world (Riaz et al., 2014) where the consequences of disability are grave due to loss of employment and hence livelihood—not just for the patient, but the whole family. The life expectancy of diabetic patients with a foot ulcer (14.4 years) or Charcot foot (13.9 years) is markedly reduced compared to a normative U.K. population (van Baal, Hubbard, Game, et al., 2010). Moreover, by the time, a patient with diabetes has developed a foot ulcer their 5-year mortality is greater than lymphoma (14%), breast (10.6%), prostate (10%), and bladder (22%) cancer (Armstrong, Wrobel, & Robbins, 2007; Marshall, Webb, Hall, et al., 2016) and is much higher than colon cancer (11%), non-Hodgkin's (31%), and Hodgkin's lymphoma (14%) (Al-Hamadani, Habermann, Cerhan, et al., 2015). Indeed, it is comparable to feared cancers such as lung cancer (48%) (Field, Duffy, Baldwin, et al., 2016; Malvezzi, Carioli, Bertuccio, et al., 2016) (Fig. 1). Furthermore, according to the World Health Organization and a recent systematic subnational analysis for the Global Burden of Disease in China, the three leading causes of death in the world are ischemic heart disease, stroke, and chronic obstructive pulmonary disease (COPD) (Zhou, Wang, Zhu, et al., 2016). The 5-year mortality after a myocardial infarct (28%) (Bata, Gregor, Wolf, et al., 2006), stroke (41.7%) (Hankey, Jamrozik,

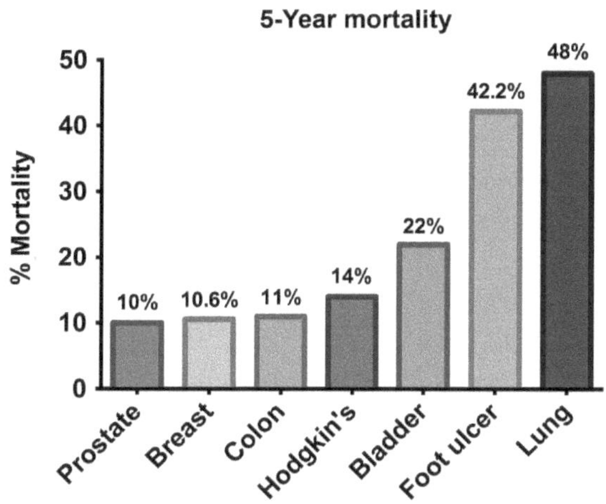

Fig. 1 Five-year mortality of common cancers compared to diabetic patients with foot ulceration.

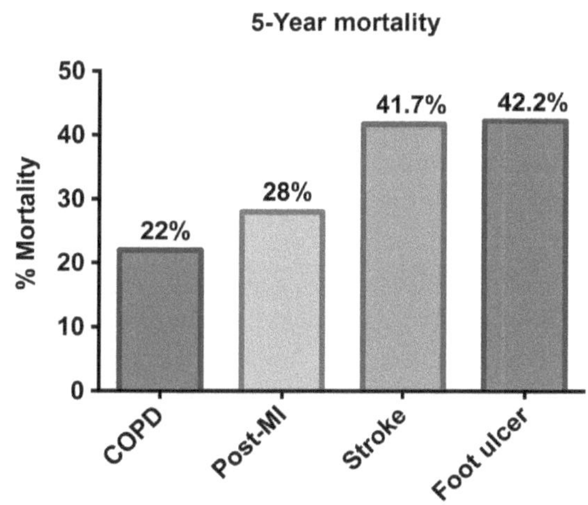

Fig. 2 Five-year mortality for the three leading causes of death in the world compared to patients with diabetic foot ulceration.

Broadhurst, et al., 2000), and COPD (22%) (Martinez, Mannino, Jaimes, et al., 2015) are, in each case, lower than in patients with diabetes and a foot ulcer (Walsh, Hoffstad, Sullivan, & Margolis, 2015) (Fig. 2).

It is instructive to compare the consequences of diabetic neuropathy with other complications of diabetes. A recent population-based survey of 414,523 people with diabetes found that 20,737 (5%) developed foot ulcers and that the 5-year mortality of these patients was 42.2%. Even after controlling for major risk factors for mortality in this population the fully adjusted hazard ratio (HR) for death were 2.48, suggesting a unique and as yet unidentified relationship between diabetic neuropathy and mortality (Walsh et al., 2015). In contrast, in a prospective (7.9 years of follow-up) study of 2237 patients with type 1 diabetes, the age- and sex-adjusted HRs of all-cause mortality were 1.45 (95% CI 0.71–2.96) and 4.16 (1.96–8.84) in patients with nonproliferative and proliferative retinopathy at baseline. After adjustment for cardiovascular risk factors, the association with nonproliferative retinopathy and proliferative retinopathy became nonsignificant (van Hecke, Dekker, Stehouwer, et al., 2005). For diabetic nephropathy, the FIELD study of 9795 patients with T2DM reported that the risk of total CVD events when comparing patients with an estimated GFR (eGFR) of $\geq$90 mL min^{-1} 1.73 m^{-2} was increased (HR [95% CI] 1.14 [1.01–1.29] for eGFR 60–89 mL min^{-1} 1.73 m^{-2} and 1.59 [1.28–1.98] for eGFR 30–59 mL min^{-1} 1.73 m^{-2}; $P<0.001$). Microalbuminuria and macroalbuminuria also increased the total CVD event rate (HR 1.25 [1.01–1.54] and 1.19 [0.76–1.85] $P=0.001$ for trend) independent of eGFR ($\geq$90 mL min^{-1} 1.73 m^{-2}), respectively. However, in multivariable analysis, 77% of the effect of eGFR and 81% of the effect of albumin:creatinine ratio were accounted for by a low HDL-cholesterol and elevated blood pressure (Drury, Ting, Zannino, et al., 2011). We readily deploy retinal screening and microalbuminuria testing to detect the earliest phases of diabetic retinopathy and nephropathy. Yet for neuropathy, we recommend the monofilament and foot examination, tests which can only identify the "high risk diabetic foot" (Mayfield & Sugarman, 2000).

In 2016 we have no FDA approved treatment for diabetic neuropathy and a graveyard of potential therapies. Many explanations, such as the wrong types of patients, severity of neuropathy, and inadequate trial duration have been suggested for the singular lack of progress in developing an effective treatment for diabetic neuropathy (Malik, 2014b). The ultimate test of the usefulness of a drug or device depends on the determination of outcomes. These have to be quantified reliably, have relevance to the underlying disease process, and should be tested in randomized clinical trials (RCTs) that use appropriate numbers and types of patients and duration of treatment. The purported gold standard end points (Tesfaye, Boulton, Dyck, et al., 2010) of symptoms and signs of diabetic neuropathy (Dyck, Overland, Low, et al., 2010) combined with neurophysiological testing (Dyck, Albers, Wolfe, et al., 2013; Litchy, Albers, Wolfe, et al., 2014) perform inadequately even in the hands of the "experts," and yet continue to be advocated as primary end points in clinical trials (Dyck, 2014).

Professional organizations continue to endorse consensus guidance, which by default only allows minor incremental change or no change at all.

The FDA will carry out due process and endorse the low-risk consensus view. The consequence of this approach has been failure after failure of clinical trials in diabetic neuropathy, so that the field has the unenviable record of a 100% failure rate. When future trials fail, and they will, investigators will readily blame "big pharma" for enrolling the wrong types of patients, too few patients, or choosing an inadequate study duration or the wrong endpoint. And thus the cycle continues.

2. WHAT CAN WE DO?

We need to learn to measure what we value, not value what we can easily measure
Marcus Aurelius

Because we have previously not been able to adequately quantify small fibers, the default position, and hence consensus, has been to ignore them. Techniques that quantify small fiber function, such as quantitative sensory testing, have been shown to be reliable (Dyck, Argyros, Russell, et al., 2014) and newer techniques that quantify small fiber structure, present intuitively attractive end points for clinical trials of diabetic neuropathy, especially for assessing early repair (Malik, 2014a). Two such techniques are the visual quantification of intraepidermal nerve fibers through skin biopsy (Fig. 3) (Lauria, Bakkers, Schmitz, et al., 2010; Lauria, Hsieh,

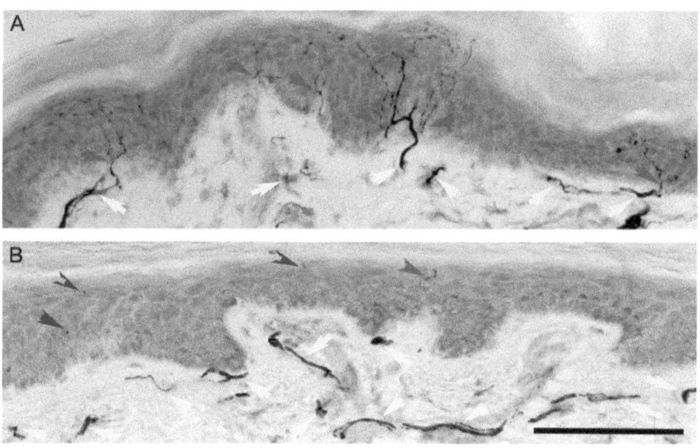

Fig. 3 Skin biopsies immunostained for neuronal marker PGP 9.5 from a healthy subject (A) and a patient with severe DPN (B). Note the almost complete depletion of IENFD (*red* (*gray* in the print version) arrows) and only very small nerve fiber profiles (*blue* (*dark* gray in the print version) arrows) scattered in the epidermis and a reduction of the subepidermal nerve plexus (*yellow* (*light gray* in the print version) arrows) in (B). Original magnification $\times$ 200, scale bar = 100 μm.

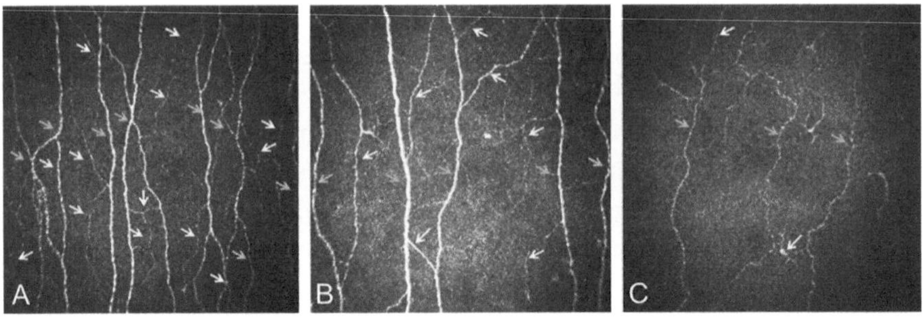

Fig. 4 Corneal confocal microscopy images of the sub-basal nerve plexus from a control (A), a patient without DPN (B), and a patient with DPN (C) showing the reduction in corneal nerves in those with DPN. *Red* (*gray* in the print version) arrows indicate main nerve fibers; *yellow* (*white* in the print version) arrows indicate branches.

Johansson, et al., 2010; Lauria, Morbin, Lombardi, et al., 2003) and corneal confocal microscopy (CCM), which allows noninvasive in vivo imaging of corneal nerves (Fig. 4) (Malik, Kallinikos, Abbott, et al., 2003). Indeed a robust body of data, which is conveniently ignored, shows that small fiber damage occurs in patients with early diabetic neuropathy (Bitirgen, Ozkagnici, Malik, et al., 2014; Malik, 2014a; Ziegler, Papanas, Zhivov, et al., 2014) and subjects with IGT (Asghar, Petropoulos, Alam, et al., 2014; Singleton, Smith, & Bromberg, 2001; Smith, Ramachandran, Tripp, et al., 2001). Of clinical relevance to the hard end point sought by the FDA (foot ulceration), an equally strong body of data shows that small cutaneous fibers release an array of neuropeptides, which are key to the process of wound healing (Ashrafi, Baguneid, & Bayat, 2015; Laverdet, Danigo, Girard, et al., 2015) and hence repair after foot ulceration.

While estimation of IENF density (IENFD) by skin biopsy offers an objective means of quantifying small fiber pathology, it is an invasive and costly procedure that does not allow for the same nerves to be visualized repeatedly (Lauria, Bakkers, et al., 2010; Lauria, Hsieh, et al., 2010; Lauria, Morbin, Lombardi, et al., 2003). Our previous NIH and JDRF funded studies have shown that CCM can quantify early small fiber axonal damage in diabetic neuropathy (Petropoulos, Alam, Fadavi, et al., 2014), reliably (Petropoulos, Manzoor, Morgan, et al., 2013), with high sensitivity and specificity (Petropoulos et al., 2014). CCM has a better sensitivity and comparable specificity for the diagnosis of diabetic neuropathy (Chen, Graham, Dabbah, et al., 2015) and correlates with IENF loss (Quattrini, Tavakoli, Jeziorska, et al., 2007). We have also shown that CCM can be deployed in children with type 1 diabetes, that corneal nerve loss is evident in young people with T1DM without retinopathy (Szalai, Deak, Modis, et al., 2016), and that it predicts incident neuropathy in type 1 diabetes (Pritchard, Edwards, Russell, et al., 2015) in those with poorer glycemic control and lower HDL (Dehghani, Pritchard, Edwards, et al., 2016). The potential of corneal nerve loss as a surrogate end point for diabetic neuropathy has been further established by showing that it detects nerve regeneration in diabetic patients undergoing simultaneous pancreas and kidney transplantation (Tavakoli, Mitu-Pretorian, Petropoulos, et al., 2013) and in patients with sarcoid (van Velzen, Heij, Niesters, et al., 2014) and diabetic (Brines, Dunne, van Velzen, et al., 2014) neuropathy treated with the novel peptide, ARA 290. These latter studies showed nerve repair within 6 months and 28 days respectively, over which time all other currently endorsed end points, such as neurological deficits, quantitative sensory testing, and neurophysiology remained unchanged. Furthermore, we have recently published a large normative reference database (Tavakoli, Ferdousi, Petropoulos, et al., 2015) and developed an automated image analysis algorithm to enable rapid and objective quantification of corneal nerve morphology (Dabbah, Graham, Petropoulos, et al., 2011). Views on the role and importance of small fibers in neuropathy and their utility in clinical trials of diabetic neuropathy have to change. Unless we incorporate a measure of small fiber damage and repair as a surrogate end point in clinical trials of diabetic neuropathy, we will continue to fulfil the self-fulfilling prophecy of failure after failure.

3. DIAGNOSTIC TESTS ARE NOT NECESSARILY GOOD SURROGATE END POINTS

An army of philosophers would not be sufficient to change the nature of error and to make it truth

Averroes

In 2010, the Toronto consensus panel separated diabetic peripheral neuropathy (DPN) into typical and atypical groups (Tesfaye et al., 2010). Typical DPN is "a symmetrical, length-dependent sensorimotor polyneuropathy attributable to metabolic, and microvessel alterations as a result of chronic hyperglycemia exposure and cardiovascular risk covariates." Atypical variants of diabetic neuropathy differ in onset, course, manifestations, associations, and putative mechanisms and are likely to be associated with pain and/or dysautonomia. It is therefore important to identify different subtypes of DPN to ensure homogeneity of patients recruited into trials.

The presentation of diabetic neuropathy can be variable and may include pain, which is reported by approximately one-third of patients with diabetes, regardless of associated neurological deficit (Abbott, Malik, van Ross, et al.,

2011). The classical description is of an unremitting burning pain that is characteristically worse at night, with a gradual distal-to-proximal progression of symptoms in a glove-and-stocking distribution. Overt motor deficits in the form of muscle wasting and weakness were typically thought to occur in more advanced neuropathy (Andersen, 2014). However, our recent study suggests that weakness and muscle atrophy may occur at an earlier stage (Almurdhi, Reeves, Bowling, et al., 2016). Measures of strength or size of muscles could potentially be incorporated as end points in clinical trials of early neuropathy, as these measures are sensitive to change and have clinical relevance.

Various tools have been validated to assess neurological impairment in diabetic neuropathy. These include the Michigan Neuropathy Screening Instrument (MNSI) and Neuropathy Disability Score (NDS). Indeed, because of the wide use and ease of administering these tools, they were incorporated into a number of intervention trials of glycemic control in type 2 diabetes. To date, they have all singularly failed to show a significant improvement in neuropathy in studies such as ACCORD and VADT trials (Calles-Escandon, Lovato, Simons-Morton, et al., 2010; Duckworth, Abraira, Moritz, et al., 2009). A Cochrane review has concluded that "improved glycemic control does not improve neuropathy in type 2 diabetes" (Callaghan, Little, Feldman, et al., 2012). Is this really the conclusion that we should be drawing? Or should we be addressing the major deficiency in these trials: namely use of the wrong end points?

Pain-specific questionnaires have also been utilized in order to quantify painful symptoms. The available instruments include the Brief Pain Inventory and the Neuropathic Pain Questionnaire (NPQ), McGill Pain Questionnaire, and visual analog scales (VAS). Additionally, neuropathy-specific quality-of-life measures, such as the NeuroQoL and the Norfolk Quality of Life Scale, may play a role in identifying patient-important factors in neuropathy (Javed, Alam, & Malik, 2015a).

4. CLINICAL TRIALS IN DIABETIC NEUROPATHY

They sent forth men to battle, But no such men return; And home, to claim their welcome, Come ashes in an urn

Aeschylus

There have been many clinical trials in subjects with diabetic neuropathy. As discussed earlier, their failure may derive from many factors ranging from incorrect choice of the presumed pathogenic mechanism to poor trial design, so that the data should be viewed with all of these caveats in mind.

4.1 Glycemic Control

The Eurodiab Insulin-Dependent Diabetes Mellitus (IDDM) study found that suboptimal glycemic control was associated with the development of DPN in a large cohort of ~3000 people with T1DM. 1172 patients with type 1 diabetes were assessed for neuropathy at baseline (1989–91) and at follow-up (1997–99), with a mean follow-up period of 7.3 ± 0.6 years. A standardized protocol for the evaluation of DPN included clinical evaluation, QST, and autonomic-function tests. A number of factors were found to be independently associated with the incidence of neuropathy including the duration of diabetes and change in HbA1c (Tesfaye, Chaturvedi, Eaton, et al., 2005). Additionally, the incidence of neuropathy was associated with potentially modifiable cardiovascular risk factors, including body mass index, smoking, and hypertension, as well as raised blood triglyceride levels (Tesfaye et al., 2005). Other observational studies have also emphasised the impact of poor glycemic control on the development and progression of DPN (Herman, Aubert, Engelgau, et al., 1998; Oyibo, Prasad, Jackson, et al., 2002).

Whether or not an improvement in glycemic control can prevent or delay progression of clinical neuropathy in patients with type 1 diabetes has been formally assessed in two key studies. The Stockholm Diabetes Intervention Study was initiated in 1982, at which time 48 patients were randomized to intensive insulin and 54 to standard insulin and followed for 7.5 years. While there was a positive effect on nerve conduction there was no change in symptoms or in vibration and thermal thresholds (Reichard, Nilsson, & Rosenqvist, 1993). In the landmark Diabetes Control and Complications Trial (DCCT), clinical neuropathy was defined as an abnormal neurologic examination and either abnormal nerve conduction in at least two peripheral nerves or unequivocally abnormal autonomic-nerve testing. Details of the DCCT trial are discussed elsewhere in this volume. Intensive therapy reduced the appearance of neuropathy after 5 years by 69% in the primary-prevention cohort and by 57% in the secondary-intervention cohort (The Diabetes Control and Complications Trial Research Group, 1993, 1995). Of note, this was a 1400-patient multicentre study and took 5 years to show an effect on patients with type 1 diabetes who had no neuropathy at baseline. The apparent success of the DCCT has been used to highlight the need for large numbers of patients without neuropathy at baseline who are followed for at least 5 years using a combination of clinical neurological examination, neurophysiology, and autonomic-function testing in clinical trials of therapies for DPN.

In type 2 diabetes, the evidence of a benefit of improved glycemic control on neuropathy is less clear—or is it? The United Kingdom Prospective Diabetes Study (Stratton, Adler, Neil, et al., 2000) reported a significant risk reduction in amputation for every 1% reduction in mean HbA1c (Stratton et al., 2000) and a lower rate of impaired VPT with intensive therapy versus standard therapy. However, this effect only became significant after 15 years, with no significant advantage observed at 3, 6, 9, and 12 years. What does this mean for trials that incorporate vibration perception threshold as a primary or secondary end point over 1–2 years?

Intensive glycemic control in type 2 diabetes has been shown to reduce microvascular complications in a 6-year randomized, prospective study (Ohkubo, Kishikawa, Araki, et al., 1995). Does this mean that the minimum duration of trials should be 6 years? Boussageon, Bejan-Angoulvant, Saadatian-Elahi, et al. (2011) conducted a meta-analysis of seven trials involving over 34,000 patients and did not find a reduction in the incidence of DPN in patients managed through intensive glycemic control. Furthermore, a Cochrane review of 17 randomized trials concluded that tight glycemic control prevented neuropathy in type 1 diabetes, driven principally by the DCCT findings, but that it had no benefit in type 2 diabetes (Callaghan, Little, et al., 2012). It should be noted that these meta-analyses have significant limitations as the targeted glucose levels, therapeutic strategies, outcome measures, trial design, and duration of follow-up differ between studies. The principal culprit for demonstrating a lack of benefit is, however, that the majority of these studies utilized crude neuropathy end points such as a monofilament and foot exam and vibration perception.

The VA Cooperative study of type 2 diabetes reported no difference in the prevalence of autonomic neuropathy between the intensive and standard therapy arms at 2 years (Duckworth et al., 2009). In contrast, the Steno-2 Trial reported that an intervention that integrated glucose control and multiple cardiovascular risk factor management reduced the prevalence of cardiac autonomic neuropathy but not unsurprisingly showed no benefit for somatic neuropathy, as assessed using vibration perception thresholds in patients with type 2 diabetes and microalbuminuria even after 7.8 years of follow-up (Gaede, Lund-Andersen, Parving, et al., 2008).

4.2 Pancreas Transplantation

The only known therapy to restore insulin secretion in response to feedback mechanisms in patients with diabetes is pancreas transplantation (Fiorina & Secchi, 2007; Tavakoli & Liong, 2012). Differing lengths of time for improvement in DPN have been reported in patients treated by pancreas transplantation. Fioretto, Steffes, Sutherland, et al. (1998) found that the reversal of DPN was only evident 10 years after transplantation. Agudo, Valls-Solé, Recasens, et al. (2002) reported an increase in action potential amplitude and conduction velocity at 3 months and 1 year after transplantation in a small study involving 26 patients. In the controlled study by Kennedy, Navarro, Goetz, et al. (1990), an improvement in sensory and motor function 12 and 24 months after pancreas transplantation was reported in a cohort of 61 patients with type 1 diabetes, but neurophysiological and autonomic-function tests did not demonstrate recovery. Navarro, Sutherland, and Kennedy (1997) reported similar findings in 115 subjects with type 1 diabetes who underwent pancreas transplantation with an improvement in composite scores assessing motor and sensory neuropathy but only a slight improvement in autonomic function after a longer 10-year follow-up. In contrast, Boucek, Havrdova, Voska, et al. (2008) assessed IENFD in 18 subjects following pancreas and kidney transplantation and found that only three patients showed an improvement, suggesting that in some patients DPN may reach a nonreversible stage. It should be noted that all these studies used different outcome measures to assess DPN, with variable follow-up periods. Hence, it is likely that the improvement in DPN reported in the literature depends entirely on the endpoint used to assess DPN. Indeed, in a contemporary cohort of 15 patients (Mehra, Tavakoli, Kallinikos, et al., 2007), we found no change in electrophysiology; quantitative sensory testing and IENFD, whereas CCM parameters improved significantly 12 months after transplantation (Tavakoli, Mitu-Pretorian, et al., 2013).

While pancreas transplantation may represent the most effective method for restoring normoglycemia, its use is limited to patients in end-stage kidney disease or, less frequently, to patients with type 1 diabetes and unpredictable hypoglycemia. This is due to the restricted and unpredictable availability of suitable organs, complications of surgery and the risks accompanying long-term immunosuppression. Recently, islet-cell transplantation has been considered as a less invasive alternative for suitable patients with diabetes. Del Carro, Fiorina, Amadio, et al. (2007) reported marked improvement in neurophysiology with no change in skin biopsies.

4.3 α-Lipoic Acid

Hyperglycemia results in increased production of reactive oxygen species due to the autooxidation of excess glucose and a failure of antioxidant mechanisms (Kaur, Pandhi, & Dutta, 2011). These oxygen free radicals mediate endothelial dysfunction by inhibiting nitric oxide, leading to ischemic nerve damage. α-Lipoic acid (ALA) is a free-radical scavenger that is suggested to alleviate this oxidative stress. It is licensed for the treatment of DPN in Germany and was found to be well tolerated and efficacious in the management of painful DPN, when administered parenterally. Indeed, Ziegler, Nowak, Kempler, et al. (2004) reported a clinically significant improvement in the symptoms of DPN after administration of a 600-mg daily dose of ALA over 3 weeks in their meta-analysis of four placebo-controlled trials. However in the more recently published NATHAN 1 study, a 4-year multicenter randomized controlled trial comparing ALA to placebo, Dyck, Norell, Tritschler, et al. (2007) found no improvement in neurophysiology, QST, and composite neuropathy scores. Unfortunately, the primary end point failed to deteriorate significantly in placebo-treated subjects, highlighting the lack of decline of placebo-treated subjects in recent trials of human diabetic neuropathy (Dyck et al., 2007). In a study of 44 participants with type 1 diabetes randomized to antioxidant treatment with allopurinol, ALA, and nicotinamide, or placebo over 24 months, there was no change in symptoms, signs, electrophysiology and intraepidermal nerve fiber density, or cardiovascular autonomic reflex testing (Pop-Busui, Stevens, Raffel, et al., 2013). In a recent meta-analysis from China, 1410 participants in 18 RCTs demonstrated a significant benefit on median and peroneal nerve conduction velocities in those given lipoic acid, prostaglandin E1, and methylcobalamin compared to prostaglandin E1 and methylcobalamin alone. However, in these studies, there were only 30–100 patients in each arm, the drugs were administered intravenously, there was no placebo group and some of the studies were <14 days duration (Jiang, Li, Wang, et al., 2015).

4.4 Aldose Reductase Inhibitors

Aldose reductase is the rate-limiting enzyme in the polyol pathway for glucose metabolism (Callaghan, Cheng, Stables, et al., 2012) and hyperglycemia increases flux through this pathway leading to reduced production of the vasodilator nitric oxide and eventually perpetuating ischemic nerve injury. Aldose reductase inhibitors (ARIs) initially showed an improvement in nerve conduction velocity and some improvement in myelinated nerve fiber density and regenerative clusters in sural nerve biopsies (Dyck, Zimmerman, Vilen, et al., 1988; Greene, Arezzo, & Brown, 1999; Sima, Bril, Nathaniel, et al., 1988), but failed consistently in subsequent phase III clinical trials (Boulton, Malik, Arezzo, et al., 2004). There are also concerns over toxicity and given the limited efficacy this has precluded their use in patients (Boulton, Kempler, Ametov, et al., 2013). Epalrestat is the only ARI which is licensed in Japan and India based on limited randomized controlled trial data (Goto, Hotta, Shigeta, et al., 1995). Indeed, a Cochrane collaboration review of 32 trials comprising 4970 participants found no overall benefit of ARIs in DPN (Chalk, Benstead, & Moore, 2007).

4.5 Vitamin B

Benfotiamine is a fat-soluble derivative of thiamine that has been shown in animal models to inhibit three major pathways implicated in oxidative stress and vascular dysfunction in diabetes: the advanced glycation end-product pathway, the hexosamine pathway, and the protein kinase C (PKC)–diacylglycerol pathway (Hammes, Du, Edelstein, et al., 2003). Although the multimodal effects of benfotiamine present an attractive option for treating DPN, clinical trials are unclear about its efficacy. A randomized placebo-controlled trial (Stracke, Gaus, Achenbach, et al., 2008) of 165 patients reported an improvement in patient-reported symptoms in the per-protocol arm of the study, although there was no improvement in the intent-to-treat group compared to placebo. Furthermore there was an improvement in symptoms but no change in deficits between the placebo and treatment arms of the study (Stracke et al., 2008). In a multicenter, randomized, double-blind, placebo-controlled trial of 214 patients with type 2 diabetes randomly assigned to 24 weeks of treatment with either L-methylfolate calcium 3 mg, methylcobalamin 2 mg, and pyridoxal-5′-phosphate 35 mg, or placebo the primary end point, not unsurprisingly, VPT did not change, while the secondary end points of the Neuropathy Total Symptom Score (NTSS-6) and Short Form 36 (SF-36), did (Fonseca, Lavery, Thethi, et al., 2013). In a recent "real-world" study, 544 patients reported a mean reduction of 35% in NTSS-6 scores after 12 weeks on a medical food product containing L-methylfolate-methylcobalamin-pyridoxal-5-phosphate (Trippe, Barrentine, Curole, et al., 2016).

4.6 Angiotensin-Converting Enzyme Inhibitors

The development and progression of nephropathy, retinopathy, and neuropathy are closely related. Angiotensin-converting enzyme (ACE) inhibitors delay progression of both nephropathy and retinopathy. We investigated the effect of ACE inhibition on diabetic neuropathy and found an improvement in neurophysiology after 12 months of treatment with the ACE inhibitor trandolapril in 41 normotensive patients with DPN, though there was no change in autonomic function compared to placebo (Malik, Williamson, Abbott, et al., 1998). In the DEMAND study, 200 patients with type 2 diabetes showed a significant reduction in the odds ratios for peripheral neuropathy at 3 years between the calcium channel blocker Manidipine and ACE inhibitor delapril vs delapril alone vs placebo: 0.45 (0.24–0.87; $P=0.017$) and 0.52 (0.27–0.99; $P=0.048$) (Ruggenenti, Lauria, Iliev, et al., 2011). Neuropathy was assessed using symptoms, deficits, autonomic function, QST, and neurophysiology.

4.7 Protein Kinase C Activation

Free radicals generated by the hyperglycemia-induced activation of PKC are thought to play a role in the pathogenesis of DPN by altering vascular permeability and causing vasoconstriction. The PKC inhibitor ruboxistaurin has been evaluated in clinical trials (Brooks, Delaney-Robinson, Molyneaux, et al., 2008; Vinik, Bril, Kempler, et al., 2005). A recent systematic review (Bansal, Badhan, Gudala, et al., 2013) of six randomized controlled trials concluded that ruboxistaurin offered no benefit in the treatment of DPN.

4.8 C-Peptide

C-peptide deficiency is an important contributing factor to the characteristic functional and structural abnormalities of peripheral nerves (Sima, 2003). C-peptide binds to cell membranes, resulting in stimulation of endothelial nitric oxide synthase (eNOS) and Na^+, K^+-ATPase (Wahren, Ekberg, & Jornvall, 2007). In the Joslin 50-Year Medallist Study (Sun, Keenan, Cavallerano, et al., 2011), protection from complications (retinopathy, nephropathy, and neuropathy) was thought to be due to the presence of enriched protective factors against microvascular complications. In two double-blind, placebo-controlled studies in type 1 DM patients, C-peptide replacement or placebo was given with the patients' regular insulin therapy (Ekberg, Brismar, Johansson, et al., 2003, 2007). Sensory NCV assessed in the sural nerve showed a significant improvement. A recent large phase III trial has reported no significant improvement in nerve conduction parameters compared to placebo, although it should be noted that the placebo arm showed an unexpected increase in nerve conduction that paralleled the anticipated treatment effect. As C-peptide, but not placebo, improved vibration perception, the authors suggested that large fiber conduction velocity did not adequately reflect large fiber-mediated sensory function (Wahren, Foyt, Daniels, et al., 2016).

4.9 Actovegin

Actovegin is a hemoderivative produced from calf blood by ultrafiltration that exerts insulin-like activity, stimulating glucose transport, pyruvate dehydrogenase, and glucose oxidation (Ziegler, Movsesyan, Mankovsky, et al., 2009). In a multicenter randomized placebo-controlled double-blind trial, 567 patients with type 2 diabetes received 20 intravenous infusions of Actovegin (2000 mg day^{-1}) ($n=281$) or placebo ($n=286$) once daily followed by three tablets of Actovegin (1800 mg day^{-1}) or placebo three times daily for 140 days. TSS ($P=0.0003$), VPT ($P=0.017$), and NIS-LL sensory function ($P=0.021$) improved significantly (Ziegler et al., 2009). However, this is the only trial of Actovegin and comprehensive clinical trials are needed to confirm its benefits in the treatment of DPN (Boulton et al., 2013).

4.10 Disease Modification

Erratic blood glucose control has been associated with DPN and it has been proposed that stable glycemic control may be essential for treating neuropathy. Previously (Callaghan, Little, et al., 2012), it has been demonstrated that continuous subcutaneous insulin infusion (CSII) may achieve near-normal glycemia in people with type 1 diabetes and also lead to an improvement in neurophysiology and painful symptoms. In a recent study by Azmi, Ferdousi, Petropoulos, et al. (2015), there was evidence of corneal nerve regeneration but no impact on QST or neurophysiology in the CSII cohort compared to subjects on a multiple daily injection insulin regimen. This was despite comparable HbA1c over 2 years of follow-up.

Dyslipidemia has been linked with DPN and Wiggin, Sullivan, Pop-Busui, et al. (2009) showed that elevated triglyceride levels correlate with a reduction in myelinated fiber density. Fried, Forrest, Ellis, et al. (2001) have suggested that lowering blood lipid levels may prevent or alleviate the symptoms of DPN. Smith et al. reported that a decrease in triglyceride levels in a cohort of patients with impaired glucose tolerance managed through lifestyle interventions was associated with an increase in IENFD. Furthermore, a recent 12-week double-blind, placebo-controlled study of rosuvastatin showed that there was an improvement in neurological deficits, symptoms, and nerve conduction parameters of DPN after 12 weeks (Hernández-Ojeda, Román-Pintos, Rodríguez-Carrízalez, et al., 2014). The significant reduction in minor amputations in the FIELD study for patients with T2DM receiving fenofibrate points to an important role for triglycerides or alternative mechanisms of benefit for the PPARα agonists (Rajamani, Colman, Li, et al., 2009).

4.11 Nerve Growth Factor

Nerve growth factor (NGF) was first isolated in 1952 by Rita Levi-Montalcini and identified as a protein that induces the growth of nerves, for which she received the Nobel prize in Physiology or Medicine in 1986 (Shelton, 2014). Tanezumab, is a monoclonal antibody, completely humanized against NGF, that binds to NGF with high affinity and selectivity, thereby blocking the NGF–TrKA interaction, inhibiting signaling through sensory and sympathetic neurones and pain (Hirose, Kuroda, & Murata, 2016). One could argue that it is potentially one of the few successes to be born out of the failure of the NGF trials in diabetic neuropathy (Apfel, Kessler, Adornato, et al., 1998). In the pivotal phase III study of recombinant human NGF (rhNGF), which enrolled 1019 patients with diabetic neuropathy, the primary end point was NIS-LL (neurological exam assessing muscle strength, reflexes, and sensory testing) and the secondary end points were vibration perception, cold perception, and heat pain thresholds (Apfel, 2002). It is not clear why one would evaluate large myelinated fiber function to assess the efficacy of a drug that targets small fibers. Not surprisingly, the study failed and perhaps prematurely or ill advisedly, Genentech stopped further trials. However, during development of rhNGF, one of the troublesome side effects of treatment by injection was small fiber-mediated neuropathic pain. This resulted in a massive dose deescalation to amounts which in hindsight were probably nonefficacious (Shelton, 2014). In a recent efficacy and safety study, a single 20 mg dose of subcutaneous Tanezumab showed a significant reduction in pain scores in patients with diabetic neuropathy (Bramson, Herrmann, Carey, et al., 2015). A logical concern for such studies is that an anti-NGF approach could potentially destroy c-fibers. In a safety study of patients with osteoarthritis who were administered Tanezumab: "Neurological safety was evaluated via a composite score made up of nerve conduction attributes and heart rate variability with deep breathing and intraepidermal nerve fiber density." Choosing to evaluate nerve conduction, a large myelinated fiber measure and indeed heart rate variability, a measure of parasympathetic function as a safety measure for a drug, which might harm unmyelinated c-fibers or sympathetic fibers appears to be looking in the wrong place. Indeed, although not significant at 24 weeks, IENFD was reduced in the group given 10 mg of Tanezumab (Brown, Herrmann, Goldstein, et al., 2014). It has been suggested that the development of postural hypotension will be an adequate marker of harm, but this will only occur with severe loss of sympathetic fibers. This appears to be a clear case of where one should employ sensitive, fiber-type-specific end points such as skin biopsy and CCM to detect early c-fiber degeneration.

4.12 Other Growth Factors

Verheyen, Peeraer, Lambrechts, et al. (2013) have suggested that targeting pathogenetic pathways implicated in DPN by utilizing vascular endothelial growth factor (VEGF) and VEGF-derived peptides may be an option for future therapies. However, a recent clinical trial of VEGF in DPN was stopped due to lack of efficacy (Ropper, Gorson, Gooch, et al., 2009). Hepatocyte growth factor (HGF) is a multifunctional, mesenchyme-derived cytokine with potent neurotrophic, angiogenic, and antiapoptotic effects. These combined neurotrophic and angiogenic properties of HGF ought to make it an ideal candidate as a disease-modifying treatment of diabetic neuropathy. Yet two clinical trials have been undertaken with pain as a primary end point. In small numbers of patients, both show efficacy after intramuscular injections of plasmid DNA expressing two different isoforms of HGF (VM202). In the first study 36 patients showed a significant reduction in mean VAS from baseline by 47.2% ($P=0.002$) at 6 months and by 44.1% ($P=0.005$) at 12 months (Ajroud-Driss, Christiansen, Allen, et al., 2013). In a larger follow-up study of 84 patients receiving 8 mg VM202 per leg, there was a significant reduction in the mean pain score at 3 but not at 6 and 9 months and there was no improvement in the MNSI, monofilament testing, or IENFD at 6 months (Kessler, Smith, Cha, et al., 2015). These disparate data sets raise the question of whether to pursue pain as a primary end point, or to consider alternative end points to show disease modification? Viromed, the sponsors of VM202, have chosen to pursue a further trial with pain as a primary end point. Is this not a case of the tail wagging the dog?

4.13 Erythropoietin

Erythropoietin (EPO) is produced in situ by cells under stress and has been found to antagonize the production of proinflammatory molecules and promote tissue healing. While EPO has been found to ameliorate experimental DPN, its use is limited by serious adverse effects, in particular increased thrombotic risk (Brines et al., 2014). ARA 290, a nonhematopoietic peptide designed from the structure of EPO, selectively interacts with the EPO receptors that mediate tissue protection. In a recent phase II trial to evaluate the activity of ARA 290 in type 2 diabetes and DPN, ARA 290 (4 mg) or placebo were self-administered subcutaneously daily for 28 days and the subjects followed for an additional month without further treatment. Neuropathic symptoms were found to improve significantly in the ARA 290 group. Additionally, a significant improvement in corneal nerve morphology was found using CCM in patients randomized to the ARA 290 arm of the trial relative to placebo. ARA 290 remains in development for DPN and type 2 diabetes (Brines et al., 2014).

4.14 Vitamin D

There is an association between vitamin D deficiency and pain in the general population (Hirani, 2012). A mechanistic link between vitamin D and pain has been supported by a recent study reporting that nociceptive calcitonin gene-related peptide (CGRP)-positive neurons have a distinct vitamin D phenotype with hormonally regulated ligand and receptor levels (Tague & Smith, 2011). Vitamin D deficiency results in increased numbers of axons containing CGRP. In cultured neurons, vitamin D receptor (VDR) expression is increased in growth cones and sprouting appears to be regulated by VDR-mediated rapid response signaling pathways (Tague, Clarke, Winter, et al., 2011). NGF is known to be depleted in experimental diabetes (Hellweg, Wohrle, Hartung, et al., 1991) and in a study of patients with diabetic neuropathy, NGF immunostaining on skin keratinocytes correlated with skin axon–reflex vasodilation, a measure of small fiber neuropathy (Anand, Terenghi, Warner, et al., 1996). In an experimental study, NGF expression was maintained in sciatic nerves of diabetic animals treated with a vitamin D analog (CB1093). Similarly Tacalcitol, active vitamin D3, induces NGF production in human epidermal keratinocytes (Fukuoka, Sakurai, Ohta, et al., 2001). Treatment with vitamin D3 has been shown to reduce demyelination in a cuprizone experimental model of demyelination (Wergeland, Torkildsen, Myhr, et al., 2011) and in a separate spinal cord compression model it has been shown to induce axonal regeneration (Bianco, Gueye, Marqueste, et al., 2011).

Several recent observational studies in patients with diabetes have demonstrated a significant association between vitamin D deficiency and paraesthesia and numbness (Soderstrom, Johnson, Diaz, et al., 2012), neurological deficits and electrophysiology (Alamdari et al., 2015; Shehab, Al-Jarallah, Mojiminiyi, et al., 2012), and parasympathetic dysfunction (Maser, Lenhard, & Pohlig, 2015). Furthermore, a recent systematic review and meta-analysis of 1484 patients with type 2 diabetes demonstrated a highly significant association (odds ratio 2.68) between vitamin D deficiency and the development of DPN (Lv, Zhao, Gong, et al., 2014). A more detailed study using electrophysiology and Douleur Neuropathique 4 (DN4) scores has shown that serum levels of vitamin D are significantly reduced while serum vitamin D-binding protein (VDBP) and vitamin D receptor (VDR) levels are comparable between diabetic patients with and without peripheral neuropathy (Celikbilek, Gocmen, Tanik, et al., 2015). Recently, the FIELD multinational study has shown that vitamin D deficiency was present in 50% of 9795 patients with type 2 diabetes and it predicted microvascular outcomes (Herrmann, Sullivan, Veillard, et al., 2015).

Few studies have investigated the potential therapeutic benefits of vitamin D therapy. An open-label prospective study in 51 patients with type 2 diabetes and painful neuropathy showed that ~2000 IU of cholecalciferol daily for 3 months resulted in an ~50% decrease in the Visual Analog Score (VAS) (Lee & Chen, 2008). More recently, in a placebo-controlled study of 112 patients with type 2 diabetes randomized to 50,000 IU of Cholecalciferol once weekly for 8 weeks, there was a significant increase in 25OHD and an improvement in the Neuropathy Symptom Score, but no change in the Neurological Disability Score or neurophysiology (Shehab, Al-Jarallah, Abdella, et al., 2015). We have recently shown a significant improvement in three separate measures of painful diabetic neuropathy following the administration of a single intramuscular dose of 600,000 IU of vitamin D_3 (Basit et al., 2016).

4.15 Angiotensin Axis

The angiotensin II type 2 receptor (AT_2R) axis has been found to play a substantial role in promoting nociceptive signaling by stimulating hyperexcitability and persistent ectopic firing of first-order sensory neurones. Rice et al. have demonstrated that EMA401, a novel small molecule AT_2R antagonist, provides significant pain relief in postherpetic neuralgia in a novel phase II randomized, placebo-controlled trial of 183 patients (Rice, Dworkin, McCarthy, et al., 2014). This drug remains in development as a therapeutic agent for neuropathic pain of different etiologies.

5. CAN WE EVER SUCCEED?

Every generation needs a new revolution

Thomas Jefferson

A number of different therapeutic approaches that target the various pathogenetic mechanisms of diabetic neuropathy have been the subject of clinical trials (Javed, Alam, & Malik, 2015b). These treatments aim to favorably impact the underlying pathophysiological abnormalities encountered in DPN by targeting different elements in the pathways leading to neurovascular dysfunction (Fig. 5). While numerous pathogenetic treatments have shown promise in experimental and early phase II studies, a recurring theme has been the lack of "translation" in phase III clinical trials. One may question the relevance of experimental findings, particularly short-term studies in animal models, which predominantly show functional and hence easily reversible deficits, rather than structural pathology. To address this, representatives from NIDDK, JDRF, and the Diabetic Neuropathy Study Group of EASD met to develop a set of consensus criteria for the phenotyping of rodent models of diabetic neuropathy (Biessels, Bril, Calcutt, et al., 2014) in relation to:

- Different rodent models of diabetes (type 1/2 diabetes, IGT/prediabetes).
- Characteristic alterations in nerve structure.
- Electrophysiological assessments of nerve function.
- Behavioral assessments of nerve function.
- The role of biomarkers in disease phenotyping.

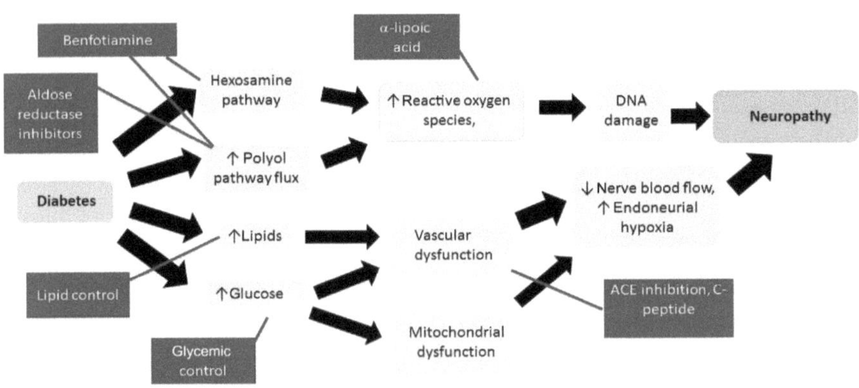

Fig. 5 Integrating pathophysiology of diabetic neuropathy and the sites of action of pathogenetic drugs (*ARIs*, aldose reductase inhibitors; *ACE*, angiotensin-converting enzyme; *AGE*, advanced glycation end products).

It was agreed that neuropathy in rodents should be defined as the presence of statistically different values between diabetic and control animals in 2 of 3 assessments:

- Nocifensive behaviour (paw withdrawal response to heat or von Frey monofilament).
- Nerve conduction velocities (sensory and motor).
- Nerve structure (nerve axon count, epidermal nerve fiber density, corneal nerve fiber density).

The salient fact remains that at present there are no FDA approved medications for the treatment of diabetic neuropathy (Malik, 2014b). One has to ask why no pathogenetic treatment for DPN has achieved regulatory approval. This is not a new question and indeed was addressed following the huge investment, and subsequent serial failure, of the ARI program during the 1980–90s, where virtually every large pharma house invested multimillions in large phase III clinical trials, all of which were deemed as failed. A Cochrane review found thirty-two randomized controlled trials of ARIs involving 879 treated participants and 909 controls with adequate data. There was no overall significant difference between the treated and control groups in neurological function (13 studies), nerve conduction parameters (27 studies), or foot ulceration (1 study) (Chalk et al., 2007). The authors concluded: "Any future clinical trials of ARIs should be restricted to compounds proven to have substantial biological or preclinical advantages over previously tested agents." Surely that is the wrong message!

Pfeifer and Schumer (1995) reviewed the outcomes and limitations of previous clinical trials and suggested future directions. They noted that there was improvement of nerve function associated with some short-term clinical trials addressing a number of possible etiologic pathways (improved glucose control, high myoinositol diet, ARIs, gamma-linolenic acid). An improvement in nerve morphometry had been demonstrated in some short-term clinical trials with ARIs (Dyck et al., 1988; Greene et al., 1999; Sima et al., 1988), perhaps attesting to the potential for using morphological measures as end points in clinical trials of diabetic neuropathy. However, the use of serial sural nerve biopsies is not something that can be advocated for clinical trials of diabetic neuropathy, given their highly invasive nature.

In 2002, Ziegler and Luft suggested that, until the mid-1990s, trials were hampered by a generally poor design, short follow-up, and by being limited to patients with advanced DPN (Ziegler & Luft, 2002). Therefore, an adequately designed RCT in diabetic neuropathy must consider:

- The type and stage of neuropathy.
- Homogeneity of the study population (type 1/2/LADA).
- Relevant surrogate end points (not just neurophysiology).
- Intermediate clinical end points.
- Ultimate clinical outcomes (foot ulceration and amputation).
- Quality of life.

The trial should also take into account the natural history of the disease in order to adequately power the study in terms of the sample size and study duration. Furthermore, the reproducibility of the surrogate and intermediate end points and their clinical relevance and applicability to all patients with diabetes needs to be considered. They (Ziegler & Luft, 2002) further suggested that trials should focus preferentially on:

- Patients with mild or moderate neuropathy.
- Be undertaken over 3–5 years.
- Aim to slow or prevent, rather than reverse diabetic neuropathy.
- Use end points that have clinical and prognostic significance.

Tesfaye, Tandan, Bastyr, et al. (2007) reported on changes in signs, symptoms, and quantitative vibration testing in the placebo-treated arms of two randomized controlled trials of ruboxistaurin in DPN. They suggested that to demonstrate deterioration in these measures in any placebo-treated DPN group, studies of >12 months were needed. In the same year, Dyck et al. examined the challenges of selecting appropriate end points for clinical trials by examining data from the placebo-treated groups of two large intervention trials as well as the Rochester Diabetic Neuropathy Study (Dyck et al., 2007). They concluded that studies were failing to demonstrate efficacy due to:

- A strong placebo effect for symptoms and signs.
- Measurement noise.
- Because DPN may progresses more gradually than previously thought due to concomitant treatment with ACE inhibitors (Malik et al., 1998; Reja, Tesfaye, Harris, et al., 1995) and lipid-lowering (Calles-Escandon et al., 2010; Davis, Yeap, Davis, et al., 2008; Rajamani et al., 2009) therapies.

Despite these analyses, we continue to enroll small numbers of heterogeneous patients in trials of short duration (1–2 years maximum), using the wrong end points such as pain for a trial of a disease modifier or large fiber nerve conduction when the underlying benefit is to small fibers. Why does this occur? Because investors want a quick return for their money and have the naïve and sometimes ill-advised hope that a quick 12–20-week pain trial will be positive for a drug that is supposed to repair nerves, or at the very least limit further degeneration. Trials are also constrained by current FDA guidance, which has not changed for ~20 years, and continues to endorse symptoms and signs, QST, and neurophysiology. In my opinion, future clinical trials in patients with diabetic neuropathy must minimally:

- Enrol patients with mild or moderate neuropathy.
- Account for the progression rate of neuropathy in the placebo cohort.
- Use surrogate end points of small fiber repair, where pertinent to the therapeutic approach.
- Operate in a regulatory environment that accepts small fiber repair as a desirable primary end point.

We can continue to go from failure to failure, or we can have the courage to question how we diagnose and conduct clinical trials in diabetic neuropathy in order to make a difference for future generations of patients with diabetic neuropathy. Many potentially efficacious therapies may well have been aborted prematurely and need to be reconsidered using the correct end points.

REFERENCES

Abbott, C. A., Carrington, A. L., Ashe, H., et al. (2002). The north-west diabetes foot care study: Incidence of, and risk factors for, new diabetic foot ulceration in a community-based patient cohort. *Diabetic Medicine*, *19*, 377–384.

Abbott, C. A., Malik, R. A., van Ross, E. R., et al. (2011). Prevalence and characteristics of painful diabetic neuropathy in a large community-based diabetic population in the U.K. *Diabetes Care*, *34*, 2220–2224.

Agudo, R., Valls-Solé, J., Recasens, M., et al. (2002). Effects of kidney and pancreas transplantation on neurophysiological rates of polyneuropathy and autonomous cardiac function in diabetic patients with end-stage renal disease. *Medicina Clínica (Barcelona)*, *118*, 534–538.

Ajroud-Driss, S., Christiansen, M., Allen, J. A., et al. (2013). Phase 1/2 open-label dose-escalation study of plasmid DNA expressing two isoforms of hepatocyte growth factor in patients with painful diabetic peripheral neuropathy. *Molecular Therapy*, *21*, 1279–1286.

Alamdari, A., Mozafari, R., Tafakhori, A., et al. (2015). An inverse association between serum vitamin D levels with the presence and severity of impaired nerve conduction velocity and large fiber peripheral neuropathy in diabetic subjects. *Neurological Sciences*, *36*, 1121–1126.

Al-Hamadani, M., Habermann, T. M., Cerhan, J. R., et al. (2015). Non-Hodgkin lymphoma subtype distribution, geodemographic patterns, and survival in the US: A longitudinal analysis of the National Cancer Data Base from 1998 to 2011. *American Journal of Hematology*, *90*, 790–795.

Almurdhi, M. M., Reeves, N. D., Bowling, F. L., et al. (2016). Reduced lower-limb muscle strength and volume in patients with type 2 diabetes in relation to neuropathy, intramuscular fat, and vitamin D levels. *Diabetes Care*, *39*, 441–447.

Anand, P., Terenghi, G., Warner, G., et al. (1996). The role of endogenous nerve growth factor in human diabetic neuropathy. *Nature Medicine*, *2*, 703–707.

Andersen, H. (2014). Motor neuropathy. *Handbook of Clinical Neurology, 126,* 81–95.
Apfel, S. C. (2002). Nerve growth factor for the treatment of diabetic neuropathy: What went wrong, what went right, and what does the future hold? *International Review of Neurobiology, 50,* 393–413.
Apfel, S. C., Kessler, J. A., Adornato, B. T., et al. (1998). Recombinant human nerve growth factor in the treatment of diabetic polyneuropathy. NGF Study Group. *Neurology, 51,* 695–702.
Armstrong, D. G., Wrobel, J., & Robbins, J. M. (2007). Guest editorial: Are diabetes-related wounds and amputations worse than cancer? *International Wound Journal, 4,* 286–287.
Asghar, O., Petropoulos, I. N., Alam, U., et al. (2014). Corneal confocal microscopy detects neuropathy in subjects with impaired glucose tolerance. *Diabetes Care, 37,* 2643–2646.
Ashrafi, M., Baguneid, M., & Bayat, A. (2015). The role of neuromediators and innervation in cutaneous wound healing. *Acta Dermato-Venereologica.* http://dx.doi.org/10.2340/00015555-2321[Epub ahead of print].
Azmi, S., Ferdousi, M., Petropoulos, I. N., et al. (2015). Corneal confocal microscopy shows an improvement in small-fiber neuropathy in subjects with type 1 diabetes on continuous subcutaneous insulin infusion compared with multiple daily injection. *Diabetes Care, 38,* e3–e4.
Bansal, D., Badhan, Y., Gudala, K., et al. (2013). Ruboxistaurin for the treatment of diabetic peripheral neuropathy: A systematic review of randomized clinical trials. *Diabetes & Metabolism Journal, 37,* 375–384.
Basit, A., Basit, K. A., Fawwad, A., Shaheen, F., Fatima, N., Petropoulos, I. N., et al. (2016). Vitamin D for the treatment of painful diabetic neuropathy. *BMJ Open Diabetes Research and Care, 4*(1), e000148. http://dx.doi.org/10.1136/bmjdrc-2015-000148.
Bata, I. R., Gregor, R. D., Wolf, H. K., et al. (2006). Trends in five-year survival of patients discharged after acute myocardial infarction. *The Canadian Journal of Cardiology, 22,* 399–404.
Bianco, J., Gueye, Y., Marqueste, T., et al. (2011). Vitamin D(3) improves respiratory adjustment to fatigue and H-reflex responses in paraplegic adult rats. *Neuroscience, 188,* 182–192.
Biessels, G. J., Bril, V., Calcutt, N. A., et al. (2014). Phenotyping animal models of diabetic neuropathy: A consensus statement of the diabetic neuropathy study group of the EASD (Neurodiab). *Journal of the Peripheral Nervous System, 19,* 77–87.
Bitirgen, G., Ozkagnici, A., Malik, R. A., et al. (2014). Corneal nerve fibre damage precedes diabetic retinopathy in patients with type 2 diabetes mellitus. *Diabetic Medicine, 31,* 431–438.
Boucek, P., Havrdova, T., Voska, L., et al. (2008). Epidermal innervation in type 1 diabetic patients: A 2.5-year prospective study after simultaneous pancreas/kidney transplantation. *Diabetes Care, 31,* 1611–1612.
Boulton, A. J., Kempler, P., Ametov, A., et al. (2013). Whither pathogenetic treatments for diabetic polyneuropathy? *Diabetes/Metabolism Research and Reviews, 29,* 327–333.
Boulton, A. J. M., Malik, R. A., Arezzo, J. C., et al. (2004). Diabetic somatic neuropathies. *Diabetes Care, 27,* 1458–1486.
Boulton, A. J., Vileikyte, L., Ragnarson-Tennvall, G., et al. (2005). The global burden of diabetic foot disease. *Lancet, 366,* 1719–1724.
Boussageon, R., Bejan-Angoulvant, T., Saadatian-Elahi, M., et al. (2011). Effect of intensive glucose lowering treatment on all cause mortality, cardiovascular death, and microvascular events in type 2 diabetes: Meta-analysis of randomised controlled trials. *BMJ, 343,* d4169.
Bramson, C., Herrmann, D. N., Carey, W., et al. (2015). Exploring the role of tanezumab as a novel treatment for the relief of neuropathic pain. *Pain Medicine, 16,* 1163–1176.

Brines, M., Dunne, A. N., van Velzen, M., et al. (2014). ARA 290, a nonerythropoietic peptide engineered from erythropoietin, improves metabolic control and neuropathic symptoms in patients with type 2 diabetes. *Molecular Medicine*, *20*, 658.

Brooks, B., Delaney-Robinson, C., Molyneaux, L., et al. (2008). Endothelial and neural regulation of skin microvascular blood flow in patients with diabetic peripheral neuropathy: Effect of treatment with the isoform-specific protein kinase C beta inhibitor, ruboxistaurin. *Journal of Diabetes and its Complications*, *22*, 88–95.

Brown, M. T., Herrmann, D. N., Goldstein, M., et al. (2014). Nerve safety of tanezumab, a nerve growth factor inhibitor for pain treatment. *Journal of the Neurological Sciences*, *345*, 139–147.

Callaghan, B. C., Cheng, H. T., Stables, C. L., et al. (2012). Diabetic neuropathy: Clinical manifestations and current treatments. *Lancet Neurology*, *11*, 521–534.

Callaghan, B. C., Little, A. A., Feldman, E. L., et al. (2012). Enhanced glucose control for preventing and treating diabetic neuropathy. *Cochrane Database of Systematic Reviews*, *6*, CD007543.

Calles-Escandon, J., Lovato, L. C., Simons-Morton, D. G., et al. (2010). Effect of intensive compared with standard glycemia treatment strategies on mortality by baseline subgroup characteristics: The Action to Control Cardiovascular Risk in Diabetes (ACCORD) trial. *Diabetes Care*, *33*, 721–727.

Celikbilek, A., Gocmen, A. Y., Tanik, N., et al. (2015). Decreased serum vitamin D levels are associated with diabetic peripheral neuropathy in a rural area of Turkey. *Acta Neurologica Belgica*, *115*, 47–52.

Chalk, C., Benstead, T. J., & Moore, F. (2007). Aldose reductase inhibitors for the treatment of diabetic polyneuropathy. *Cochrane Database of Systematic Reviews*, CD004572.

Chen, X., Graham, J., Dabbah, M. A., et al. (2015). Small nerve fiber quantification in the diagnosis of diabetic sensorimotor polyneuropathy: Comparing corneal confocal microscopy with intraepidermal nerve fiber density. *Diabetes Care*, *38*, 1138–1144.

Dabbah, M., Graham, J., Petropoulos, I., et al. (2011). Automatic analysis of diabetic peripheral neuropathy using multi-scale quantitative morphology of nerve fibres in corneal confocal microscopy imaging. *Medical Image Analysis*, *15*, 738–747.

Davis, T. M., Yeap, B. B., Davis, W. A., et al. (2008). Lipid-lowering therapy and peripheral sensory neuropathy in type 2 diabetes: The Fremantle Diabetes Study. *Diabetologia*, *51*, 562–566.

Dehghani, C., Pritchard, N., Edwards, K., et al. (2016). Risk factors associated with corneal nerve alteration in type 1 diabetes in the absence of neuropathy: A longitudinal In Vivo corneal confocal microscopy study. *Cornea*, (in press).

Del Carro, U., Fiorina, P., Amadio, S., et al. (2007). Evaluation of polyneuropathy markers in type 1 diabetic kidney transplant patients and effects of islet transplantation: Neurophysiological and skin biopsy longitudinal analysis. *Diabetes Care*, *30*, 3063–3069.

Drury, P. L., Ting, R., Zannino, D., et al. (2011). Estimated glomerular filtration rate and albuminuria are independent predictors of cardiovascular events and death in type 2 diabetes mellitus: The Fenofibrate Intervention and Event Lowering in Diabetes (FIELD) study. *Diabetologia*, *54*, 32–43.

Duckworth, W., Abraira, C., Moritz, T., et al. (2009). Glucose control and vascular complications in veterans with type 2 diabetes. *The New England Journal of Medicine*, *360*, 129–139.

Dyck, P. J. (2014). Methodology for conduct of epidemiologic surveys and randomized controlled trials of diabetic polyneuropathy. *Handbook of Clinical Neurology*, *126*, 335–338.

Dyck, P. J., Albers, J. W., Wolfe, J., et al. (2013). A trial of proficiency of nerve conduction: Greater standardization still needed. *Muscle & Nerve*, *48*, 369–374.

Dyck, P. J., Argyros, B., Russell, J. W., et al. (2014). Multicenter trial of the proficiency of smart quantitative sensation tests. *Muscle & Nerve*, *49*, 645–653.

Dyck, P. J., Norell, J. E., Tritschler, H., et al. (2007). Challenges in design of multicenter trials: End points assessed longitudinally for change and monotonicity. *Diabetes Care, 30*, 2619–2625.

Dyck, P. J., Overland, C. J., Low, P. A., et al. (2010). Signs and symptoms versus nerve conduction studies to diagnose diabetic sensorimotor polyneuropathy: Cl vs. NPhys trial. *Muscle & Nerve, 42*, 157–164.

Dyck, P. J., Zimmerman, B. R., Vilen, T. H., et al. (1988). Nerve glucose, fructose, sorbitol, myo-inositol, and fiber degeneration and regeneration in diabetic neuropathy. *The New England Journal of Medicine, 319*, 542–548.

Ekberg, K., Brismar, T., Johansson, B. L., et al. (2003). Amelioration of sensory nerve dysfunction by C-peptide in patients with type 1 diabetes. *Diabetes, 52*, 536–541.

Ekberg, K., Brismar, T., Johansson, B. L., et al. (2007). C-Peptide replacement therapy and sensory nerve function in type 1 diabetic neuropathy. *Diabetes Care, 30*, 71–76.

Field, J. K., Duffy, S. W., Baldwin, D. R., et al. (2016). UK Lung Cancer RCT Pilot Screening Trial: Baseline findings from the screening arm provide evidence for the potential implementation of lung cancer screening. *Thorax, 71*, 161–170.

Fioretto, P., Steffes, M. W., Sutherland, D. E., et al. (1998). Reversal of lesions of diabetic nephropathy after pancreas transplantation. *The New England Journal of Medicine, 339*, 69–75.

Fiorina, P., & Secchi, A. (2007). Pancreatic islet cell transplant for treatment of diabetes. *Endocrinology and Metabolism Clinics of North America, 36*, 999–1013. ix.

Fonseca, V. A., Lavery, L. A., Thethi, T. K., et al. (2013). Metanx in type 2 diabetes with peripheral neuropathy: A randomized trial. *The American Journal of Medicine, 126*, 141–149.

Fried, L. F., Forrest, K. Y., Ellis, D., et al. (2001). Lipid modulation in insulin-dependent diabetes mellitus: Effect on microvascular outcomes. *Journal of Diabetes and its Complications, 15*, 113–119.

Fukuoka, M., Sakurai, K., Ohta, T., et al. (2001). Tacalcitol, an active vitamin D3, induces nerve growth factor production in human epidermal keratinocytes. *Skin Pharmacology and Applied Skin Physiology, 14*, 226–233.

Gaede, P., Lund-Andersen, H., Parving, H. H., et al. (2008). Effect of a multifactorial intervention on mortality in type 2 diabetes. *The New England Journal of Medicine, 358*, 580–591.

Goto, Y., Hotta, N., Shigeta, Y., et al. (1995). Effects of an aldose reductase inhibitor, epalrestat, on diabetic neuropathy. Clinical benefit and indication for the drug assessed from the results of a placebo-controlled double-blind study. *Biomedicine & Pharmacotherapy, 49*, 269–277.

Greene, D. A., Arezzo, J. C., & Brown, M. B. (1999). Effect of aldose reductase inhibition on nerve conduction and morphometry in diabetic neuropathy. Zenarestat Study Group. *Neurology, 53*, 580–591.

Hammes, H. P., Du, X., Edelstein, D., et al. (2003). Benfotiamine blocks three major pathways of hyperglycemic damage and prevents experimental diabetic retinopathy. *Nature Medicine, 9*, 294–299.

Hankey, G. J., Jamrozik, K., Broadhurst, R. J., et al. (2000). Five-year survival after first-ever stroke and related prognostic factors in the Perth Community Stroke Study. *Stroke, 31*, 2080–2086.

Hellweg, R., Wohrle, M., Hartung, H. D., et al. (1991). Diabetes mellitus-associated decrease in nerve growth factor levels is reversed by allogeneic pancreatic islet transplantation. *Neuroscience Letters, 125*, 1–4.

Herman, W. H., Aubert, R. E., Engelgau, M. M., et al. (1998). Diabetes mellitus in Egypt: Glycaemic control and microvascular and neuropathic complications. *Diabetic Medicine, 15*, 1045–1051.

Hernández-Ojeda, J., Román-Pintos, L. M., Rodríguez-Carrízalez, A. D., et al. (2014). Effect of rosuvastatin on diabetic polyneuropathy: A randomized, double-blind, placebo-controlled Phase IIa study. *Diabetes, Metabolic Syndrome and Obesity: Targets and Therapy*, *7*, 401–407.

Herrmann, M., Sullivan, D. R., Veillard, A. S., et al. (2015). Serum 25-hydroxyvitamin D: A predictor of macrovascular and microvascular complications in patients with type 2 diabetes. *Diabetes Care*, *38*, 521–528.

Hirani, V. (2012). Vitamin D status and pain: Analysis from the Health Survey for England among English adults aged 65 years and over. *The British Journal of Nutrition*, *107*, 1080–1084.

Hirose, M., Kuroda, Y., & Murata, E. (2016). NGF/TrkA signaling as a therapeutic target for pain. *Pain Practice*, *16*, 175–182.

Javed, S., Alam, U., & Malik, R. A. (2015a). Burning through the pain: Treatments for diabetic neuropathy. *Diabetes, Obesity & Metabolism*, *17*, 1115–1125.

Javed, S., Alam, U., & Malik, R. A. (2015b). Treating diabetic neuropathy: Present strategies and emerging solutions. *The Review of Diabetic Studies*, *12*, 63–83.

Jiang, D. Q., Li, M. X., Wang, Y., et al. (2015). Effects of prostaglandin E1 plus methylcobalamin alone and in combination with lipoic acid on nerve conduction velocity in patients with diabetic peripheral neuropathy: A meta-analysis. *Neuroscience Letters*, *594*, 23–29.

Kaur, S., Pandhi, P., & Dutta, P. (2011). Painful diabetic neuropathy: An update. *Annals of Neurosciences*, *18*, 168–175.

Kennedy, W. R., Navarro, X., Goetz, F. C., et al. (1990). Effects of pancreatic transplantation on diabetic neuropathy. *The New England Journal of Medicine*, *322*, 1031–1037.

Kessler, J. A., Smith, A. G., Cha, B. S., et al. (2015). Double-blind, placebo-controlled study of HGF gene therapy in diabetic neuropathy. *Annals of Clinical and Translational Neurology*, *2*, 465–478.

Lauria, G., Bakkers, M., Schmitz, C., et al. (2010). Intraepidermal nerve fiber density at the distal leg: A worldwide normative reference study. *Journal of the Peripheral Nervous System*, *15*, 202–207.

Lauria, G., Hsieh, S. T., Johansson, O., et al. (2010). European Federation of Neurological Societies/Peripheral Nerve Society Guideline on the use of skin biopsy in the diagnosis of small fiber neuropathy. Report of a joint task force of the European Federation of Neurological Societies and the Peripheral Nerve Society. *European Journal of Neurology*, *17*, 903–912. e944-909.

Lauria, G., Morbin, M., Lombardi, R., et al. (2003). Axonal swellings predict the degeneration of epidermal nerve fibers in painful neuropathies. *Neurology*, *61*, 631–636.

Laverdet, B., Danigo, A., Girard, D., et al. (2015). Skin innervation: Important roles during normal and pathological cutaneous repair. *Histology and Histopathology*, *30*, 875–892.

Lee, P., & Chen, R. (2008). Vitamin D as an analgesic for patients with type 2 diabetes and neuropathic pain. *Archives of Internal Medicine*, *168*, 771–772.

Litchy, W. J., Albers, J. W., Wolfe, J., et al. (2014). Proficiency of nerve conduction using standard methods and reference values (Cl. NPhys Trial 4). *Muscle & Nerve*, *50*, 900–908.

Lv, W. S., Zhao, W. J., Gong, S. L., et al. (2014). Serum 25-hydroxyvitamin D levels and peripheral neuropathy in patients with type 2 diabetes: a systematic review and meta-analysis. *Journal of Endocrinological Investigation*.

Malik, R. A. (2014a). Which test for diagnosing early human diabetic neuropathy? *Diabetes*, *63*, 2206–2208.

Malik, R. A. (2014b). Why are there no good treatments for diabetic neuropathy? *The Lancet. Diabetes & Endocrinology*, *2*, 607–609.

Malik, R. A., Kallinikos, P., Abbott, C. A., et al. (2003). Corneal confocal microscopy: A non-invasive surrogate of nerve fibre damage and repair in diabetic patients. *Diabetologia, 46*, 683–688.

Malik, R. A., Williamson, S., Abbott, C., et al. (1998). Effect of angiotensin-converting-enzyme (ACE) inhibitor trandolapril on human diabetic neuropathy: Randomised double-blind controlled trial. *Lancet, 352*, 1978–1981.

Malvezzi, M., Carioli, G., Bertuccio, P., et al. (2016). European cancer mortality predictions for the year 2016 with focus on leukaemias. *Annals of Oncology*.

Marshall, D. C., Webb, T. E., Hall, R. A., et al. (2016). Trends in UK regional cancer mortality 1991–2007. *British Journal of Cancer, 114*, 340–347.

Martinez, C. H., Mannino, D. M., Jaimes, F. A., et al. (2015). Undiagnosed obstructive lung disease in the United States. Associated factors and long-term mortality. *Annals of the American Thoracic Society, 12*, 1788–1795.

Maser, R. E., Lenhard, M. J., & Pohlig, R. T. (2015). Vitamin D insufficiency is associated with reduced parasympathetic nerve fiber function in type 2 diabetes. *Endocrine Practice, 21*, 174–181.

Mayfield, J. A., & Sugarman, J. R. (2000). The use of the Semmes-Weinstein monofilament and other threshold tests for preventing foot ulceration and amputation in persons with diabetes. *The Journal of Family Practice, 49*, S17–S29.

Mehra, S., Tavakoli, M., Kallinikos, P. A., et al. (2007). Corneal confocal microscopy detects early nerve regeneration after pancreas transplantation in patients with type 1 diabetes. *Diabetes Care, 30*, 2608–2612.

Navarro, X., Sutherland, D. E., & Kennedy, W. R. (1997). Long-term effects of pancreatic transplantation on diabetic neuropathy. *Annals of Neurology, 42*, 727–736.

Ohkubo, Y., Kishikawa, H., Araki, E., et al. (1995). Intensive insulin therapy prevents the progression of diabetic microvascular complications in Japanese patients with non-insulin-dependent diabetes mellitus: A randomized prospective 6-year study. *Diabetes Research and Clinical Practice, 28*, 103–117.

Oyibo, S. O., Prasad, Y. D., Jackson, N. J., et al. (2002). The relationship between blood glucose excursions and painful diabetic peripheral neuropathy: A pilot study. *Diabetic Medicine, 19*, 870–873.

Petropoulos, I. N., Alam, U., Fadavi, H., et al. (2014). Rapid automated diagnosis of diabetic peripheral neuropathy with in vivo corneal confocal microscopy. *Investigative Ophthalmology & Visual Science, 55*, 2071–2078.

Petropoulos, I. N., Manzoor, T., Morgan, P., et al. (2013). Repeatability of in vivo corneal confocal microscopy to quantify corneal nerve morphology. *Cornea, 32*, e83–e89.

Pfeifer, M. A., & Schumer, M. P. (1995). Clinical trials of diabetic neuropathy: Past, present, and future. *Diabetes, 44*, 1355–1361.

Pop-Busui, R., Stevens, M. J., Raffel, D. M., et al. (2013). Effects of triple antioxidant therapy on measures of cardiovascular autonomic neuropathy and on myocardial blood flow in type 1 diabetes: A randomised controlled trial. *Diabetologia, 56*, 1835–1844.

Pritchard, N., Edwards, K., Russell, A. W., et al. (2015). Corneal confocal microscopy predicts 4-year incident peripheral neuropathy in type 1 diabetes. *Diabetes Care, 38*, 671–675.

Quattrini, C., Tavakoli, M., Jeziorska, M., et al. (2007). Surrogate markers of small fiber damage in human diabetic neuropathy. *Diabetes, 56*, 2148–2154.

Rajamani, K., Colman, P. G., Li, L. P., et al. (2009). Effect of fenofibrate on amputation events in people with type 2 diabetes mellitus (FIELD study): A prespecified analysis of a randomised controlled trial. *Lancet, 373*, 1780–1788.

Reichard, P., Nilsson, B. Y., & Rosenqvist, U. (1993). The effect of long-term intensified insulin treatment on the development of microvascular complications of diabetes mellitus. *The New England Journal of Medicine, 329*, 304–309.

Reja, A., Tesfaye, S., Harris, N. D., et al. (1995). Is ACE inhibition with lisinopril helpful in diabetic neuropathy? *Diabetic Medicine*, *12*, 307–309.

Riaz, M., Miyan, Z., Zaidi, S. I., Alvi, S. F., Fawwad, A., Ahmadani, M. Y., et al. (2014). Characteristics of a large cohort of patients with diabetes having at-risk feet and outcomes in patients with foot ulceration referred to a tertiary care diabetes unit. *International Wound Journal*. http://dx.doi.org/10.1111/iwj.12289[Epub ahead of print].

Rice, A. S., Dworkin, R. H., McCarthy, T. D., et al. (2014). EMA401, an orally administered highly selective angiotensin II type 2 receptor antagonist, as a novel treatment for postherpetic neuralgia: A randomised, double-blind, placebo-controlled phase 2 clinical trial. *Lancet*, *383*, 1637–1647.

Ropper, A. H., Gorson, K. C., Gooch, C. L., et al. (2009). Vascular endothelial growth factor gene transfer for diabetic polyneuropathy: A randomized, double-blinded trial. *Annals of Neurology*, *65*, 386–393.

Ruggenenti, P., Lauria, G., Iliev, I. P., et al. (2011). Effects of manidipine and delapril in hypertensive patients with type 2 diabetes mellitus: The delapril and manidipine for nephroprotection in diabetes (DEMAND) randomized clinical trial. *Hypertension*, *58*, 776–783.

Shehab, D., Al-Jarallah, K., Abdella, N., et al. (2015). Prospective evaluation of the effect of short-term oral vitamin d supplementation on peripheral neuropathy in type 2 diabetes mellitus. *Medical Principles and Practice*, *24*, 250–256.

Shehab, D., Al-Jarallah, K., Mojiminiyi, O. A., et al. (2012). Does Vitamin D deficiency play a role in peripheral neuropathy in Type 2 diabetes? *Diabetic Medicine*, *29*, 43–49.

Shelton, D. (2014). Development of nerve growth factor (NGF) inhibition as a strategy for treatment of pain. *Journal of the Peripheral Nervous System*, *19*(Suppl. 2), S12–S13.

Sima, A. A. (2003). New insights into the metabolic and molecular basis for diabetic neuropathy. *Cellular and Molecular Life Sciences*, *60*, 2445–2464.

Sima, A. A., Bril, V., Nathaniel, V., et al. (1988). Regeneration and repair of myelinated fibers in sural-nerve biopsy specimens from patients with diabetic neuropathy treated with sorbinil. *The New England Journal of Medicine*, *319*, 548–555.

Singleton, J. R., Smith, A. G., & Bromberg, M. B. (2001). Increased prevalence of impaired glucose tolerance in patients with painful sensory neuropathy. *Diabetes Care*, *24*, 1448–1453.

Smith, A. G., Ramachandran, P., Tripp, S., et al. (2001). Epidermal nerve innervation in impaired glucose tolerance and diabetes-associated neuropathy. *Neurology*, *57*, 1701–1704.

Soderstrom, L. H., Johnson, S. P., Diaz, V. A., et al. (2012). Association between vitamin D and diabetic neuropathy in a nationally representative sample: Results from 2001–2004 NHANES. *Diabetic Medicine*, *29*, 50–55.

Stracke, H., Gaus, W., Achenbach, U., et al. (2008). Benfotiamine in diabetic polyneuropathy (BENDIP): Results of a randomised, double blind, placebo-controlled clinical study. *Experimental and Clinical Endocrinology & Diabetes*, *116*, 600–605.

Stratton, I. M., Adler, A. I., Neil, H. A., et al. (2000). Association of glycaemia with macrovascular and microvascular complications of type 2 diabetes (UKPDS 35): Prospective observational study. *BMJ*, *321*, 405–412.

Sun, J. K., Keenan, H. A., Cavallerano, J. D., et al. (2011). Protection from retinopathy and other complications in patients with type 1 diabetes of extreme duration: The Joslin 50-Year medalist study. *Diabetes Care*, *34*, 968–974.

Szalai, E., Deak, E., Modis, L., Jr., et al. (2016). Early corneal cellular and nerve fiber pathology in young patients with type 1 diabetes mellitus identified using corneal confocal microscopy. *Investigative Ophthalmology & Visual Science*, *57*, 853–858.

Tague, S. E., Clarke, G. L., Winter, M. K., et al. (2011). Vitamin D deficiency promotes skeletal muscle hypersensitivity and sensory hyperinnervation. *The Journal of Neuroscience*, *31*, 13728–13738.

Tague, S. E., & Smith, P. G. (2011). Vitamin D receptor and enzyme expression in dorsal root ganglia of adult female rats: Modulation by ovarian hormones. *Journal of Chemical Neuroanatomy*, *41*, 1–12.

Tavakoli, M., Ferdousi, M., Petropoulos, I. N., et al. (2015). Normative values for corneal nerve morphology assessed using corneal confocal microscopy: A multinational normative data set. *Diabetes Care*, *38*, 838–843.

Tavakoli, A., & Liong, S. (2012). Pancreatic transplant in diabetes. *Advances in Experimental Medicine and Biology*, *771*, 420–437.

Tavakoli, M., Mitu-Pretorian, M., Petropoulos, I. N., et al. (2013). Corneal confocal microscopy detects early nerve regeneration in diabetic neuropathy after simultaneous pancreas and kidney transplantation. *Diabetes*, *62*, 254–260.

Tesfaye, S., Boulton, A. J., Dyck, P. J., et al. (2010). Diabetic neuropathies: Update on definitions, diagnostic criteria, estimation of severity, and treatments. *Diabetes Care*, *33*, 2285–2293.

Tesfaye, S., Chaturvedi, N., Eaton, S. E. M., et al. (2005). Vascular risk factors and diabetic neuropathy. *New England Journal of Medicine*, *352*, 341–350.

Tesfaye, S., Tandan, R., Bastyr, E. J., 3rd., et al. (2007). Factors that impact symptomatic diabetic peripheral neuropathy in placebo-administered patients from two 1-year clinical trials. *Diabetes Care*, *30*, 2626–2632.

The Diabetes Control and Complications Trial Research Group. (1993). The effect of intensive treatment of diabetes on the development and progression of long-term complications in insulin-dependent diabetes mellitus. *The New England Journal of Medicine*, *329*, 977–986.

The Diabetes Control and Complications Trial Research Group. (1995). The effect of intensive diabetes therapy on the development and progression of neuropathy. *Annals of Internal Medicine*, *122*, 561–568.

Trippe, B. S., Barrentine, L. W., Curole, M. V., et al. (2016). Nutritional management of patients with diabetic peripheral neuropathy with L-methylfolate-methylcobalamin-pyridoxal-5-phosphate: Results of a real-world patient experience trial. *Current Medical Research and Opinion*, *32*, 219–227.

van Baal, J., Hubbard, R., Game, F., et al. (2010). Mortality associated with acute Charcot foot and neuropathic foot ulceration. *Diabetes Care*, *33*, 1086–1089.

van Hecke, M. V., Dekker, J. M., Stehouwer, C. D., et al. (2005). Diabetic retinopathy is associated with mortality and cardiovascular disease incidence: The EURODIAB prospective complications study. *Diabetes Care*, *28*, 1383–1389.

van Velzen, M., Heij, L., Niesters, M., et al. (2014). ARA 290 for treatment of small fiber neuropathy in sarcoidosis. *Expert Opinion on Investigational Drugs*, *23*, 541–550.

Verheyen, A., Peeraer, E., Lambrechts, D., et al. (2013). Therapeutic potential of VEGF and VEGF-derived peptide in peripheral neuropathies. *Neuroscience*, *244*, 77–89.

Vinik, A. I., Bril, V., Kempler, P., et al. (2005). Treatment of symptomatic diabetic peripheral neuropathy with the protein kinase C beta-inhibitor ruboxistaurin mesylate during a 1-year, randomized, placebo-controlled, double-blind clinical trial. *Clinical Therapeutics*, *27*, 1164–1180.

Wahren, J., Ekberg, K., & Jornvall, H. (2007). C-peptide is a bioactive peptide. *Diabetologia*, *50*, 503–509.

Wahren, J., Foyt, H., Daniels, M., et al. (2016). Long-acting C-peptide and neuropathy in Type 1 diabetes: A 12-month clinical trial. *Diabetes Care*, (in press).

Walsh, J. W., Hoffstad, O. J., Sullivan, M. O., & Margolis, D. J. (2015). Association of diabetic foot ulcer and death in a population-based cohort from the United Kingdom. *Diabetic Medicine*. http://dx.doi.org/10.1111/dme.13054.

Wergeland, S., Torkildsen, O., Myhr, K. M., et al. (2011). Dietary vitamin D3 supplements reduce demyelination in the cuprizone model. *PloS One*, *6*, e26262.

Wiggin, T. D., Sullivan, K. A., Pop-Busui, R., et al. (2009). Elevated triglycerides correlate with progression of diabetic neuropathy. *Diabetes*, *58*, 1634–1640.

Zhou, M., Wang, H., Zhu, J., et al. (2016). Cause-specific mortality for 240 causes in China during 1990–2013: A systematic subnational analysis for the Global Burden of Disease Study 2013. *Lancet*, *387*, 251–272.

Ziegler, D., & Luft, D. (2002). Clinical trials for drugs against diabetic neuropathy: Can we combine scientific needs with clinical practicalities? *International Review of Neurobiology*, *50*, 431–463.

Ziegler, D., Movsesyan, L., Mankovsky, B., et al. (2009). Treatment of symptomatic polyneuropathy with actovegin in type 2 diabetic patients. *Diabetes Care*, *32*, 1479–1484.

Ziegler, D., Nowak, H., Kempler, P., et al. (2004). Treatment of symptomatic diabetic polyneuropathy with the antioxidant alpha-lipoic acid: A meta-analysis. *Diabetic Medicine*, *21*, 114–121.

Ziegler, D., Papanas, N., Zhivov, A., et al. (2014). Early detection of nerve fiber loss by corneal confocal microscopy and skin biopsy in recently diagnosed type 2 diabetes. *Diabetes*, *63*, 2454–2463.

INDEX

Note: Page numbers followed by "*f*" indicate figures, and "*t*" indicate tables.

T

V

Z

CONTENTS OF RECENT VOLUMES

Volume 39

Volume 40

Volume 41

Volume 42

Volume 43

Volume 44

Volume 45

Volume 46

Volume 47

Volume 48

Volume 49

Volume 50

Volume 51

Volume 52

Volume 53

Volume 54

Volume 55

Volume 56

Volume 57

Volume 58

Volume 59

Volume 60

Volume 61

Volume 62

Volume 63

Volume 64

Volume 65

Volume 66

Volume 67

Volume 68

Volume 69

Volume 70

Volume 71

Volume 72

Volume 73

Volume 74

Volume 75

Volume 76

Volume 77

Volume 78

Volume 79

Volume 80

Volume 81

Volume 82

Volume 83

Volume 84

Volume 85

Volume 86

Volume 87

Volume 88

Volume 89

Volume 90

Volume 91

Volume 92

Volume 93

Volume 94

Volume 95

Volume 96

Volume 97

Volume 98

Volume 99

Volume 100

Volume 101

Volume 102

Volume 103

Volume 104

Volume 105

Volume 106

Volume 107

Volume 108

Volume 109

Volume 110

Volume 111

Volume 112

Volume 113

Volume 114

Volume 115

Volume 116

Volume 117

Volume 118

Volume 119

Volume 120

Volume 121

Volume 122

Volume 123

Volume 124

Volume 125

Volume 126

MIX
Papier | Fördert
gute Waldnutzung
FSC® C083411

Kolibri 360 GmbH
Ferdinand-Jühlke-Straße 7
99095 Erfurt, Deutschland
produktsicherheit@kolibri360.de